THE NEW YORK CITY

CONTEST PROBLEM BOOK

Problems and Solutions from the
New York City Interscholastic Mathematics League
1975–1984

Mark E. Saul
Bronx High School of Science/Bronxville Schools

Gilbert W. Kessler
Canarsie High School

Sheila Krilov
Hunter College Campus School

Lawrence Zimmerman
Brooklyn Technical High School

DALE SEYMOUR PUBLICATIONS

ISBN 0-86651-307-8

Order Number DS01614

abcdefghij-MA-8932109876

CONTENTS

PREFACE

The New York City Interscholastic Mathematics League is the oldest and largest local organization of its type in the United States. Records exist of competitions given by the League since the early 1920s, and there are indications that these contests were the continuation of an even earlier tradition. The contests have contributed to the unique position that New York City holds in mathematics education today. In national and international contests, New York City high school students have shown a high degree of talent, interest, and achievement. Many of these students have gone on to make their careers in mathematics, and some of them have contributed problems that appear in the present book. Thus the New York City IML contests have been instrumental in developing a mathematical problem-solving community in New York City, which in turn has made its contribution to the field on a national level.

The questions in this book are from the Senior A division contests. This division is open to teams of students at or below their senior year in high school. A series of five contests is given each term. Each contest consists of three pairs of questions, with a time limit of 9-12 minutes for each pair. A Senior B division is organized similarly, with somewhat easier questions. The junior division is open to teams of students at or below their junior year in high school; three contests are offered each term. The IML also runs eighth and ninth year contests and a series of elementary school competitions.

The IML has always been an organization of classroom teachers. Math teams are often coached by teachers on their own time, using their own talents and interests in developing materials for team practice. For this reason, it is a particularly difficult task to acknowledge all the people whose work appears in this book. Many teachers and former students have contributed one or two problems, or one or two ideas that later became problems. In every case, the problems and solutions were subject to much editing, so that the original problem is often unrecognizable in its final form. Nonetheless, it seems appropriate to mention a few individuals who have been instrumental in creating the contests of the last ten years.

Steven R. Conrad, formerly of Benjamin Cardozo High School and President of the League from 1968 to 1978, is responsible in large measure for the scope and format of the contests as they are held today. Irwin Kaufman, Director of Technology for the New York City Board of Education, Joel Arougheti of the High School of Art and Design, and Richard Geller of Stuyvesant High School, also contributed much time and effort to the organization of the League.

Of the many contributors of problems, Harold Shapiro and Peter Ungar of the Courant Institute of Mathematical Sciences, New York University, deserve special mention, as does Murray Klamkin of the University of Alberta.

In the final preparation of the contest, Robert Saenger, Sherrill Mirsky, John Reutershan, and Werner Weingartner of the Bronx High School of Science, and Harry Ruderman of Hunter College High School devoted much time and effort. Maria De Salvio of the Bureau of Curriculum and Instruction and Miriam Refkin, now retired from New York City School District 15, helped enormously with the myriad organizational details. Finally, the late Harry Sitomer was responsible for many original problems and solutions.

Perhaps the most important, though least tangible, contribution has been made by the math team coaches of the New York City high schools and by their students. Without the teachers' work and the students' interest, the contests would have been meaningless. As graduates, the following students later made direct contributions to the contests: David Karr, Brian Sheppard, and Benji Fisher of the Bronx High School of Science; David Wolland of Hunter College High School; Ashfaq Munshi and David Zagorski of Stuyvesant High School; Alan Edelman of Canarsie High School; Mark Prysant of South Shore High School; and Nicomedes Alonso of Long Island City High School.

All of these individuals, and many others, have made the New York City Interscholastic Mathematics League a major force in the field of mathematics education in New York City and throughout the country.

Mark E. Saul
President, NYCIML

SUGGESTIONS FOR USING THIS BOOK

The problems in this book are presented, for the most part, exactly as they appeared in the original contests. In a very few cases, there were errors in the statement of the problem, and these have been corrected.

The percentage of contestants who correctly solved the problem is indicated below the problem number for many of the contests. This information may help in anticipating the level of difficulty of a given problem (a task that is surprisingly difficult, even after much experience). References to items in Appendix A (Selected Theorems) are given as decimal numbers, while references to the bibliography are made by giving the author's name (and the book's title, if appropriate).

The two appendices and the index are provided to help organize the material contained in the problems. Appendix A is cross-referenced. A search for related problems can begin with a topic in the appendix, by following the references given to problems using that topic. Or it can begin with a problem, by following the reference given in the problem to the appendix. The index groups problems in the same general area. We have tried to list problems under more than one category. The process of classifying a problem is one step toward solving it, and a hasty classification may be a step in the wrong direction. Some of the most interesting problems resist classification. We have grouped these under the heading "Difficult to Classify."

The annotated bibliography is intended to help bridge the gap between short-answer contest problems and mathematics in its more natural setting. A contest problem at its best will stimulate the problem solver to do further investigation. The bibliography provides a set of starting points for this process.

Neither the appendices, nor the index, nor the bibliography are meant as complete surveys of their domains. Many interesting theorems and formulas are not mentioned in the appendix, simply because they are not used in the problems. Similarly, the bibliography is by no means exhaustive. It offers only a small part of the rich literature that has developed in problem solving and recreational mathematics. It is intended to furnish an introduction to the tradition on which the New York City contests are based.

PROBLEMS

Problems—Spring 1975

1. (19%) The positive integer n, when divided by 3, 4, 5, 6, and 7, leaves remainders of 2, 3, 4, 5, and 6, respectively. Find the least possible value of n.

2. (15%) Find all positive numbers x that satisfy

$$(2 + \log_{10}x)^3 + (-1 + \log_{10}x)^3 = (1 + \log_{10}x^2)^3.$$

3. (3%) Triangles I and II are coplanar, and each point of Triangle II is interior to and 2 units from Triangle I. If the side-lengths of Triangle I are 13, 14, and 15, find the area of the region that is the intersection of the interior of Triangle I and the exterior of Triangle II.

4. (23%) There are two times between 5 A.M. and 6 A.M. when the hands of an accurate clock are perpendicular. Exactly how many minutes must elapse between these two times?

5. (9%) Quadrilateral $ABCD$ has right angles at B and D. If $ABCD$ is kite shaped with $AB = AD = 20$ and $BC = CD = 15$, find the length of a radius of the circle inscribed in $ABCD$.

6. (55%) How many real values of x satisfy $\sqrt{x^2 - 4} - 2\sqrt{x^2 - 1} = x$?

7. (1%) The two-digit base ten numeral $AB (A \neq 0)$ represents the positive integer x. Find every integer x for which the sum of the digits in the product kx is equal to $A + B$ for every integral value of k from 1 to 9 inclusive.

8. (31%) Find the coordinates of that point on the circle with equation $(x - 6)^2 + (y - 5)^2 = 25$ that is nearest the point (-2,11).

9. (20%) Each of a set of four congruent circles is internally tangent to a fifth circle and externally tangent to exactly two other circles in the set. If the radius-length of each of the congruent circles is $m(\sqrt{2} - 1)$, find the radius-length of the fifth circle in terms of m.

10. (47%) If $a < b$, find the ordered pair of positive integers (a,b) that satisfies

$$\sqrt{10 + \sqrt{84}} = \sqrt{a} + \sqrt{b}\ .$$

11. (8%) The length of each side of square $ABCD$ is 12. Points E, F, G, and H are the trisection points of $\overline{AB}$, $\overline{BC}$, $\overline{CD}$, and $\overline{DA}$, nearer A, B, C, and D, respectively. Find the area of the region bounded by $\overline{AF}$, $\overline{BG}$, $\overline{CH}$, and $\overline{DE}$.

12. (32%) Find the value of the infinite continued fraction

$$\cfrac{1}{2 + \cfrac{1}{2 + \cdots}},$$

where the dots indicate an endless repetition of the indicated pattern.

13. (52%) Find all real values of x that satisfy

$$\frac{x^3 - x^2 - x + 1}{x^3 - x^2 + x - 1} = 0.$$

14. (46%) If $A > 0$ and if $(x + 1)(x + 2)(x + 3)(x + 4) + 1 = (Ax^2 + Bx + C)^2$, find the ordered triple (A,B,C).

15. (29%) Circle O has radius-length 3, and point A, coplanar with the circle, is 5 units from the center O. One tangent from A to the circle intersects the circle at point T. If X is a point on $\overline{AT}$ such that $\overline{XO}$ intersects the circle at a point Y with $XA = XY$, find the length of $\overline{XA}$.

16. (39%) If $a > 0$, $b > 0$, $a \neq b$, and $\dfrac{a\sqrt{b} + b\sqrt{a}}{a\sqrt{b} - b\sqrt{a}} - \dfrac{a\sqrt{b} - b\sqrt{a}}{a\sqrt{b} + b\sqrt{a}} = \sqrt{ab}$, write an equation expressing a explicitly in terms of b.

17. (27%) The ratio of the width of a rectangle to its length equals the ratio of its length to half its perimeter. If the area of the rectangle is $50(\sqrt{5} - 1)$ and the width of the rectangle is $k(\sqrt{5} - 1)$, compute the value of k.

18. (11%) Let q, r, and s be three distinct real numbers such that $(x - q)(x - r) = x^2 + ax + bc$ and $(x - q)(x - s) = x^2 + bx + ac$. If a, b, and c are real numbers such that $abc(a - b) \neq 0$, write an equation expressing c explicitly in terms of a and b.

19. (36%) Find all real values of x that satisfy $|x| + 3 - |x + 3| = 6$.

20. (8%) Find all positive real values of x that satisfy

$$\frac{1}{x + \sqrt{x}} + \frac{1}{x - \sqrt{x}} \leq 1.$$

21. (35%) The length of $\overline{AB}$ is 12, and X is a variable point on $\overline{AB}$. Squares $AXCD$ and $XBEF$ are drawn in a plane on the same side of $\overline{AB}$. The centers of these squares are Y and Z, respectively, and W is the midpoint of $\overline{YZ}$. As X moves from A to B, find the length of the path traced out by point W.

22. (7%) Find all real values of x that satisfy

$$\frac{3x^3 - 2x^2 + x + 1}{3x^3 - 2x^2 - x - 1} = \frac{3x^3 - 2x^2 + 5x - 13}{3x^3 - 2x^2 - 5x + 13}.$$

23. (39%) In ΔABC, perpendiculars from A to the bisectors of angle B and angle C meet the bisectors in D and E, respectively. The line through D and E intersects $\overline{AC}$ at X and $\overline{AB}$ at Y. What fractional part of the area of ΔABC is the area of region $XYBC$?

24. (20%) Find all real values of x that satisfy $x^4 - 4x^3 + 5x^2 - 4x + 1 = 0$.

25. (31%) If A and B are digits and the base ten numeral $30AB5$ can be expressed as the product $225n$, find all positive integral values of n.

26. (25%) If $a = \sqrt{.16}$, $b = \sqrt[3]{.0639}$, $c = \sqrt[6]{.0041}$, and $d = (.2)^2$, arrange a, b, c, and d in increasing order.

27. (18%) In ΔABC, C' is on $\overline{AB}$ such that $AC':C'B = 1:2$, and B' is on $\overline{AC}$ such that $AB':B'C = 3:4$. If $\overline{BB'} \cap \overline{CC'} = \{P\}$ and if $\{A'\} = \overrightarrow{AP} \cap \overline{BC}$, find $AP:PA'$.

28. (18%) The equation of line m is $3x + 4y = 12$ in a rectangular coordinate system. If the line whose equation is $3x + 4y = k$ is 2 units from m, find the two possible values of k.

29. (26%) In isosceles right triangle ABC, M is the midpoint of hypotenuse $\overline{AB}$. An equilateral triangle has one vertex on $\overline{AC}$, one on $\overline{BC}$, and one at M. If $AB = 24$ and if the side-length of the equilateral triangle is $k(\sqrt{3} - 1)$, find k.

30. (10%) If $x^2 + y^2 + z^2 = a^2$ and $x + y + z = x^3 + y^3 + z^3 = a$, write an equation expressing xyz explicitly in terms of a.

Problems—Fall 1975

1. (9%) Let $[x]$ denote the greatest integer n such that $n \le x$. Let $f(x) = [x/12\frac{1}{2}] \cdot [-12\frac{1}{2}/x]$. If $0 < x < 90$, then the range of f consists of k elements. Find the value of k.

2. (15%) ABC and CBA are, respectively, the base nine and base seven numerals for the same positive integer. Express this integer using a base ten numeral.

3. (45%) Find all ordered pairs of integers (x,y) that satisfy $x^2 + y^2 \leq 25$ and $y = x - 3$.

4. (22%) Some people agree to share in the cost of buying a boat. If ten of them later decide not to buy in, each of those remaining would have to chip in one dollar more. If the sole payment actually occurs after an additional fifteen people drop out, each of those ultimately remaining has to pay two dollars more than he would have had to pay had only the first ten dropped out. How many people originally agreed to buy the boat?

5. (34%) The lengths of the sides of a triangle are 6, 8, and 10. Find the area of the larger of the two triangles into which the bisector of the larger acute angle partitions the original triangle.

6. (16%) The roots of $x^3 + 2px^2 - px + 10 = 0$ are integral and form an arithmetic progression. Find the value of p.

7. (44%) A train, x meters long, traveling at a constant speed, takes 20 seconds from the time it first enters a tunnel 300 meters long until the time it completely emerges from the tunnel. One of the stationary ceiling lights in the tunnel is directly above the train for 10 seconds. Find the value of x.

8. (14%) If a, b, and c are different numbers and if $a^3 + 3a + 14 = 0$, $b^3 + 3b + 14 = 0$, and $c^3 + 3c + 14 = 0$, find the value of $\frac{1}{a} + \frac{1}{b} + \frac{1}{c}$.

9. (40%) If the lengths of the bases of a trapezoid inscribed in a circle are 10 and 20 and the length of one of the legs is $\sqrt{89}$, find the length of one diagonal.

10. (8%) If i represents the imaginary unit, find all ordered pairs of real numbers (a,b) that satisfy

$$a + bi = x + \frac{1}{x} \text{ and } x^3 + \frac{1}{x^3} = 110.$$

11. (46%) The nth term of a sequence is represented by a_n. If $a_1 = 1$, $a_2 = 3$, $a_3 = 5$, and if $a_n = a_{n-1} + a_{n-2} - a_{n-3}$ for all $n > 3$, find the sum of the first thirty terms of the sequence.

12. (45%) In ΔABC, D is on $\overline{AB}$ and E is on $\overline{BC}$. Let $\overline{CD} \cap \overline{AE} = \{K\}$ and let $\overrightarrow{BK} \cap \overline{AC} = \{F\}$. If $AK:KE = 3:2$ and $BK:KF = 4:1$, find $CK:KD$.

13. (16%) The sum of the ages of a father, a mother, and a daughter is 73. When the father will be twice as old as the daughter, the sum of their two ages will be 132. Write an equation expressing the mother's present age, m, in terms of the daughter's present age, d.

14. (9%) The bases $\overline{AB}$ and $\overline{CD}$ of isosceles trapezoid $ABCD$ are 12 units apart, $AB = 10$, and $CD = 8$. Point R is on the axis of symmetry of the trapezoid so that angle CRB measures 90°. Find the possible distances from R to AB.

15. (36%) The distances from a point on the circumcircle of an equilateral triangle to the two nearest vertices are 3 and 6. Find the distance from the point to the furthest vertex.

16. (16%) If $a_1 = X$, $a_2 = X^{a_1}$, and, in general, $a_n = X^{a_{n-1}}$, calculate $a_{1,000,000}$ correct to the nearest thousandth when $X = \left(\frac{4}{3}\right)^{3/4}$.

17. (80%) On hypotenuse $\overline{AB}$ of right triangle ABC, D is the point for which $CB = BD$. If angle B measures 40°, find the measure of angle ACD.

18. (25%) A two-person game is played in which each player, in turn, removes 1, 2, 3, 4, 5, 6, or 7 toothpicks from a common pile, until the pile is exhausted. The player taking the last toothpick LOSES. If the starting pile contains 1000 toothpicks, how many toothpicks must the first player take on his first turn to guarantee a win with perfect subsequent play?

19. (20%) Find all integral values of x for which $\dfrac{12(x^2 - 4x + 3)}{x^3 - 3x^2 - x + 3}$ has a positive integral value.

20. (39%) A piece of paper in the shape of an equilateral triangle ABC has $AB = 8$. When A is folded over to any point on $\overline{BC}$, a crease of length x is formed which joins some point of $\overline{AC}$ to some point of $\overline{AB}$. As A is moved along $\overline{BC}$ from B to C, x varies. If the greatest value of x is M and the least value of x is m, find the ordered pair (M,m).

21. (5%) The coordinates of the vertices of pentagon $ABCDE$ are $A(0,0)$, $B(11,0)$, $C(11,2)$, $D(6,2)$ and $E(0,8)$. The line whose equation is $x = k$ partitions this pentagonal region into two regions of equal area. If $k = a + b\sqrt{6}$, find the ordered pair of integers (a,b).

22. (09%) Two men, A and B, start at the same point, walk in opposite directions, and reach their respective destinations, X and Y, each in one hour. Had they exchanged destinations from the start, A would have arrived at Y 35 minutes after B would have arrived at X. Find the ratio of A's speed to B's speed.

23. (30%) Find the integer x for which $x^5 = 656{,}356{,}768$.

24. (22%) The length of each side of ΔABC is 6. If X is the trisection point of $\overline{CA}$ nearer C and if median $\overline{AM}$ intersects $\overline{BX}$ at U, then $MU = k\sqrt{3}$. Find k.

25. (41%) A unit fraction is a fraction of the form $\frac{1}{n}$, where n is an integer greater than 1. Find the two largest unit fractions, each of which is the square of a rational number, whose sum is also the square of a rational number.

26. (17%) If $a \neq 0$ and if $x + y = a$ and $x^3 + y^3 = b$, write an equation expressing $x^2 + y^2$ explicitly in terms of a and b.

27. (23%) Find all ordered pairs of real numbers (a,b) for which $3\sqrt{x - 2y} + \dfrac{3}{\sqrt{x - 2y}} = 10$ and $x = ay + b$.

28. (2%) A piece of paper in the shape of an equilateral triangle ABC has $AB = 15$. When A is folded over to the point D on $\overline{BC}$ for which $BD = 3$, a crease is formed along a line that joins a point on $\overline{AB}$ to a point on $\overline{AC}$. The length of the crease is $\frac{1}{2}\sqrt{n}$. Find n.

29. (50%) The length of diameter $\overline{AB}$ of a circle is 12, and its trisection points are C and D. Semicircles are drawn on $\overline{AC}$ and $\overline{AD}$ as diameters on the same side of $\overline{AB}$, and on $\overline{BD}$ and $\overline{BC}$ as diameters, on the other side of $\overline{AB}$. These four semicircles partition the circle into three regions. If the area of the middle region is $k\pi$, find k.

30. (59%) In the five-digit base ten numeral $ABCDE$ $(A \neq 0)$, different letters do not necessarily represent different digits. If this numeral is the fourth power of an integer and if $A + C + E = B + D$, find the digit C.

Problems—Spring 1976

1. Find the sum of the infinite series whose n^{th} term is $\dfrac{7^{n-1}}{10^n}$.

2. In ΔABC, angle C measures 60° and angle $A >$ angle B. The bisector of angle C intersects $\overline{AB}$ at E. If CE is a mean proportional between AE and EB, find the value of AE/AC.

3. In ΔABC, D is on $\overline{AB}$ so that $AD:DB = 1:2$, and G is on $\overline{CD}$ so that $CG:GD = 3:2$. If $\overrightarrow{BG}$ intersects $\overline{AC}$ at F, find $BG:GF$.

4. In base fifty, the integer x is represented by the numeral CC and x^3 is represented by the numeral $ABBA$. If $C > 0$, express all possible values of B in base ten.

5. In ΔABC, angle A measures 120°, $AB + BC = 21$, and $AC + BC = 20$. Find BC.

6. One of the roots of

$$\frac{x^2 + 1}{x} + \frac{x}{x^2 + 1} = \frac{29}{10} \quad \text{is} \quad \frac{1 + \sqrt{k}}{5},$$

where k is a negative integer. Find k.

7. (57%) If 100 more than the sum of n consecutive integers is equal to the sum of the next n consecutive integers, find n.

8. (15%) External squares are drawn on each side of a rhombus, and the centers of these squares are joined to form convex quadrilateral Q. If the length of each side of the rhombus is 6 and if one of its angles has a degree-measure of 30, find the area of quadrilateral Q.

9. (17%) A lattice point in a rectangular coordinate plane is a point both of whose coordinates are integers. How many lattice points are there on the circle that is the graph of the equation $x^2 + y^2 = 625$?

10. (18%) A proper divisor of a positive integer is a positive integral divisor of the integer, other than the integer itself. Find the largest proper divisor of 1,030,301.

11. (28%) The points of intersection of the graphs of $xy = 20$ and $x^2 + y^2 = 41$ are joined to form a convex quadrilateral. Find the area of the quadrilateral.

12. (47%) A merchant bought some oranges at the rate of 3 for 16 cents. He bought twice as many oranges at the rate of 4 for 21 cents. To make a profit of 20%, based on his investment, he sold them all at the rate of 3 for k cents. Find k.

13. (86%) A club found that it could achieve a membership ratio of 2 adults for each minor either by inducting 24 adults or by expelling x minors. Find x.

14. (44%) The length of each side of ΔABC is 10, and a circle of radius-length 3 is tangent to $\overline{AB}$ and to $\overline{AC}$. If the distance from the center of the circle to $\overline{BC}$ is $a\sqrt{3} - b$, find the ordered pair of rational numbers (a,b).

15. (55%) In ΔABC, D is on $\overline{BC}$ so that $BD:DC = 3:2$, and E is on $\overline{AD}$ so that $AE:ED = 5:6$. If $\overline{BE}$ intersects $\overline{AC}$ at F, find $BE:EF$.

16. (29%) Find the sum of the infinite series $a_1 + a_2 + a_3 + \ldots + a_n + \ldots$, where, for m a positive integer,

$$a_n = \begin{cases} \dfrac{1}{2^{n-1}} & \text{if } n = 4m - 3 \text{ or } n = 4m - 2 \\[2ex] \dfrac{-1}{2^{n-1}} & \text{if } n = 4m - 1 \text{ or } n = 4m. \end{cases}$$

17. (35%) Let a, b, and c be positive integers such that a is the cube of an integer, $c = b + 1$, and $a^2 + b^2 = c^2$. Find the least possible value of c.

18. (13%) In rectangle $ABCD$, $AB = 6$ and $BC = 8$. Equilateral triangles ADE and DCF are drawn on the exterior of the rectangle. If the area of ΔBEF is $a\sqrt{3} + b$, find the ordered pair of rational numbers (a,b).

19. (78%) An integer is represented by a two-digit base ten numeral. If three times the sum of its digits is added to the integer, the result is the original integer with digits reversed. Find all such positive integers.

20. (32%) In a triangle, the lengths of the three medians are 9, 12, and 15. Find the length of the side to which the longest median is drawn.

21. (28%) The points $A(-1,4)$ and $B(2,-3)$ are in a rectangular coordinate system. Point C is on $\overline{AB}$ with $AC:CB = 3:4$. Find the coordinates of point C.

22. (32%) Find all ordered triples of real numbers (x,y,z) that satisfy

$$\sqrt{x - y + z} = \sqrt{x} - \sqrt{y} + \sqrt{z},$$
$$x + y + z = 8, \text{ and}$$
$$x - y + z = 4.$$

23. (54%) Let $\overline{AB}$ be a leg of the right triangle of least perimeter whose sides have integral lengths, whose hypotenuse is one unit longer than AB, and in which $AB > 100$. Find AB.

24. (18%) In an arithmetic progression whose n^{th} term is a_n, $a_1 = 6$, $a_2 = 8$, and $a_3 = 10$. A second sequence, whose n^{th} term is b_n, has $b_1 = 3$ and, for $n > 1$, $b_n = b_{n-1} + a_{n-1}$. Write an equation expressing b_n explicitly in terms of n.

25. The expression $\sqrt{10 + \sqrt{10 + \sqrt{10 + \ldots}}}$, where the dots indicate an infinite repetition of the indicated pattern of operations, can be expressed in the form $\frac{a + \sqrt{b}}{c}$, where a, b, and c are integers, no two of which have a common prime factor. Find the ordered triple (a,b,c).

26. In a rectangular coordinate system, find the area of the region bounded by the graph of $|x| + |y - 1| = 2$.

27. The altitude and the median from vertex A of ΔABC are 4 and 5 units long, respectively. The angle determined by $\overline{AB}$ and the median is bisected by the altitude. Find AC.

28. The squares of three positive integers are in arithmetic progression, and the third integer is 12 more than the first. Find the three integers.

29. A rectangular sheet of paper, $ABCD$, is folded and creased so that $\overline{AD}$ lies along $\overline{DC}$. The crease formed determines a point E on $\overline{AB}$, and it is found the E is the trisection point of $\overline{AB}$ nearer B. Find $EC:BC$.

30. Find all real values of x that satisfy $(16x^2 - 9)^3 + (9x^2 - 16)^3 = (25x^2 - 25)^3$.

(Due to unusual circumstances, the six problems that follow served as replacements for the six problems immediately preceding.)

31. (23%) If $\sin x + \cos x = -\frac{1}{5}$ and if $\frac{3}{4}\pi \leq x \leq \pi$, find the value of $\cos 2x$.

32. (28%) Find all ordered triples of real numbers (x,y,z) that satisfy

$$yz + xz = 13,$$
$$xz + xy = 25,$$
$$xy + yz = 20.$$

33. (37%) In honor of the economic freeze in the state of New York, government officials are rumored to have put forth a plan to institute one of two new types of thermometers. On these new scales, °F represents degrees Ford and °C represents degrees Carey. It is known that 40°F = 25°C, that 280°F = 125°C, and that degrees Ford is a linear function of degrees Carey. Write an equation expressing °F explicitly in terms of °C.

34. (13%) In ΔABC, $AB = 5$, $BC = 6$, and $AC = 7$. Points P on $\overline{AB}$, Q on $\overline{BC}$, and R on $\overline{AC}$ are located so that $AP = 2$, $BQ = 2$, and $CR = 3$. If the area of ΔABC is x, then the area of ΔPQR is $kx/35$. Find the numerical value of k.

35. (41%) In the sequence 6, x, y, 16, the first three terms form an arithmetic sequence and the last three terms form a geometric sequence. Find all possible ordered pairs (x,y).

36. (38%) The parabola that is the graph of the equation $y = ax^2 + bx + c$ passes through the points $(-1,-11)$, $(1,1)$, and $(2,4)$. Find the coordinates of the vertex of this parabola.

Problems—Fall 1976

1. (35%) For all nonzero real numbers x, the function f is defined by $[f(1 + x)]^{1/x} = k$. If y is a nonzero real number, find the value of $[f(1 + \frac{4}{y})]^{8y}$ in terms of k.

2. (33%) In the complex domain, find all solutions of $(x + 1)^5 = x^5 + 1$.

3. (17%) If $\tan\left(\frac{1}{4}\pi + x\right) = A \sec 2x + B \tan 2x$ is an identity for all real values of x for which both sides of the equation are defined, find the ordered pair of rational numbers (A,B).

4. (19%) In acute triangle ABC, $AC = 14$ and the length of the altitude from B is 12. A square has two of its vertices on $\overline{AC}$ and its other two vertices on $\overline{AB}$ and $\overline{BC}$, respectively. Find the length of a side of this square.

5. (52%) In ΔABC, D is the point on $\overline{BC}$ for which $\overline{AD}$ bisects median $\overline{BE}$. Find $BD:DC$.

6. (8%) Find all ordered triples of real numbers (x,y,z) that satisfy $x + yz = 6$, $y + xz = 6$, and $z + xy = 6$.

7. (55%) Find the maximum area of a triangle in which the length of one side is 6 and the sum of the lengths of the other two sides is 10.

8. (27%) Of all ordered triples of positive integers (x,y,z) that satisfy $2^{3x} + 2^{4y} = 2^{5z}$, find the ordered triple with the smallest value of z.

9. (73%) Find the positive integer whose cube exceeds its square by 4624.

10. (21%) The length of a radius of a circle is $3 + 2\sqrt{3}$. Three congruent circles are drawn in the interior of the original circle, each internally tangent to the original circle and externally tangent to the others. Find the length of a radius of one of the three congruent circles.

11. (45%) If $\log(\log[\log(\log x)]) = 0$, where the base of each logarithm is 10, then $x = 10^k$. Find the positive integer k.

12. (16%) If $x \neq \frac{k\pi}{2}$, where k is an integer, then $\frac{\cot^2 x - \tan^2 x}{2 + \cot^2 x + \tan^2 x} + 2\sin^2 x$ has a minimum value of m and a maximum value of M, for x a real number. Find the ordered pair (m,M).

13. (49%) In ΔABC, D is on $\overline{AB}$ such that $AD:DB = 3:2$ and E is on $\overline{BC}$ such that $BE:EC = 3:2$. If $\overrightarrow{DE}$ intersects $\overrightarrow{AC}$ at F, find $DE:EF$.

14. (19%) Find the value of $\tan x$ if $\sin x + \cos x = \frac{1}{5}$ and $\frac{1}{2}\pi < x < \pi$.

15. (44%) Find the sum of the infinite series $\frac{2}{1} + \frac{1}{3} + \frac{2}{9} + \frac{1}{27} + \frac{2}{81} + \frac{1}{243} + \ldots$ $= \sum_{n=0}^{\infty} \frac{\frac{3}{2} + \frac{1}{2}(-1)^n}{3^n}$.

16. (53%) Two jugs each have a capacity of x gallons. One is filled with wine and the other is filled with water. Two gallons are taken from each jug and are then transferred into the other, after which each jug has its contents thoroughly mixed. If two gallons were now taken from each jug and then transferred into the other, the amount of wine in each jug would be the same. Find x.

17. (63%) Bases $\overline{AB}$ and $\overline{CD}$ of trapezoid $ABCD$ are each perpendicular to leg $\overline{AD}$. If $CD = 6$, $AB = 8$, and $AD = 10$, find the average of the areas of all noncongruent triangles two of whose vertices are A and D and whose third vertex is a point on $\overline{BC}$.

18. (45%) Find all solutions of $x(x^3 - 1) - 6(x^2 + x + 1) = 0$.

19. (27%) In base x, $\frac{1}{5}$ and $.\overline{17}$ are numerals for the same number. Find x.

20. (47%) Half the perimeter of a rectangle exceeds the length of one of its diagonals by one-third the length of the longer side. Find the tangent of the angle formed by its longer side and one of its diagonals.

21. (37%) While B is riding a bicycle from Here to There, C is driving a car from There to Here, each at a steady rate along the same road. They start at the same time and, after passing each other, B takes 25 times as long to complete the journey as C. Find the ratio of the speed of the bicycle to the speed of the car.

22. (20%) If a square and a regular hexagon have equal areas, then the ratio of their respective perimeters is $\sqrt[4]{k}:1$. Find k.

23. (46%) A pair of twin primes is a pair of primes that are consecutive odd integers. Find the largest integer that is a divisor of the sum of the two elements in every pair of twin primes, if these two elements are each primes greater than 3.

24. (21%) If the length of a circumradius of regular octagon $ABCDEFGH$ is 4, the area of the quadrilateral determined by lines $\overleftrightarrow{AB}$, $\overleftrightarrow{CD}$, $\overleftrightarrow{EF}$, and $\overleftrightarrow{GH}$ is $k(2 + \sqrt{2})$. Find k.

25. (39%) Find the value of $\cos\left(\text{Arcsin}\,\frac{4}{5} + \text{Arctan}\,\frac{5}{12}\right)$, where Arc denotes principal value.

26. (12%) A man lives in a building located five blocks south and five blocks west of the building in which his girlfriend lives. Thus, in driving to his girlfriend's home each evening, he drives ten blocks. All the streets are in a rectangular pattern and all are available to him for driving. In how many different ways can he drive from his home to his girlfriend's home, driving only ten blocks?

27. (50%) If $c > a > 0$ and if $a - b + c = 0$, find the larger root of $ax^2 + bx + c = 0$.

28. (32%) A box, in the shape of a rectangular solid, has a length of 12, a width of 8, and a height of 8. What is the length, along the outside surface of the box, of the shortest path that can be drawn between two of the vertices of the box which do not lie in the same face of the box?

29. (33%) The ancient Greeks used $\frac{7}{5}$ as an approximation of $\sqrt{2}$. The ratio of the absolute value of the approximation error to $\sqrt{2}$ is $\left(1 - \frac{\sqrt{k}}{10}\right):1$. Find k.

30. (18%) The sides of a right triangle all have integral lengths, and the sum of the lengths of one of the legs and the hypotenuse is 49. Find all possible lengths for the other leg.

Problems—Spring 1977

1. (18%) In a triangle, segments are drawn from one vertex to the trisection points of the opposite side. A median drawn from a second vertex is divided, by these segments, into the continued ratio $x:y:z$. If $x \geq y \geq z$, find $x:y:z$.

2. (32%) An integer is chosen at random from $\{x \mid 100 < x < 300\}$. Find the probability that this integer is divisible by 7 or 11.

3. (28%) For all real numbers x and y, the function f satisfies $f(xy) = f(x) \cdot f(y)$ and $f(0) \neq 0$. Find the numerical value of $f(1977)$.

4. (3%) If $0 \leq x \leq \pi$ and if $\sin \frac{1}{2}x = \sqrt{1 + \sin x} - \sqrt{1 - \sin x}$, find all possible values of $\tan x$.

5. (25%) Find all ordered pairs of real numbers (x,y) that satisfy $x + y + \sqrt{x + y} = 56$ and $x - y + \sqrt{x - y} = 30$.

6. (36%) In the binomial expansion of $(x + y)^3$, the sum of the two middle terms is equal to the sum of the first and last terms. If $x + y \neq 0$, find the numerical value of the ratio $(x - y)^2 : xy$.

7. (22%) Find the ordered pair of positive numbers (x,y) that satisfies $x - y = xy = x^2 - y^2$.

8. (15%) If $\sin^6\theta + \cos^6\theta = \frac{2}{3}$, find all possible values of $\sin 2\theta$.

9. (68%) For all $x > 0$ and all $y > 0$, the function f satisfies $f(xy) = f(x) + f(y)$. If $f(2) = a$ and $f(3) = b$, find the value of $f(72)$ explicitly in terms of a and b.

10. (47%) Find the ordered pair of rational numbers (x,y) that satisfies $\sqrt{\frac{21}{4} + 3\sqrt{3}} = x + \sqrt{y}$.

11. (35%) The three-digit base x numeral $7y3$ is twice the three-digit base x numeral $3y7$. If x is a positive integer, express the sum $x + y$ as a base ten numeral.

12. (33%) In an infinite sequence of regular hexagons, each hexagon except the first is formed by connecting the midpoints of the sides of the previous hexagon. If the perimeter of the first hexagon is 12, find the sum of the perimeters of all the hexagons.

13. (15%) If Arc denotes principal value, find all real values of x that satisfy $\text{Arctan } 2x + \text{Arctan } 3x = \frac{1}{4}\pi$.

14. (51%) Two different-sized vertical poles are staked into level ground. The two laser beams from the top of each pole to the ground-level base of the other pole cross at a point 24 meters above ground. If the top of the shorter pole is 40 meters above ground and the top of the taller pole is x meters above ground, find x.

15. (80%) Let f and g be functions defined by $f(x) = 9x + 1$ and $g(x) = x^2$. Find all values of x that satisfy $f[g(x)] = g[f(x)]$.

16. (28%) Every expression of the form $a^2b^2 + b^2c^2 + c^2d^2 + d^2a^2$ can be expressed as the sum of two squares in at least two different ways. Find any one of the three possible ordered pairs of positive integers (x,y), with $x > y$, that satisfies $x^2 + y^2 = 44^2 \cdot 10^2 + 10^2 \cdot 33^2 + 33^2 \cdot 5^2 + 5^2 \cdot 44^2$.

17. (46%) If $xy \neq 0$, express in simplest form the value of

$$\left(x + \frac{1}{x}\right)^2 + \left(y + \frac{1}{y}\right)^2 + \left(xy + \frac{1}{xy}\right)^2 - \left(x + \frac{1}{x}\right)\left(y + \frac{1}{y}\right)\left(xy + \frac{1}{xy}\right).$$

18. (65%) When the system $3x - 4y < 6$ and $6x + 4y = 15$ is solved, all solutions (x,y) have the form $\left(\frac{7 - r}{3}, \frac{1 + s}{4}\right)$. Find the ratio $r:s$.

19. (62%) Find the value of x that satisfies $\log_x 25 - \log_x 4 = \log_x \sqrt{x}$.

20. (11%) Find all ordered pairs of positive integers (x,y) that satisfy $x^2 + x + 29 = y^2$.

21. (35%) Find all ordered pairs of real numbers (x,y) that satisfy $x^3 - y^3 = 19$ and $x^2y - xy^2 = 6$.

22. (25%) A circle is inscribed in a 3-4-5 triangle. A segment is drawn from the smaller acute angle to the point of tangency on the opposite side. This segment is divided in the ratio $p:q$ by the segment drawn from the larger acute angle to the point of tangency on its opposite side. If $p > q$, find $p:q$.

23. (45%) If Arc denotes principal value and if $x = \text{Arcsin}\,\frac{3}{5} + \text{Arccos}\,\frac{15}{17}$, find the value of $\sin x$.

24. (57%) If $x > 1$, the equation $\sqrt{x + \sqrt{2x - 1}} - \sqrt{x - \sqrt{2x - 1}} = \sqrt{k}$ is an identity. Find, in simplest form, the value of k.

25. (54%) Let N be the base x numeral representing the number $\frac{x^5 - 1}{x - 1}$, where x is an integer greater than 1. Find N.

26. (12%) A grain of sand, when at the point (x,y) of a coordinate plane, is allowed to be moved to the point $(x + 1,y)$ or to the point $(x,y + 1)$, but not both. If the grain starts at (0,0) and moves to (4,4), find the probability that it passes through (2,2).

27. (75%) Traveling at respective rates of a and b kilometers per hour, A and B head directly towards each other across a distance of 120 kilometers. If both start at 10:30 A.M., they will meet at noon. If A starts at 10 A.M. and B starts at 11 A.M., they will also meet at noon. Find the ordered pair (a,b).

28. (38%) In ΔABC, the lines containing the bisector of angle B and the bisector of an exterior angle at C intersect at Q. A line, through Q, drawn parallel to $\overline{BC}$, intersects $\overline{AC}$ at E and $\overline{AB}$ at D. If $BD = 8$ and $EC = 6$, find ED.

29. (68%) Let x be a nonintegral positive number and let $[x]$ denote the greatest integer n such that $n \leq x$. Find the value of $([x] + [-x])^5$.

30. (22%) If $0 < x < \pi$, find all values of x that satisfy $\sin 5x + \sin 3x = 0$.

Problems—Fall 1977

1. (33%) Find all ordered pairs of real numbers (x,y) that satisfy $2x^2 - 2xy + y^2 = 2$ and $3x^2 + 2xy - y^2 = 3$.

2. (3%) In parallelogram $ABCD$, angle A is acute and $AB = 5$. Point E is on $\overleftrightarrow{AD}$ with $AE = 4$ and $BE = 3$. A line through B, perpendicular to $\overleftrightarrow{CD}$, intersects $\overleftrightarrow{CD}$ at F. If $BF = 5$, find EF.

3. (14%) A convex quadrilateral is inscribed in a circle so that one of its sides is a diameter of the circle. The lengths of the two sides of the quadrilateral which have exactly one endpoint on this diameter are 7 and 20. If the length of the longest side of this quadrilateral is 25, find the length of the fourth side of the quadrilateral.

4. (22%) Two drivers, A and B, were 225 kilometers apart. They traveled towards each other at the same constant speed of x km per hour, with A having had a head start of 30 minutes. Upon meeting, each continued to the other's starting point at a constant speed of $x - 10$ km per hour. If A completed the entire trip in five hours, find x.

5. (11%) The polynomials $x^3 + ax^2 + 9x + 6$ and $x^3 + bx^2 + 6x + 3$ have a common quadratic factor over the set of polynomials with integral coefficients. Find the ordered pair of integers (a,b).

6. (49%) Alphonse and Beauregard each roll one fair cubical die, numbered from 1 through 6. Find the probability that Alphonse rolls a higher number than Beauregard.

7. (23%) Find all possible third terms of a five-term geometric progression of real numbers, the sum of whose terms is sixteen times the sum of their reciprocals.

8. (32%) The lengths of the sides of a right triangle are the integers a, b, and c, these integers having no common prime factor. If $a < b < c$ and $(c - a):b = 4:5$, find the ordered triple (a,b,c).

9. (22%) If $\cos 3x + \cos x = 0$, find all possible values of $\cos x$.

10. (3%) Find all ordered pairs of real numbers (x,y) that satisfy

$$(2x - y)^3 + (x - 2y)^3 = 27(x - y)^3 \text{ and } \sqrt{\frac{x + 1}{y}} + \sqrt{\frac{y}{x + 1}} = \frac{5}{2}.$$

11. (55%) For every real number x, $[x]$ denotes the greatest integer less than or equal to x. Find all values of x in the interval $1 \leq x < 2$ that satisfy $[x]^2 = [x^2]$.

12. (6%) Two coplanar right triangles, whose interiors have no points in common, share a common hypotenuse whose length is 25. If the lengths of the legs of these two triangles are all integers, find all possible distances between the vertices of their right angles.

13. The equation $x^3 - x^2 + mx + 1 = 0$ has two equal real roots, distinct from the third root. Find the value of m.

14. If x, y, z, a, b, and c are nonzero real numbers satisfying $(4x^2 + 9y^2 + z^2)(a^2 + b^2 + c^2) = (2ax + 3by + cz)^2$, find the continued ratio $x:y:z$ in terms of a, b, and c.

15. Two distinct numbers are randomly selected from among the first 64 positive integral perfect squares. Find the probability that these two numbers are both cubes of integers.

16. The diagonals of a parallelogram partition it into four triangles, one of which has G as its centroid (intersection of its medians). If the area of the parallelogram is 180, find the largest possible area of a triangle one of whose vertices is G and one of whose sides is a side of the parallelogram.

17. Express in simplest form, the numerical value of $\dfrac{\sin 40° + \sin 80°}{\cos 40° + \cos 80°}$.

18. If $2^{22} = 4{,}194{,}304$, find the sum

$$1 \cdot 2 + 2 \cdot 2^2 + 3 \cdot 2^3 + \ldots + 21 \cdot 2^{21} = \sum_{n=1}^{21} n \cdot 2^n.$$

19. (83%) Find the positive value of x that satisfies $4^{\log_2 x} + x^2 = 8$.

20. (25%) If $n > 1$, find the two smallest integral values of n for which $x^2 + x + 1$ is a factor of $(x + 1)^n - x^n - 1$, over the set of polynomials with integral coefficients.

21. (37%) The sum of two of the three roots of $x^3 + ax^2 + bx + c = 0$, where a, b, and c are real constants, is zero. Write an equation expressing c explicitly in terms of a and b.

22. (1%) There are two circles that pass through (1,9) and (8,8) and are tangent to the x-axis. Find the lengths of their radii.

23. (19%) In an isosceles triangle, the lengths of the base and the median to one leg are equal. Find the cosine of the vertex angle.

24. (17%) A function f is defined on the positive integers by $f(1) = 0$ and $f(n) = 3f(n - 1) + 1$ for all $n > 1$. Express the numerical value of $f(1003) - f(1001)$ in the form $(p^r)(q^s)$, where p and q are positive primes and r and s are integers.

25. (53%) The lengths of the sides of a triangle are 13, 37, and 40. Find the length of the altitude to the longest side.

26. (32%) If $a + b + c = 0$ and $a^3 + b^3 + c^3 = 216$, find the value of abc.

27. (47%) A detachment of z marchers was arranged in a square formation in which there were as many marchers in each row as there were rows. At the last minute, 64 marchers were removed from the detachment, and those who remained could again be arranged into a square formation. If $z > 64$, find all possible values of z.

28. (21%) If $\sin^2 15°$ is one root of $x^2 + bx + c = 0$, find the ordered pair of rational numbers (b,c).

29. (40%) With seamstress X making x dresses per day and seamstress Y making y dresses per day, X requires 5 more days than does Y to make 60 dresses. If each seamstress were to increase her daily output by 1 dress, X would need only 3 days more than Y to make 60 dresses. Find the ordered pair (x,y).

30. Suppose f and g are functions respectively defined by $f(x) = x + 2$ and $g(x) = x$. Find all $x > -2$ for which

$$3^{g(x) \cdot \log_3 f(x)} = f(x).$$

Problems—Spring 1978

1. Three different numbers are randomly selected from among the first ten positive integers. Find the probability that the sum of these three numbers is 15.

2. A circle is inscribed in a 3-4-5 triangle. A second circle, interior to the triangle, is tangent to the first circle and to both sides of the larger acute angle of the triangle. If the length of a radius of the second circle is r, find the numerical value of $\dfrac{1 + r}{1 - r}$.

3. The lengths of the sides of a triangle are 4, 13, and 15. Find the sine of the angle opposite the side whose length is 13.

4. In base eight, the four-digit numeral $BBCC$ is the square of the two-digit numeral AA. Find the ordered triple of digits (A,B,C).

5. Find the length of the line segments whose endpoints have polar coordinates $(7,40°)$ and $(15,100°)$.

6. Find the first of 100 consecutive odd integers whose sum is 100^{100}.

7. Find the real value of x that satisfies $\log_3 x + \log_9 x + \log_{81} x = 7$.

8. In arranging the ordered pairs of positive integers thusly: (1,1), (1,2), (2,1), (1,3), (2,2), (3,1), (1,4), ..., if two ordered pairs have a different element-sum, the one with the smaller element-sum comes first. If they have the same element-sum, the one with the smaller first element comes first. In this arrangement, find the 1978th ordered pair.

9. Find all ordered pairs of rational numbers (x,y) that satisfy $6x(2x + 1) + 1 = y^3$. (You may want to use this special case of Fermat's Last Theorem: If $a^3 + b^3 = c^3$ has a solution in integers, then $abc = 0$.)

10. In a right triangle whose legs have integral lengths a and b and whose hypotenuse has integral length c, if $a = 3$ and $c = 5$, then $a + c = 8$, a perfect cube. Find the least value of c greater than 5 for which a and c are relatively prime odd integers whose sum is the cube of an integer.

11. The longer base of an isosceles trapezoid is a chord of a circle, and the shorter base is tangent to the circle. If the length of one leg is 5 and the lengths of the bases are 24 and 18, find the area of the circle.

12. Find the simplified numerical value of $\tan 20° + \tan 40° + \sqrt{3} \tan 20° \tan 40°$.

13. If $\sin 2x = \frac{24}{25}$, find the value of $\sin^4 x + \cos^4 x$.

14. For all complex numbers x, the polynomial function P is defined by $P(x + 2) - P(x) = 6$ and $P(0) = 2$. Write an equation expressing $P(x)$ explicitly in terms of x.

15. All the circles in an infinite chain of successively tangent circles of decreasing size are tangent to both sides of a 60° angle. If the length of a radius of the largest of the circles in the chain is 105, find the sum of the lengths of the radii of all the circles in the chain.

16. The sum of the squares of the first and fourth terms of an arithmetic progression is 200, while the sum of the squares of the second and third terms is 136. Find the product of these four terms.

17. If a, b, and c are positive numbers, find the numerical value of the product $\log_a b^2 \cdot \log_b c^2 \cdot \log_c a^2$.

18. Find the numerical value of b for which the length of the path from $A(0,2)$ to $B(b,0)$ to $C(c,10)$ to $D(5,9)$ will be a minimum.

19. In an isosceles triangle with a 30° vertex angle, the perpendicular bisector of one leg divides the other leg into the ratio $k:2$. If $k \geq 2$, find k.

20. The function f is defined by $f(1) = 1$ and $f(n) = f(n - 1) + 2n$ for every positive integer $n \geq 2$. Write an equation expressing $f(n)$ explicitly in terms of n.

21. After witnessing a four-car "Demolition Derby," a young man constructed the alphametic $(4)(NEW) = DENT$, in which NEW is odd and $T = 4$. If NEW and $DENT$ respectively represent three-digit and four-digit base ten numerals (with digits N, E, W and D, E, N, T, respectively), find $N + E + W$.

22. Find the fundamental period of the function defined by $f(x) = \sin^4 x + \cos^4 x$, where x is a real number.

23. The value of t \$2 bills and f \$5 bills is \$87. If all ordered pairs (t,f) are equally likely, find the probability that both t and f are prime numbers.

24. The decimal expansion of $\sum_{n-1}^{\infty} \frac{2^n}{10^{2n}}$ begins .02040816. Find the product of the next four digits of this expansion.

25. In angle ABC, angle $A = 45$ and angle $C = 30$. If altitude $\overline{BH}$ intersects median $\overline{AM}$ at P, then $AP:PM = 1:k$. Find k.

26. Find all ordered pairs of positive integers (x,y) that satisfy $x^2 + y^2 - xy = 49$.

27. Two students toss a fair coin. Whenever it lands heads up, the first student wins one dollar. If not, (s)he loses one dollar. Find the probability that the first student's net winnings after five tosses is exactly one dollar.

28. In terms of n, the sum of the squares of the first n positive integers is $\frac{n^3}{3} + \frac{n^2}{2} + \frac{n}{6}$. If the sum of the squares of the first n positive odd integers is $an^3 + bn^2 + cn + d$, find the ordered 4-tuple (a,b,c,d).

29. If x is a real number, find the minimum value of $(x + 8)(x - 8)(x + 6)(x - 6)$.

30. A circle is inscribed in a right triangle the lengths of whose legs are 30 and 40. Find the length of the line segment whose endpoints are the vertex of the right angle and the point of tangency, on the hypotenuse, of the inscribed circle.

Problems—Fall 1978

1. (49%) In the isosceles triangle, the length of the base is 30 and the length of an altitude to one of the legs is 24. Find the length of one of the legs.

2. (47%) Find all integral values of x that satisfy

$$\frac{2}{x + \sqrt{2 - x^2}} + \frac{2}{x - \sqrt{2 - x^2}} = x.$$

3. (33%) When 16 is subtracted from a three-digit number (greater than 100) and the resulting difference is divided by 2, the result is a three-digit number (greater than 100) whose digits are those of the original number, but in reverse order. If the sum of the three digits is 20, find the original number.

4. (7%) The degree-measure of angle ABC is 120, and $AB = 1$. Lying on $\overline{BC}$ are the points $B_1, B_2, B_3, \ldots, B_n$ such that the distance from A to B_k is $\sqrt{k}$. If the distance from B_k to B_{3k} is 2, find k.

5. (52%) Which of the following integers is the largest?

$$5^{3^2},\quad 2^{3^5},\quad 3^{5^2}$$

6. (1%) A fair coin is tossed 17 times. Determine the probability of obtaining at least 11 consecutive heads.

7. (39%) A middle-aged man, noting that his present age was a prime number, observed that the next occurrence of this kind for him was exactly as far away as was the most recent one of this kind. The last time he could have made such an observation was when he was five years old. How old is the man (in years)?

8. (22%) In a sequence whose first term is 1, the nth term, a_n, is recursively defined by $a_n = 3a_{n-1} + 1$, $n > 1$. Find a_{1978}.

9. (47%) Find all values of x that satisfy

$$6\left(x + \frac{1}{x}\right)^2 - 35\left(x + \frac{1}{x}\right) + 50 = 0.$$

10. (5%) Find the numerical value of $\sin 40° \sin 80° + \sin 80° \sin 160° + \sin 160° \sin 320°$.

11. (49%) If $\log_{10} 3 = a$ and $\log_{10} 7 = b$, find the value of $\log_7 9$ in terms of a and b.

12. (16%) Point C lies on a circle one of whose diameters is $\overline{AB}$. The bisector of angle CAB intersects $\overline{BC}$ at D and intersects the circle at E. If $BD = 25$ and $CD = 7$, find BE.

13. (46%) Find the least positive integral value of z^3 for which there exist different positive integers x, y, and z that satisfy $x^2 + y^2 = z^3$.

14. (29%) Find the value of x that satisfies $\log_3 x = (-2 + \log_2 100)(\log_3 \sqrt{2})$.

15. (89%) For all real numbers x and y, the function f has the property that $f(x + y) = f(x) + f(y)$. If $f(1) = 3$, find the value of $f(10)$.

16. (81%) In a triangle, the degree-measure of one angle is 60 more than that of another. The ratio of the lengths of the sides opposite these two angles is 2:1. Find the degree-measure of the largest angle of this triangle.

17. (29%) Mr. Smith left on a trip very early one morning. Not wishing to wake Mrs. Smith, Mr. Smith packed in the dark. He had socks that were alike except for color, and his socks came in six different colors. Find the least number of socks he would have had to pack to be sure of getting at least four matching pairs.

18. (25%) Consider eight points that are equally spaced on a circle in which the length of a radius is 1. Find the product of the distances from one of the points to each of the other seven points.

19. (39%) The points with rectangular coordinates (1,1), (3,5), and (-1,4) are three vertices of a parallelogram, Find all possible ordered pairs that could represent the coordinates of the fourth vertex of this parallelogram.

20. (13%) In calculating his income tax, a teacher found that he could calculate his federal tax as 20% of his income after state taxes were deducted, and he could calculate his state tax as 10% of his income after federal taxes were deducted. If the teacher's income before taxes was $9800, how much federal tax did he pay (in dollars)?

21. (19%) Using only multiplication, find the least number of multiplications required to calculate x^{31}, given the real number x.

22. (55%) For any fixed real numbers a, b, and c, find the maximum number of different real solutions for x in the equation

$$\sqrt[3]{x - a} + \sqrt[3]{x - b} = \sqrt[3]{x - c}.$$

23. (59%) A lattice point in a rectangular coordinate plane is defined as a point both of whose coordinates are integers. Find the number of lattice points in the region defined by $|x| + |y| < 5$.

24. (43%) Find all values of a for which the set of simultaneous equations $x + y + z = a$, $x^2 + y^2 + z^2 = a$, $x^3 + y^3 + z^3 = a$, $x^4 + y^4 + z^4 = a$ is consistent (i.e., has a common solution).

25. (53%) If $[k]$ denotes the greatest integer $\leq k$, then the set of all real numbers x that satisfy $[x^2 - (x - 1)^2] = 10$ is $\{x | a \leq x < b\}$. Find the ordered pair (a,b).

26. (30%) If $a < 0$, the tangents from $(0,a)$ to the parabola $y = x^2$ are perpendicular. Find a.

27. (23%) Find all real numbers x such that: $(1 - \sqrt{x^2 - 1} - \sqrt{x^2 + 2x + 1})$

$$+ \frac{1}{(1 - \sqrt{x^2 - 1} - \sqrt{x^2 + 2x + 1})} = 2.$$

28. (65%) In convex quadrilateral $ABCD$, the area of triangle DAB is 1, that of triangle ABC is 6, and that of triangle CDA is 2. Find the area of triangle BCD.

29. (97%) Find three distinct positive integers whose sum equals the product of the largest two of them.

30. (45%) If $(1 + \cos 60° + i \sin 60°)^{12} = a + bi$ where a and b are real numbers and i is the imaginary unit, find the value of b.

Problems—Spring 1979

1. Find the numerical value of $(\log_2 3)(\log_9 4)$, expressing your answer in simplest form.

2. The lengths of the four sides of a trapezoid are 2, 3, $\sqrt{7}$, and $2\sqrt{7}$, with the last two being the lengths of the parallel sides. From a point P within or on the trapezoid, the four perpendiculars $\overline{PA}$, $\overline{PB}$, $\overline{PC}$, and $\overline{PD}$ are drawn to the respective sides of the trapezoid (where one or more of these perpendiculars may have a length of 0). Determine the maximum value of $PA + PB + PC + PD$.

3. On a trip to the supermarket, a shopper bought 25 items, none of which cost more than \$1.10. The sales tax on the total purchase was 65¢. The bill was incorrectly calculated to be \$28.97. If the cashier's error on any one item was less than 10¢, at least how many errors were made by the cashier?

4. Find the degree-measure of the least positive angle that satisfies $\sin 20° + \sin 40° = \sin \theta$.

5. In the cube determined by $1 \le x \le 2$, $1 \le y \le 2$, $1 \le z \le 2$, determine the maximum numerical value of the function f defined by $f(x,y,z) = xyz - 3yz + 2x - 5$.

6. Joe was asking \$1000 for his car, but Moe offered only \$800. Joe scoffed at the offer; but to keep the bargaining going, he offered to narrow the distance between their two offers by 10% by saying that he would accept \$980. Moe said that he too would be willing to narrow the gap by 10% and offered \$818. This bargaining process continued until the two offers, rounded down to the nearest cent, were the same. When the bargaining ceased, to what price were both offers rounded down?

7. The graph of a quadratic function passes through the points with coordinates $(-1,-2)$, $(1,0)$, $(2,7)$, and $(3,y)$. Find y.

8. Find, in degrees, the least positive value of $x + y$ for which $2 \sec 2x = \tan y + \cot y$.

9. Twenty recruits stand in a long column, one behind the other. At a signal from the sergeant, the recruits standing in places 10 and 20 step forward into places 1 and 2 respectively and the others step back. Find the least number of these maneuvers that must occur before the recruit originally in front will be in front once again.

10. There are eight positive integers n for which $2^n + 1$ is a divisor of $2^{210} - 1$. Four of these divisors are $2^1 + 1$, $2^3 + 1$, $2^5 + 1$, and $2^7 + 1$. Find the other four.

11. A man gets 24 miles per gallon on regular gas costing 60¢ per gallon. He can get more miles per gallon by using high test gas costing 70¢ per gallon. How many more miles per gallon must he get with the high test gas to make it cost the same as regular gas, on the basis of miles he can get per dollar's worth of gasoline?

12. Find the largest number of diagonals that, except for their end-points, can lie in the exterior of a simple (non-self-intersecting) planar polygon of 12 sides.

13. Find the ordered pair of numbers (x,y) that satisfies

$$\frac{x}{\sqrt{x} + \sqrt{y}} = 18 \text{ and } \frac{y}{\sqrt{x} + \sqrt{y}} = 2.$$

14. Let b be the largest number such that a rectangle with sides of lengths 1 and b will fit into a unit square so that the sides of the rectangle are not parallel to the sides of the square. Find b.

15. In a polar coordinate system, the vertices of a triangle are $(6,30°)$, $(2,90°)$, and $(4,150°)$. Find the area of this triangle.

16. If $(\sqrt{2} + 1)^{99} = N + r$, where N is a positive integer and $0 < r < 1$, determine (in simplest form) the numerical value of $(N + r)r$.

17. The real roots of $2x^3 - 7x^2 + kx - 2 = 0$ are in geometric progression. Find these three roots.

18. A regular 30-gon (polygon with 30 sides) and a regular 60-gon have congruent radii but different areas. The ratio of the area of the larger polygon to that of the smaller can be expressed as $k \cdot f(\theta)$, where f is a circular (trigonometric) function. Find this ratio.

19. On bus line A, the buses go every 8 minutes, precisely. On bus line B, the buses go every 15 minutes, precisely. Pat is a 10-minute walk from the nearest stop of line A and a 5-minute walk from the nearest stop of line B. To which of these two stops (A or B) should Pat walk to minimize the expected time that must pass before Pat can board a bus?

20. Determine the ordered triple of nonzero numbers (p,q,r) for which p, q, and r are the three roots of $x^3 - px^2 + qx - r = 0$.

21. If $\sin x + \cos x = 1/2$, find the numerical value of $\sin^3 x + \cos^3 x$.

22. In triangle ABC, $CB = 6$. Point E is on $\overline{AC}$ so that angle $CEB = 80°$. Point D is on AB so that angle $BDC = 100°$. The bisectors of angles ECD and DBE intersect at F. Find the numerical value of $CF^2 + FB^2$.

23. If n is a positive integer, find (in terms of n) all values of x greater than 1 that satisfy

$$4x^{1/(2n)} - x^{1/n} - 3 = 0.$$

24. In the canonical (accepted) ten-pin configuration in bowling illustrated below, how many different equilateral triangles are there with vertices at the pins?

25. In a simple code, each letter of the alphabet is assigned its numerical position in the alphabet. A one-word message was received in this code, but was lost. All that the operator remembered was that the message had the form x, $x + 7$, $x + 6$, $x + 5$, that the second letter was a vowel, and that the word was an English word. What was this one-word message?

26. Points O and P on the graph of $x^2 = 20y - 100$ are the centers of circles tangent to both axes. These circles intersect at $(0,k)$. Find k.

27. The expression $\max(a,b)$ means the larger of a and b. If $x > y > z$, then $x + y + z - \max(x,y) - \max(x,z) - \max(y,z) + \max(x,y,z)$ always equals one of x, y, or z. Determine which one.

28. In a circle, two chords of lengths 4 and 11 respectively subtend central angles whose degree-measures are in the ratio 1 to 3 respectively. Determine the length of a radius of the circle.

29. Consider all positive integers between 3000 and 4000 which, when divided by any of the first three primes, leave a remainder of 1. Find the sum of all these integers.

30. Suppose the probability that the Yankees will beat the Mets in a game of baseball is 2/5. If these teams were to meet in a World Series, what would be the probability that the Mets would win by four games to two? (For those who may not know, there are no ties in the game of baseball, and a World Series ends when one team has won four games).

Problems—Fall 1979

1. (17%) The three positive integers 3, 36, and x are such that their arithmetic mean exceeds their geometric mean by 13. Find the value of x.

2. (42%) The number N, represented by the decimal numeral ABC, is divided by the decimal numeral AC. The quotient is 9 and the remainder is 0. What is the maximum value possible for N?

3. (28%) If $\log 80 = a$ and $\log 45 = b$, find $\log 36$ in terms of a and b.

4. (16%) In equilateral triangle ABC, points D, E, and F are on $\overline{AB}$, $\overline{BC}$, and $\overline{CA}$, respectively, with $AD = BE = CF = 1$ and $DB = EC = FA = \sqrt{3}$. The area of triangle DEF is $a + b\sqrt{3}$. Find the ordered pair of rational numbers (a,b).

5. (26%) If $\sin A = \frac{24}{25}$ and $90° < A < 180°$, find $\sin \frac{A}{2}$.

6. (11%) Find all real numbers x that satisfy the equation $(\sqrt{3x+1} + \sqrt{2-x}) + (\sqrt{3x+1} + \sqrt{2-x})^2 + (\sqrt{3x+1} + \sqrt{2-x})^3 = 14$.

7. (67%) The sum of the first six terms of an arithmetic progression whose first term is 1 is equal to the sum of the first six terms of the geometric progression beginning 1, 2, 4, ... Find the sixth term of the arithmetic progression.

8. (19%) If a permutation of the digits 1, 2, 3, 4, 5 is written at random, what is the probability that exactly two of the digits will occur in their correct places? (The digit n is in its correct place if it occupies the nth position in the permutation.)

9. (52%) If $S = \log\frac{1}{2} + \log\frac{2}{3} + \log\frac{3}{4} + \ldots + \log\frac{n}{n+1} + \ldots + \log\frac{99}{100}$, where the base of the logarithms is 10, find the integral value of S in simplest form.

10. (32%) The slopes of lines L_1 and L_2 are $\frac{1}{7}$ and 1. Find the slope of the line that bisects the acute angle formed by lines L_1 and L_2.

11. (33%) If $A + B = 225°$, and $(\cot A)(\cot B) \neq 0$, find the numerical value of $\frac{1 + \cot A}{\cot A} \cdot \frac{1 + \cot B}{\cot B}$.

12. (31%) Write an equation expressing c explicitly in terms of a and b if $a = x + y$, $b = x^2 + y^2$, and $c = x^3 + y^3$.

13. (27%) The largest power of 7 that divides 343! is 7^x. Find the value of x. (For positive integral n, $n! = n(n - 1)(n - 2) \ldots (3)(2)(1)$.)

14. (57%) How many real values of x satisfy the equation $\tan[\text{Arctan } x + \text{Arctan}(x^2)] = x$?

15. (47%) The two legs of an isosceles triangle have length y. What is the maximum area that the triangle can have? Express your answer in terms of y.

16. (13%) Find all values of x for which there is no ordered pair (x,y) satisfying the following simultaneous equations:

$$\frac{x^2 - 2xy - 3y^2}{x - 2y + 1} = 0$$

$$\frac{x^2 - 4xy + 3y^2}{2x - 3y - 2} = 0 .$$

17. (13%) The dimensions of a rectangle are x and y, where x and y are integers. Its area is divided into unit squares by lines parallel to the sides of the rectangle. The number of unit squares touching the sides of the rectangle is equal to the number of unit squares not touching the sides. Find all possible values of the product xy.

18. (10%) When a certain polynomial is divided by $x - 2$, the remainder is 2. When the polynomial is divided by $x + 2$, the remainder is -2. What is the remainder when the polynomial is divided by $x^2 - 4$?

19. (29%) In how many zeroes does the quantity 127! end?

20. (34%) Find all values of x that satisfy the equation:

$$\frac{\sqrt{x} + \sqrt{\frac{1}{x}} - \sqrt{2}}{\sqrt{x} + \sqrt{\frac{1}{x}}} = \frac{\sqrt{x} + \sqrt{\frac{1}{x}}}{\sqrt{x} + \sqrt{\frac{1}{x}} + 3\sqrt{2}}$$

21. (26%) Find n if $n > 2$ and $(n!)^2(n^2 - 2)! = \frac{5}{6}(n^2)!((n - 2)!)^2$.

22. (32%) If $0° < x < 90°$ and $\tan^2 x - 4 \tan x + 1 = 0$, find the numerical value of $\sin 2x$ in simplest form.

23. (37%) Find all ordered triples (x,y,z) of real numbers that satisfy $xyz^3 = 24$, $xy^3z = 54$, and $x^3yz = 6$ simultaneously.

24. (17%) In regular hexagon $ABCDEF$, the bisector of angle AFB meets $\overline{AB}$ at G, and H is a point on the bisector of angle BFE such that $FH = FG$. Find the numerical value of the ratio of the area of triangle FGH to the area of the hexagon.

25. (32%) What is the base of the numeral system in which $\frac{1}{5} = .333\ldots$ (where the digit 3 repeats forever)?

26. (43%) Find the ordered pair of real numbers (x,y) that satisfies simultaneously $x + y + \sqrt{x + y} = 12$ and $x - y + \sqrt{x - y} = 6$.

27. (22%) Four students are about to take a test containing 100 questions of equal difficulty. The probabilities are that the first student will answer 60 questions correctly, the second will answer 50 correctly, and the others will answer 40 and 25 correctly, respectively. What is the probability that at least one student will correctly answer the first question?

28. (36%) Given that $N > 1$ and $\frac{1}{\log_2 N} + \frac{1}{\log_4 N} + \frac{1}{\log_6 N} + \frac{1}{\log_8 N} + \frac{1}{\log_{10} N} = \frac{1}{\log_x N}$, find the value of the integer x.

29. (18%) Points A, B, C, and D lie on the graph of $y^2 + 2xy + x^2 + 3x + 4y + 2 = 0$, and each has an abscissa of 1 or -1. Of these points, let A and C be the furthest apart. If $AC = \sqrt{p + q\sqrt{3}}$, where p and q are integers, find $p + q$.

30. (15%) The area of a circle circumscribed about an isosceles triangle whose legs are 1 unit long and whose base is x units long is given by the formula $A = \frac{\pi}{a + bx^2}$, where a and b are real numbers. Find the ordered pair (a,b).

Problems—Spring 1980

1. Five math league coaches, one each from the Bronx, Nassau, Westchester, Canton, and Hudson, are to be seated in a row at the speakers' table at a convention. How many possible seating arrangements are there if the Bronx and Canton coaches cannot be seated next to one another?

2. Express $x^3 + x + 2x^4 + 4x^2 + 2$ as the product of two polynomials of degree two.

3. Find all integers x such that $(\sqrt{x} - 1)\sqrt{x}(\sqrt{x} + 1) = 6$.

4. Find the sum of the (finite) series $\frac{1}{\sqrt{2} + \sqrt{1}} + \frac{1}{\sqrt{3} + \sqrt{2}} + \frac{1}{\sqrt{4} + \sqrt{3}}$ $+ \ldots + \frac{1}{\sqrt{25} + \sqrt{24}}$ whose n^{th} term is $\frac{1}{\sqrt{n + 1} + \sqrt{n}}$.

5. Solve for all real values of x, $0° \leq x \leq 90°$: $\sin 40° + \cos 40°$ $= \sqrt{1 + \sin x}$.

6. In triangle ABC, $AC = 3$, $BC = 4$, and $AB = 5$. A semicircle with its center on segment $\overline{AC}$ is tangent to $\overline{AB}$ and $\overline{BC}$. Find the radius of the semicircle.

7. After graduation exercises at North Beach High School, each senior gave a snapshot of himself (herself) to every other senior and received a snapshot in return. If 870 snapshots were exchanged, how many seniors were in the graduating class?

8. The length x of a side of an equilateral triangle is also a root of the equation $x^8 - 2x^4 - 120 = 0$. Find the area of this triangle.

9. Find the largest three-digit number (written in decimal notation) that is divisible by 22 and such that the sum of the units digit and the tens digit is 11.

10. The polynomial $ax^4 + bx^3 + cx^2 + dx + e$, where $a > 0$, has integral coefficients whose greatest common divisor is 1. Find the polynomial if its value is zero when $x = \sqrt{-7}$ and when $x = \sqrt{3}$.

11. Find the ordered pair of real numbers (x,y) that satisfies

$$\frac{x}{x^2 + y^2} = 1 \text{ and } \frac{y}{x^2 + y^2} = 2.$$

12. Three circles with centers O_1, O_2, and O_3 are externally tangent in pairs at points A, B, and C. Find the area of the triangular region ABC if the radii of the circles are 1, 2, and 3.

13. Solve for all real values of x: $\log_2 x + \frac{1}{\log_x 2} = 4$.

14. If the domain for x is all complex numbers, find the solution set of $4x^4 + 1 = 0$. Express each element of the solution set in the form $a + bi$, where a and b are rational numbers.

15. There are two circles that have their centers on the x-axis, pass through the origin, and are tangent to the circle $(x - 7)^2 + (y - 6)^2 = 25$. Find the radius of the smaller of these two circles.

16. Urn A contains 2 white balls, and urn B contains 1 white ball and 2 black balls. Two balls are drawn at random from Urn B and, without their color being noted, are placed in urn A. Then one ball is drawn from urn A. What is the probability that this ball is white?

17. Find the degree-measure of all angles A for which $-180° < A \leq 180°$ and $\cos^2 A + 2 \sin A = 1$.

18. The lengths of the sides of a triangle are in arithmetic progression. The area of the triangle is 126 square centimeters, and the radius of the inscribed circle is 3 centimeters. Find the length, in centimeters, of the shortest side of the triangle.

19. Find all positive integers n for which $n^5 - 1$ is prime.

20. A polynomial $f(x)$, with real coefficients, is of degree 3. If $f(4) = f(-4) = -4$ and $f(0) = 1$, how many real solutions are there to the equation $f(x) = 0$?

21. The arithmetic mean of two positive numbers exceeds their geometric mean by 50. By how much does the square root of the larger of the two numbers exceed the square root of the smaller?

22. $P(x)$ and $Q(x)$ are polynomials with positive integral coefficients such that $[Q(x)]^2 = [P(x)]^2 + 2x^2 + 1$. Find $Q(x)$.

23. The equation $4x^2 + ax + b = 0$ is such that one of its roots is the sum, and the other the product, of the roots of $2x^7 + x^6 - 3 = 0$. Find the ordered pair (a,b).

24. An infinite sequence of real numbers $a_1, a_2, a_3, \ldots$ has the property that for all $n > 1$, $a_1 + a_2 + a_3 + \ldots + a_n = n^2 a_n$ and and $a_1 = \frac{1}{2}$. Find the numerical value of $a_1 + a_2 + a_3 + \ldots + a_{50}$.

25. The integers from 1 through 10 are written on index cards. It is desired to color each of these cards with one of three colors (red, white, or blue), in such a way that any two cards with integers differing by less than three are different colors. In how many ways can this be done?

26. Find all ordered pairs of positive integers (x,y) that satisfy $x^3 - y^3 = 721$.

27. Solve for x: $(\log_4 (\log_3 (\log_2 x))) = 0$.

28. Chord AB of a circle is extended through B to an exterior point D, and a line is drawn from D tangent to the circle at C. If $AB = CD$ and triangle BCD has area $2\sqrt{5}$ square centimeters, then the area of triangle ABC can be expressed as $a + b\sqrt{5}$ square centimeters. Find the ordered pair (a,b).

29. One vertex of a triangle is the point $(0,6)$, and one side lies along the x-axis. The area of this triangle is divided into two equal parts by the line $y = k$. Find k.

30. A man has forgotten how many children he has. He does, however, know that if he adds the number of children to its square, squares the result, and then subtracts the fourth power of the original number, he obtains the same result as the one obtained if he adds the original number to its square, subtracts 3, and then multiplies by 12. How many children does he have?

Problems—Fall 1980

1. (83%) A telegram costs m cents for 12 words and s cents for each additional word. If $k > 12$, express in terms of m, k, and s the cost of sending a telegram of k words.

2. (57%) If $P(x + 3) = x^2 + 7x + 4$ and $P(x) = ax^2 + bx + c$, find the ordered triple (a,b,c).

3. (87%) In a basketball tournament, each of six teams played five games: one game against each of the other five teams. No game ended in a tie, and, at the conclusion of the tournament, no two teams had won the same number of games. How many games were won by the team that won the tournament?

4. (54%) In parallelogram $ABCD$, $\overline{AC}$ and $\overline{BD}$ are diagonals. If $BC = 7$, $AB = 8$, and angle $C = 60°$, find the value of $AC^2 - BD^2$.

5. (47%) It is known that any odd number greater than 7 can be expressed as the sum of three odd primes, not necessarily distinct. Find the smallest odd number that can be expressed as the sum of three odd primes in three different ways (1 is not considered a prime).

6. (52%) Find the numerical value of $\log_{1/125} 25\sqrt[3]{25}$.

7. (35%) When the polynomial $f(x)$ is divided by $x - a$, the quotient is $Q(x)$ and the remainder is R. When $f(x)$ is divided by $3(x - a)$, the quotient is $g \cdot Q(x)$ and the remainder is $h \cdot R$, where g and h (as well as a and R) are constants. Find the ordered pair (g,h).

8. (7%) In isosceles triangle ABC, E is the midpoint of base $\overline{BC}$. Points F and G are on line segments $\overline{AB}$ and $\overline{AC}$, respectively, and EFG is an equilateral triangle. If $BC = 6$ and $AB = 5$, then $FG = (-24/11)(a + b\sqrt{3})$.
Find the ordered pair (a,b) of rational numbers.

9. (47%) The roots of the equation $x^2 - px + q = 0$ are a and b. The roots of the equation $x^2 - mx + n = 0$ are $a + 2/b$ and $b + 2/a$. Write an equation expressing n explicitly in terms of q only.

10. (59%) The sum of n consecutive positive odd integers, starting with the least one, a, is 105. Find the smallest possible value of a.

11. (30%) Find the ordered pair of positive integers (x,y) for which $\log(\log(xy)) = \log(\log x) + \log(\log y)$. (All logarithms have base ten.)

12. (20%) A median of a triangle is equal in length to the geometric mean of the lengths of the sides that include it. If these two sides are 7 and 10, find the length of the side of the triangle to which the median is drawn.

13. (37%) In triangle ABC, D is the midpoint of $\overline{BC}$ and E is the midpoint of $\overline{AD}$. Ray $\overrightarrow{BE}$ intersects $\overline{AC}$ at F. Find the numerical value of $\frac{FE}{EB} + \frac{AF}{FC}$.

14. (42%) Find the greatest common factor of all numbers of the form $2^n \cdot 3^{2n} - 1$, where n is an integer greater than 1.

15. (35%) The integer m is greater than 7 and is the first of five consecutive integers that are, respectively, multiples of 3, 4, 5, 6, and 7. Find the smallest such m.

16. (42%) In acute triangle ABC, $AB = c$, $BC = a$, $CA = b$, and $ac = 2b$.
Find the numerical value of $\frac{\cos A}{a} + \frac{\cos C}{c}$.

17. (66%) The angles of a triangle are in the ratio 3:4:5. Find the numerical value of the ratio of the sine of the smallest angle to the sine of the next smallest angle.

18. (18%) $\overline{AB}$ is the diameter of a circle, and $ABCD$ is an inscribed trapezoid. If the areas of triangles ABC and ACD are 150 and 120 square units, respectively, find the length of a leg of the trapezoid.

19. (40%) If $\sin x = \frac{1}{\tan x}$, find the numerical value of $\cos x$.

20. (35%) Runners A and B run a race from point P to point Q and back again. B starts out running at 9 miles per hour and A at 12 miles per hour. As B is running towards point Q he meets A, who is already returning to point P. As soon as they meet, B increases his speed, and the race ends in a tie. By how much did B increase his speed?

21. (35%) Find the largest positive integer q such that q is not the square of an integer and $\sqrt{10 + \sqrt{q}}$ can be expressed as $\sqrt{x} + \sqrt{y}$, where x and y are positive integers, $x \neq y$.

22. (16%) In trapezoid $ABCD$, base $\overline{AD}$ is twice as long as base $\overline{BC}$. E is the midpoint of leg $\overline{AB}$ and F is a point on leg $\overline{CD}$. Ray $\overrightarrow{EF}$ intersects ray $\overrightarrow{AD}$ in G and $AD = DG$. Find the ratio $CF:FD$.

23. (31%) If the logarithm is taken to the base ten and x denotes radian measure, for how many positive values of x does $\log_{10}x = \sin x$?

24. (54%) The four jacks, four queens, and four kings of an ordinary deck of cards are removed and thoroughly shuffled. The first three cards are turned face up one at a time in an experiment to detect extrasensory perception. A subject in another room announces that the first card is a jack, the second is a queen, and the third is a king. Assuming he is guessing, what is the probability that he is correct?

25. (25%) In triangle ABC, $AB = 13$, $BC = 14$, $CA = 14$, and $\overline{DC}$ and $\overline{EA}$ are the altitudes from C and A, respectively. Find DE.

26. (9%) The equation $x^3 + ax^2 + bx + c = 0$ has the root $\sqrt[3]{\sqrt{28} + 6} - \sqrt[3]{\sqrt{28} - 6}$. If a, b, and c are integers, find the ordered triple (a,b,c).

27. (42%) If $0 \leq x \leq \pi/2$, find all values of x for which $\cos 3x \cdot \cos x + \sin 3x \cdot \sin x = \cos \frac{3\pi}{5}$.

28. (59%) Urn A contains 8 balls, each either black or white. Urn B contains 3 black balls and 2 white balls. Each morning, a ball is drawn at random from urn A and placed in urn B. Every evening a ball is chosen from urn B. Then the balls are arranged as before for the next morning's drawing. The evening ball is white 3/8 of the time. How many white balls are in urn A?

29. (25%) If the reciprocal of $(x + y)$ is equal to the sum of the reciprocal of x and the reciprocal of y, find all possible numerical values of $\frac{x}{y}$.

30. (11%) Find the numerical value of $(\sin 10°)(\sin 50°)(\sin 70°)$.

Problems—Spring 1981

1. (72%) If the ratio of a to b is r and $a \neq b$, express, in terms of r, the ratio of $a + b$ to $a - b$.

2. (23%) In triangle ABC, points D and E lie on $\overline{CA}$ and $\overline{CB}$, respectively, so that $\overline{DE}$ is parallel to $\overline{BA}$. $\overline{DE}$ divides the triangle into two regions such that the ratio of the area of the smaller region to that of the larger equals the ratio of the area of the larger region to that of triangle ABC. If DEC is the smaller region, find the ratio of the area of triangle DEC to that of triangle ABC.

3. (75%) Find all ordered pairs (p,q) of real numbers for which $|p| = |q| = 1$ and the equation $x^2 + px + q = 0$ has only real roots.

4. (25%) A circle of radius 1 is externally tangent to seven smaller congruent circles. Each of these smaller circles is externally tangent to two of the other smaller circles. If $a = \sin \pi/7$, express the radius of one of these smaller circles in terms of a.

5. (46%) If $a + \frac{1}{a} = 3$, find the absolute value of $a - \frac{1}{a}$.

6. (27%) A man lost in the woods decides to flip a coin, walking one mile east each time it appears heads and one mile west each time it appears tails. He flips the coin six times and walks accordingly. Find the probability that he will be at least two miles from his starting point at the end of this procedure.

7. (66%) After a hard math test, each of the 35 students in a math class got a peek at the teacher's grade book. No student saw all of the grades, and no student was able to see his own grade. Each student noticed exactly ten failures among the grades (s)he saw. What is the smallest number of failures that could have occurred in the class?

8. (47%) If the radius of the inscribed circle of a regular heptagon (seven-sided figure) is 1, find the distance from the center of the inscribed circle to one of the longest diagonals. If $b = \sin \pi/14$, express your answer in terms of b.

9. (17%) Let $A = \dfrac{x^r}{1 - x^r}$, $B = \dfrac{x^s}{1 - x^s}$ and $C = \dfrac{x^{r+s}}{1 - x^{r+s}}$, where $(1 - x^r)(1 - x^s)(1 - x^{r+s}) \neq 0$. Write an equation expressing C explicitly in terms of A and B.

10. (52%) Gilbert is eight years younger than Irwin. If Irwin were one year older, he would be exactly n times as old as Gilbert is now, where n is a positive integer. Find all possible integral values of Gilbert's age.

11. (54%) If $(a + bi)^2 = b + ai$, where $i = \sqrt{-1}$, find the ordered pair of positive numbers (a,b).

12. (46%) Four points are coplanar but not collinear. What is the minimum number of midpoints of line segments determined by these four points?

13. (43%) If $0 < y < x$, and $a = x - y$, $b = xy$, $c = x + y$, write an equation expressing c explicitly in terms of a and b.

14. (19%) At 10 minutes past noon, the hands of a clock make an angle of $x°$. After how many minutes will the hands again make an angle of $x°$?

15. (60%) Four points lie along a line. What is the minimum number of midpoints of line segments determined by these four points?

16. (26%) A binary sequence of length K is an ordered k-tuple whose entries are all 0's or 1's. A block of length m is a sequence of m consecutive 0's in a binary sequence that is neither preceded nor followed by another 0. Find the number of binary sequences of length 5 that contain blocks of length 2.

17. (27%) In a certain game with four players, the odds against Amy winning are 4 to 5, against Beth winning are 5 to 2, against Carl winning are 8 to 1, and against Debbie winning are x to 1. Exactly one of these players will win the game. Find x.

18. (13%) What is the minimum value of the sum of the squares of the distances of a point to the vertices of a triangle whose sides are 3, 4, and 5?

19. (69%) Find all real numbers x such that $\sqrt{x} + \sqrt[4]{x} = 12$.

20. (23%) Equilateral triangle ABC is inscribed in a circle. Point P lies on this circle, and $PB = 8$. If $AB = 7$ and $PA < PC$, find the ordered pair (PA,PC).

21. (57%) Find the shortest altitude of a triangle whose sides are 17, 25, and 28.

22. (41%) A bridge is supported by some old cables, each of which has 1/3 probability of remaining intact for an hour. As long as one cable lasts, the bridge will hold. Rigid safety rules require that there be at least an 80% probability that the bridge stand for an hour. What is the minimum number of cables that must be on the bridge for it to pass the safety rules?

23. (29%) A regular tetrahedron is a solid with four faces, each of which is an equilateral triangle. From each vertex of such a figure, a line is drawn to the centroid (intersection of the medians) of the opposite face. These four segments meet in a point, which thus divides each segment into a larger part and a smaller part. Find the ratio of the larger part to the smaller part.

24. (13%) A rectangular solid block of wood is 6" high, M" wide, and N" long, where $M > N$ are positive integers. The outside surface of the block is painted, and the block is then cut into $6MN$ unit cubes. Exactly half of these unit cubes have no paint at all on them. Find the least possible value of N.

25. (44%) The ratio of John's age to Mary's age is now r. If $1 \leq r < 2$, express in terms of r the ratio of John's age to Mary's age when John was as old as Mary is now.

26. (17%) Point P lies on minor arc AB of the circle circumscribing square $ABCD$. If $AB = 5$ and $PA = 4$, find the length of $\overline{PD}$.

27. (59%) The degree of the polynomial $f(x)$ is greater than three, and $f(x^2) = (f(x))^2$ for all real x. Find the coefficient of x^3 in the polynomial $f(x)$.

28. (8%) Each of four unit spheres is externally tangent to the other three. A larger sphere is tangent to all four of these. If the radius of the large sphere is $\frac{1}{2}(a + \sqrt{b})$, find the ordered pair (a,b) of rational numbers.

29. (34%) A particle starts at (0,0) and moves at each step one unit in either the positive x or y direction. The probability that it will move in the x direction is p. If the probability that the particle passes through the point (3,1) is 16 times as great as the probability that it passes through the point (1,3), find p.

30. (22%) In trapezoid $ABCD$, shorter base $CD = 2$, leg $AD = 6$, and angles CBA and DAB are both 60°. The intersection point of $\overline{AC}$ and $\overline{BD}$ is E. F and G are on segments $\overline{AC}$ and $\overline{BD}$, respectively, such that $AF = CE$ and $BG = DE$. Find the length of FG.

Problems—Fall 1981

1. (84%) Two students went to buy Munchie candy bars. Each student had some money, but one student found that he needed 7¢ more to buy one bar and the other found she needed 2¢ more. They pooled their money, but still did not have enough. How many cents does one Munchie bar cost?

2. (81%) Find the smallest value of y if, for all real x, $y = 2^x + \dfrac{1}{2^x}$.

3. (63%) Find the degree-measure of an acute angle formed by two diagonals of a regular pentagon that do not share a common endpoint.

4. (29%) A polynomial is said to be prime if it has integral coefficients and if no polynomial of lower degree (and integral coefficients) divides it without a remainder. Find all prime quadratic factors of the polynomial $x^6 - 64$.

5. (85%) In parallelogram $ABCD$, $AB = CD = 8$. A line intersecting AB at E and CD at F cuts the parallelogram into two polygons with equal area. If $AE = 3$, find DF.

6. (38%) If, for all real x, $xf(x) + f(1 - x) = x^2 + 2$, write an equation expressing $f(x)$ explicitly in terms of x.

7. (53%) In right triangle ABC, CM is the median to hypotenuse AB. If angle $A = 60°$ and $AB = 12$, find the distance from point B to line $\overleftrightarrow{CM}$.

8. (15%) If $2(4^x) + 6^x = 9^x$ and $x = \log_{2/3}a$, find the numerical value of a.

9. (36%) The number 1000027 has exactly one prime factor that is larger than 100 but less than 1000. Find this prime factor.

10. (48%) The Gnostic Gnomes spend their lives writing positive integers on tiny slips of paper. The slip on which the integer N is written weighs $1/N$ grams. The Grand Gnomon takes all the slips containing numbers of the form 10^k, k a positive integer, and puts them in a sack. He will of course never finish, but what limit will the weight of the sack's contents approach as the gnomes write more and more integers?

11. (82%) Twelve consecutive integers are added together. What is the remainder when the sum is divided by 4?

12. (43%) The area common to the circles $(x - 2)^2 + (y - 2)^2 = 25$ and $(x - 2)^2 + (y - 6)^2 = 25$ is divided into two equal parts by the line $14x + 3y = k$. Find k.

13. (31%) In right triangle ABC, leg $AC = \sec 15°$ and leg $BC = \csc 15°$. Find the numerical value of the length of hypotenuse AB.

14. (41%) In a game played by the Elves of Elfland, a positive integer is picked at random. The probability of picking the number n is given by $\frac{1}{2^n}$. In this game, what is the probability that the positive integer picked will be divisible by 3?

15. (69%) The complex number $a + bi$, when squared, gives $-3 + 4i$. If $a > 0$, find the ordered pair of real numbers (a,b).

16. (26%) Find the radius of the circle in which chord AB, measuring 2 centimeters, and chord AC, measuring $5\sqrt{3}$ centimeters, meet at an angle of 30°.

17. (29%) If Arctan denotes principal value, solve for x.

$$\text{Arctan}\ \frac{3}{5} + \text{Arctan}\ \frac{3}{4} = \text{Arctan}\ x.$$

18. (10%) The integer 999,999,995,904 may be factored as $a^{16} \cdot b^2 \cdot c \cdot d \cdot e \cdot f$, where a, b, c, d, e, and f are primes and $a < b < c < d < e < f$. Compute the value of f.

19. (85%) In rectangle $ABCD$, $AB = 8$ and $AD = 6$. A parallelogram has each of its vertices on a different side of the rectangle, and each side of the parallelogram is parallel to one of the rectangle's diagonals. Find the perimeter of the parallelogram.

20. (40%) An ordinary, six-sided die is thrown five times. What is the probability that the number showing on each throw is higher than the one that was showing on the previous throw?

21. (94%) A computer is programmed so that it takes as input a positive integer and prints out its common English name, then counts the number of characters (spaces and hyphens excluded) in that name. For example, given 492, it prints "four hundred ninety-two," then "20." A programmer accidentally "loops" the machine, so that it receives as input the letter-count it just put out. Eventually the machine will settle down to printing the same number over and over. What is that number?

22. (28%) In acute triangle ABC, $\sin A = \frac{4}{5}$ and $\cos B = \frac{5}{13}$. Find $\tan C$.

23. (39%) Find all positive integral values of x for which $\sqrt{x + 1} + \sqrt{x + 34}$ is also an integer.

24. (29%) If $x = 20$, then $x^2 + x - 16$ is a multiple of 101. Find another positive integer $N < 100$ such that $N^2 + N - 16$ is a multiple of 101.

25. (65%) A piece of gold weighs 6 grams and one of silver weighs 12 grams. Equal pieces (by weight) of each metal are cut off, and each cut-off piece is alloyed with the remainder of the <u>other</u> metal. If the resulting alloys have identical percentages of gold, how many grams of gold were cut from the original piece?

26. (15%) The bisector of the right angle of a right triangle divides the hypotenuse into two segments whose ratio is 2:5. The altitude to the hypotenuse of this triangle also divides it into two segments. Find the ratio of the smaller of these segments to the larger.

27. (41%) The sequence t_n of "triangular" numbers is defined for integers $n \geq 1$ by $t_n = \frac{1}{2}n(n+1)$. How many of the first 100 "triangular" numbers are multiples of 3?

28. (55%) For certain integers x, one can find another integer n so that $nx = n + 10x$. Find the largest such integer x.

29. (48%) Find the degree-measure of all angles θ such that $0° < \theta < 90°$ and $\sin\theta + \sin 2\theta = \cos\theta + \cos 2\theta$.

30. (19%) In acute triangle ABC, $AB = 5$, $AC = 7$, and angle $BAC = 60°$. A point X is selected on arc BC of the circle circumscribing ABC, and perpendiculars $\overline{XN}$ and $\overline{XK}$ are dropped to lines AB and AC, respectively. Find the largest possible value for the length of $\overline{NK}$.

Problems—Spring 1982

1. The units digit of the difference of the squares of two integers (written in base ten notation) is 2. What is the units digit of the sum of these two squares?

2. Find the sum of all natural numbers that divide 3087 without remainder.

3. If John gets 97 on the next math test, his average will be 90. If he gets 73, his average will be 87. How many tests has John already taken?

4. Points X, Y, Z are chosen on sides AB, BC, AC respectively, of triangle ABC. If $AX:XB = BY:YC = CZ:ZA = 1:4$, find the ratio of the area of triangle XYZ to that of triangle ABC.

5. A girl bought some 1¢, 2¢, and 5¢ stamps, for a total value of \$1.00. She bought at least one of each kind, and got ten times as many 1¢ stamps as 2¢ stamps. How many 5¢ stamps did she buy?

6. If the ordered pair (x,y) of nonzero real numbers satisfies simultaneously $x^2 + 2xy + y^2 - 2x - 2y = 0$ and $x^2 - y^2 - 2x + 2y = 0$, find the numerical value of $x + y$.

7. Of five lines, no two are parallel and no three are coincident. Into how many bounded regions do they divide the plane?

8. The number $500 = 2^2 \cdot 5^3$ has no prime factors other than 2 and 5. Some other such numbers are 2, 16, 10, and 1 (this last has no prime factors at all, so certainly none other than 2 or 5). Find the sum of all such numbers less than or equal to 500.

9. Point X is on side CD of parallelogram $ABCD$. Angle ADC is acute and $XC = CB$, $XB = BA$, and $XA = AD$. Find the measure of angle ADX.

10. The polynomial $x^4 + 64$ can be expressed as $[P(x)]^2 - [Q(x)]^2$, where $P(x)$ and $Q(x)$ are polynomials with positive integer coefficients, and the degree of $P(x)$ is larger than that of $Q(x)$. Find $P(x)$.

11. Find the numerical value of $\dfrac{\cos 2b}{\sin^2 2b(\cot^2 b - \tan^2 b)}$ for $0 < b < \dfrac{\pi}{2}$, $b \neq \dfrac{\pi}{4}$.

12. Points X, Y, and Z are chosen on sides AB, BC, CA, respectively, of triangle ABC, and $AX{:}XB = BY{:}YC = CZ{:}ZA < 1$. If the area of triangle XYZ is 7/25 that of triangle ABC, find the ratio $AX{:}XB$.

13. The number $\sqrt{20 + \sqrt{384}}$ can be expressed as $\sqrt{a} + \sqrt{b}$, where a and b are rational and $a < b$. Find the ordered pair (a,b).

14. In triangle ABC, angle ABC = angle $ACB = 72°$. Find (in radical form) the numerical value of $AB{:}BC$.

15. In the figure, each small square is to be colored either red, white, or blue. Each row and each column is to have three different colored squares in it. In how many different ways can the figure be colored (colorings that are reflections or rotations of one another are to be counted as different)?

16. The polynomial $P(x) = (2x^2 + 2x + 1)^2$ can be represented as $[Q(x)]^2 + [R(x)]^2$, where both $Q(x)$ and $R(x)$ are polynomials of degree ≥ 1, with positive integer coefficients, and where $Q(x)$ has the larger degree. Find $Q(x)$.

17. Points A, B, and C are consecutive vertices of a regular polygon whose side is 1 unit long. Point D is the foot of the perpendicular from C to $\overline{AB}$ (extended). If $\overline{BD} = y$ and $\overline{AC} = a$, write an equation expressing y in terms of a.

18. The sequence $n_1, n_2, n_3, n_4, \ldots$ consists of the number 1 and all those natural numbers greater than 1 whose prime factorizations consist of only 2's, 3's, and 5's. Find the infinite sum $\dfrac{1}{n_1} + \dfrac{1}{n_2} + \dfrac{1}{n_3} + \dfrac{1}{n_4} + \ldots$.

19. If the number of permutations of n objects taken r at a time is 6 times the number of combinations of n objects taken r at a time, find the numerical value of r.

20. Find the greatest integer less than $2\sqrt{5} + \sqrt{21}$.

21. A worker received a salary increase of $p\%$ one year and a decrease of $p\%$ the next year. The net result over the two years was a decrease of $x\%$. Write an equation expressing x in terms of p.

22. The number $2^{2^2} + 1$ has exactly one prime factor greater than 1000. Find it.

23. In isosceles triangle ABC, angle $ACB = 108°$. Find in radical form the numerical value of $AB:AC$.

24. If A and B are positive acute angles, $\tan A = \frac{1}{7}$, and $\sin B = \frac{1}{\sqrt{10}}$, find the degree-measure of $A + 2B$.

25. A square-free number is a natural number that is not divisible by the square of any natural number (other than 1). How many square-free numbers are there less than or equal to 100?

26. If $x = \sqrt[3]{20 + 14\sqrt{2}} + \sqrt[3]{20 - 14\sqrt{2}}$, find the numerical value of $x^3 - 6x$ in simplest form.

27. For how many value(s) of x, $0 \leq x < 2\pi$, does $\sin x + \sin 2x + \sin 3x = 0$?

28. In triangle ABC, $AB = BC = 5$. Point D is on line segment AC and $DC = 4$. If $BD = 4$, find AD.

29. The probability of getting exactly k heads in 35 tosses of a fair coin is the same as the probability of getting exactly $k + 1$ heads in 36 tosses. Find k.

30. In isosceles triangle ABC, CD is the altitude to base AB and AE is the bisector of angle CAB. If $AE = 2CD$, find the degree-measure of angle ACB.

Problems—Fall 1982

1. (82%) Find all real x for which $x + \frac{1}{x} = \frac{10}{3}$.

2. (55%) How many triples (a,b,c) are there of positive integers with the following two properties: (1) a, b, and c are primes and (2) $a^2 - b^2 = c$?

3. (34%) If the hundreds digit of a three-digit number is increased by n, and both the tens and units digits are decreased by n, the new number formed is n times greater than the original number. Find the original number.

4. (67%) Express in simplest form the real number $\left(\sqrt[5]{\sqrt{18}+\sqrt{2}}\right)^2$.

5. (41%) Two circles of radii 6 and 2 are coplanar, and their centers are 12 units apart. Find the length of their common external tangent.

6. (5%) In how many ways can ten different marbles be placed in three different urns, if each urn must have at least one marble in it?

7. (41%) The product of two positive integers is 588000. Find the largest possible value for the greatest common divisor of the two integers.

8. (36%) Find all real x such that $\sqrt{\dfrac{x+4}{x-1}}+\sqrt{\dfrac{x-1}{x+4}}=\dfrac{5}{2}$.

9. (32%) For $x > 0$, find the smallest possible value of $x + \dfrac{5}{x}$.

10. (32%) From the corners of square *ABCD*, equal isosceles triangles are cut off, leaving a regular octagon. Find the ratio of the area of the octagon to that of the original square.

11. (40%) Two circles of radii 6 and 2 are coplanar, and their centers are 12 units apart. How far from the <u>smaller</u> circle do their common <u>internal</u> tangents meet?

12. (55%) If a, b, c are real numbers such that $a + b + c = 0$, find the largest possible value of

$$\frac{(a^2+b^2+c^2)^2}{a^4+b^4+c^4}.$$

13. (93%) How many of the numbers $2^1, 2^2, 2^3, \ldots, 2^{100}$, written in base ten notation, end in 6?

14. (52%) Find all real $x > 0$ such that $\log_x 4 + \log_4 x = \dfrac{17}{4}$.

15. (48%) For $x > 0$, find the minimum value of $\sqrt{\dfrac{(4+x)(1+x)}{x}}$.

16. (47%) Two nonintersecting circles have radii of 9 and 16. A third circle is tangent to these two and also to one of their common external tangents. If the centers of all three circles are collinear, find the radius of the third circle.

17. (82%) Twice Jack's age plus three times Jill's age equals 31. Their ages are integers, and the difference in their ages is a multiple of 4. If Jack's age is a and Jill's age is b, find the ordered pair (a,b).

18. (38%) Three sides of a triangle are in arithmetic progression. The middle side has length 2, and the angle opposite this side is 60°. Find the area of the triangle.

19. (49%) If $\tan\theta + \cot\theta = \frac{10}{3}$ and $0 < \theta < \frac{\pi}{2}$, find all possible values of $\sec\theta$.

20. (71%) If $ax + by = 8$ passes through $(2,-3)$ and $(1,4)$, it also passes through $(11,k)$. Find k.

21. (69%) For $x > 0$, find the smallest possible value of $\frac{(4 + x)(1 + x)}{x}$.

22. (42%) Find the largest prime p such that no number of the form 999...9 has a factor of p.

23. (12%) Two nonintersecting circles have radii 1 and 9, and one of their <u>common external tangents</u> is 12 units long. A third circle is <u>tangent to the first two</u> and also to one of their common external tangents. Find the radius of the third circle.

24. (61%) The three angles A, B, C of triangle ABC satisfy $\cos A \cos B + \sin A \sin B \sin C = 1$. Find the degree-measure of angle A.

25. (60%) For all a, $b > 0$, $\sqrt{\frac{a}{b}\sqrt{\frac{b}{a}\sqrt{\frac{a}{b}}}} = \left(\frac{a}{b}\right)^k$. Find the rational number k.

26. (66%) Find the numerical value of $1 + 3\left(\frac{1}{3}\right) + 5\left(\frac{1}{3}\right)^2 + 7\left(\frac{1}{3}\right)^3 + 9\left(\frac{1}{3}\right)^4 + \ldots$.

27. (63%) If x can be any real number, find the minimal value of

$$(4 + \sqrt{15})^x + (4 - \sqrt{15})^x.$$

28. (35%) How many four-digit numbers, made up only of the digits 1, 2, 3, are multiples of 9?

29. (31%) The decimal representation of the integer N consists entirely of 9's, and N is a multiple of 17. Find the smallest possible number of digits in the decimal representation of N.

30. (01%) Find all values of θ such that $0 \leq \theta < \pi$ and

$$(\sin 3\theta)(\sin 3\theta - \cos\theta) = \sin\theta\,(\sin\theta - \cos 3\theta).$$

Problems—Spring 1983

1. (91%) The integer 1234321 is a perfect square. Find its positive square root.

2. (91%) Points P, Q, R, and S are chosen on sides $\overline{AB}$, $\overline{BC}$, $\overline{CD}$, and $\overline{DA}$, respectively, of square $ABCD$ so that $AP:PB = BQ:QC = CR:RD = DS:SA = 3:1$. Find the ratio of the area of square $PQRS$ to that of $ABCD$.

3. (69%) The minute hand of a clock travels $k\pi$ radians between 1 P.M. and 2:35 P.M. Find k.

4. (48%) How many ways can a set of 6 different elements be divided into 3 subsets of 2 each?

5. (17%) The roots of the equation $x^2 - qx + p = 0$ are the squares of the roots of the equation $x^2 - px + q = 0$. Find the ordered pair of nonzero real numbers (p,q).

6. (46%) For how many natural numbers N less than or equal to 12 is $\overline{4^N + 5^N + 6^N + 7^N}$ a multiple of 11?

7. (82%) Express in simplest form the real number $\left(\sqrt[3]{\sqrt{75} - \sqrt{12}}\right)^{-2}$.

8. (29%) Line ℓ is drawn through the centroid (intersection of the medians) of triangle ABC. Points B and C are on the opposite side of line ℓ from point A. The (perpendicular) distances from A and B to ℓ are 10 and 6, respectively. Find the distance from point C to ℓ.

9. (47%) If $A = 1 + \cfrac{1}{1 + \cfrac{1}{1 + \ddots}}$ and $B = 2 + \cfrac{1}{2 + \cfrac{1}{2 + \ddots}}$

(where both fractions are assumed to converge), find the numerical value of $A + B - (\frac{1}{A} + \frac{1}{B})$.

10. (69%) If A is an acute angle and $(\sin A)(\cos A) = \frac{60}{169}$, find the numerical value of $\sin A + \cos A$.

11. (25%) A and B both represent nonzero digits. If the base ten numeral AB divides (without remainder) the base ten numeral $A0B$ (whose middle digit is zero), find all possible values for the integer AB.

12. (15%) In triangle ABC, $AB = 20$, $BC = 30$, and $\overline{BD}$ is an angle bisector (point D is on $\overline{AC}$). Point E is chosen on $\overline{BC}$ so that $\overline{DE} \| \overline{AB}$, and point K is chosen on $\overline{DC}$ so that $\overline{EX} \| \overline{BD}$. If $AD - KC = 1$, find the length of AC.

13. (36%) By using a "yardstick" that was too long, a dealer made a profit of 20% on some cloth he sold, instead of 30%. The cloth was paid for by the (linear) yard. How many inches were there in his "yard" stick?

14. (50%) Points W, X, Y, and Z are chosen on sides $\overline{AB}$, $\overline{BC}$, $\overline{CD}$, and $\overline{DA}$, respectively, of square $ABCD$ to form square $WXYZ$. If the area of square $WXYZ$ is $\frac{5}{8}$ that of $ABCD$ and $AW < WB$, find the numerical value of the ratio $AW:WB$.

15. (51%) How many fractions $\frac{a}{b}$ are there such that $0 < a < b$, $\frac{a}{b}$ is in lowest terms, and b divides 24 (note that b could equal 24)?

16. (37%) If x and y are real numbers such that $xy = 7$, find the smallest possible value of $20x^2 + 5y^2$.

17. (88%) If all numbers are written in base ten notation, how many digits are in the numeral representing 61224^2?

18. (14%) Define an interior diagonal of a three-dimensional polyhedron P to be a line segment whose endpoints are distinct vertices of P and which (except for its endpoints) lies entirely in the interior of P. Find the maximum possible number of interior diagonals if P has 10 vertices.

19. (70%) A fleeble factory was supposed to fill a certain order for fleebles in 12 days. The factory produced 25% more fleebles each day than it was supposed to, and as a result it filled the order in 10 days, producing 42 extra fleebles besides. If the same number of fleebles were produced each day, what was that number?

20. (84%) Find the value of $\text{Arcsin}\,\frac{5}{13} + \text{Arccos}\,\frac{5}{13}$, where "Arc" denotes principal value.

21. (47%) If x and y are nonnegative real numbers such that $x^2 + 4y^2 = 50$, find the largest possible value of $x + 2y$.

22. (41%) In triangle ABC, D is on side $\overline{AB}$ and E is on side $\overline{AC}$. The area of triangle ADE is half that of triangle ABC, and $AD:DB = 4:3$. Find the ratio $AE:EC$.

23. (55%) Some children bought sticks of gum for 1¢ each, rolls for 10¢ each, and candy bars for 50¢ each. They spent \$5 and got 100 items. How many sticks of gum did they buy?

24. (17%) If $r_1, r_2, \ldots, r_5$ are the distinct roots of the equation $3x^5 + 8x^4 + 3x^3 + x^2 - 4x + 1 = 0$, find the numerical value of $(1 + r_1)(1 + r_2)\ldots(1 + r_5)$.

25. (40%) Find the largest natural number A such that x and x^3 leave the same remainder when divided by A, for any integer $x > 5$.

26. (45%) Find the area of a triangle whose sides have lengths a, b, and c if $4a^2b^2 - (a^2 + b^2 - c^2)^2 = 16$.

27. (65%) Find all ordered triples of integers (a,b,c) such that $0 < a < b < c$ and $a^2 + b^2 + c^2 = 90$.

28. (24%) If $\cos x + \cos y + \cos z = \sin x + \sin y + \sin z = 0$, find the numerical value of $\cos(x - y) + \cos(y - z) + \cos(z - x)$.

29. (30%) A cube is to be colored, with six colors, so that each face is a single different color. Two colorings are considered the same if two cubes so colored can be placed next to each other so that the colorings of corresponding faces are identical. Colorings that are mirror images of each other are considered distinct. How many such distinct colorings are there?

30. (07%) A chord of a triangle is a line segment whose endpoints lie on the sides of the triangle (but not at the vertices). For a triangle whose sides are 4, 5, 6, find the length of the shortest chord that divides the triangle into two regions of equal area.

Problems—Fall 1983

1. The sequence $\{a_n\}$ is defined by $a_1 = a_4 = 1$, $a_2 = a_3 = a_5 = -1$, $a_6 = a_1a_2$, $a_7 = a_2a_3$, and in general for $k > 5$, $a_k = a_{k-5}a_{k-4}$. Find a_{1983}.

2. The numbers 1, 2, 3,..., 1000 are written in a row. Sam started at 1 and circled every 24th number in red. Janet started at 1 and circled every 15th number in blue. What is the smallest possible (positive) difference between a red number and a blue number?

3. How many five-digit numbers (in base ten notation, with the left-most digit not equal to zero) are there such that each digit is strictly greater than the sum of the digits to its right (in particular, the tens digit is larger than the units digit)?

4. If $0 < A < \pi$, $0 < B < \pi$, and $\sin A + \sin B = \cos A + \cos B$, find the numerical value of $A + B$.

5. Line XY is tangent to circle O (with center O) at X and to circle P (with center P) at Y. The radii of the circles are 5 and 8, respectively, and points O and P are both on the same side of line XY. If $XY = \sqrt{7}$, find the length of OP.

6. Find the smallest natural number N such that $\frac{N}{2}$ is a perfect square and $\frac{N}{3}$ is a perfect cube.

7. A circular path is 330 meters in circumference. A man makes a mark on the path, then walks around it several times, making a mark every 75 meters. He stops when the mark he is about to make coincides with his very first mark. When he is done, what is the shortest (positive) distance (measured along the circular path) between two of the marks?

8. In a plane, points A and B are on the same side of line L. They are each 3 cm from line L and they are 4 cm from each other. Find the radius of the circle through points A and B that is tangent to line L.

9. In right triangle ABC, leg $AC = 4$ and leg $BC = 8$. A square is drawn exterior to the triangle with $\overline{AB}$ as one side. Find the distance from C to the intersection of the diagonals of the square.

10. The sequence $\{a_i\}$ is defined as follows: $a_1 = 3^{1983}$, and for $i > 1$, a_i is the sum of the digits in the decimal representation of a_{i-1}. Find the numerical value of a_{10}.

11. In triangle ABC, $AB = 3$, $BC = 4$, $AC = 6$. If $\overline{BC}$ is extended through C to D and $BC = CD$, find AD.

12. How many ordered pairs (x,y) of positive integers are there such that both x and y are less than 100 and the expression $\log_{10}x + \log_{10}y$ has an <u>integral</u> value?

13. In triangle ABC, $\sin^2 A + \sin^2 B = 1$. Find the degree-measure of angle C.

14. The number N is represented by the base q numeral 1441. When divided by eleven, N leaves a remainder of 1. If $1 < q \leq 10$, find q.

15. A certain plane geometric figure may be made to coincide with itself if it is rotated in a plane about a fixed point P through an angle of 48°. What is the smallest positive angle through which you can be <u>sure</u> the figure may be rotated (about point P) and still coincide with itself?

16. In triangle ABC, $AC = 6$, $BC = 8$, and $AB = 10$. A line dividing the triangle into two regions of equal area is perpendicular to $\overline{AB}$ at X. Find BX.

17. If $m,n > 1$ and for all $x > 0$, $\log_n x = 3 \log_m x$, write an equation expressing m explicitly in terms of n.

18. In trapezoid $ABCD$, $AB = BC = CD = 6$ and $AD = 8$. Points P and R are chosen on line AD (on opposite sides of point A), so that there exists a point Q on the plane for which B is the midpoint of $\overline{PQ}$, C is the midpoint of $\overline{RQ}$, and $BQ = QC$. Find PQ.

19. The sequence $\{a_i\}$ is defined by setting $a_1 = 7$ and, for $i > 1$, taking a_i to be the sum of the digits in the decimal representation of $(a_{i-1})^2$. Find a_{1983}.

20. Let n be the maximum number of points that can be arranged in a plane so that of any four of them, there are three that determine an equilateral triangle. Find n.

21. Find all values of a such that the three equations $\begin{cases} ax + y = 1 \\ x + y = 2 \\ x - y = a \end{cases}$ are satisfied simultaneously by some ordered pair (x,y).

22. Points M and N are on side $\overline{AC}$ of triangle ABC, and points P and Q are on side $\overline{AB}$. The lines MP, NQ, and BC are parallel, and they divide the triangle into three regions of equal area. If $NQ = 4$, find BC.

23. In triangle ABC, m angle $ACB = 90$ and m angle $ABC = 45$. Points X, Y, Z are on sides $\overline{AC}$, $\overline{CB}$, $\overline{BA}$, respectively, so that $AX:XC = CY:YB = BZ:ZA = 2:1$. If $\overline{CZ}$ intersects $\overline{XY}$ in P, find the degree-measure of angle CPY.

24. The symbol $[x]$ represents the largest integer not exceeding x. Find all positive integral values of n for which the expression $\left[\frac{n^2}{3}\right]$ represents a prime number (1 is not prime).

25. A sequence of digits has the property that each pair of successive digits, taken in the order written, forms a decimal numeral representing a multiple of either 17 or 23. If the first digit is 9 and the sequence is finite, what is its last (possible) digit?

26. In parallelogram $ABCD$, $AB = 4$, $AD = 9$, and m angle $BAD = 30$. The circle through points A, B, and D intersects $\overline{BC}$ at B and X. Find XC.

27. A right circular cone has a base with radius 12. A plane parallel to the plane of the base cuts the cone into two equal volumes. Find, in radical form, the radius of the circle of intersection of this plane with the cone.

28. If $[x]$ denotes the greatest integer not exceeding x and $\{x\} = x - [x]$, find all ordered triples (x,y,z) of real numbers such that $x + [y] + \{z\} = 1.1$, $y + [z] + \{x\} = 2.2$, $z + [x] + \{y\} = 3.3$.

29. If $0 < x < \pi$ and $2^{\tan x} = 8^{\sin x}$, find the numerical value of $\cos x$.

30. Acute triangle ABC is inscribed in a circle. Altitudes $\overline{AM}$ and $\overline{CN}$ are extended to meet the circle again at P and Q, respectively. If $PQ:AC = 7:2$, find the numerical value of $\sin B$.

Problems—Spring 1984

1. Instead of finding twice the square of a number, a student found twice the square root of that number and got 10 as her answer. What is the correct answer?

2. Diameter $\overline{AB}$ of a circle whose center is O is extended past A to point P, and tangent segment $\overline{PT}$ is drawn. On $\overline{AO}$ as diameter, a small semicircle is drawn on the same side of $\overline{AB}$ as point T. If $\overline{OT}$ intersects the smaller semicircle at X and angle $TPB = 40°$, find the degree measure of arc $\overset{\frown}{OX}$.

3. In triangle ABC, $AC = 6$, $BC = 8$, and $AB = 10$. Squares $ACXY$ and $BCWZ$ are drawn exterior to the triangle (the triangle has no interior point in common with either of the squares). Find the distance between the centers of the two squares.

4. If $a = \log_{10}(\sqrt{13} + \sqrt{3})$, express in terms of a the value of $\log_{10}(\sqrt{13} - \sqrt{3})$.

5. The nth term, a_n, of a sequence of numbers is given by $a_1 = 1$ and $a_n = a_{n-1} + 2n$ for $n > 1$. Write an equation expressing a_n as a polynomial in n.

6. In parallelogram $PQRS$, point X is on $\overline{PQ}$ and $PX:PQ = 1:3$. Point Y is on $\overline{PS}$ and $PY:PS = 1:4$. The line XY intersects diagonal $\overline{PR}$ in Z. Find the numerical value of the ratio $PZ:PR$.

7. The number N has three digits when written in base ten notation. Its cube root is the sum of its three digits. Find N.

8. A train leaves the Jerome Avenue subway terminal every 12 minutes. These trains travel down the Jerome branch line, then down the Lexington Avenue trunk line. A train leaves the White Plains Road terminal every 6 minutes. These trains travel down the White Plains branch line, then also join the Lexington Avenue trunk line. All the trains are scheduled so that there are constant intervals between them as they run down Lexington Avenue. How many minutes are there in this interval?

9. Find the largest integer smaller than $\log_4 9 + \log_9 28$.

10. In equilateral triangle ABC, $AB = 12$. One vertex of a square is at the midpoint of $\overline{AB}$, and the two adjacent vertices are on the other two sides of the triangle. The length of a side of the square may be expressed as $p\sqrt{2} + q\sqrt{6}$ where p and q are rational numbers. Find the ordered pair (p,q).

11. If $0 < x < \frac{\pi}{4}$ and $\cos x + \sin x = \frac{5}{4}$, find the numerical value of $\cos x - \sin x$.

12. Find the least positive remainder when the integer $2^0 + 0^2 + 2^1 + 1^2 + 2^2 + 2^2 + 2^3 + 3^2 + 2^4 + 4^2 + \ldots + 2^{100} + 100^2$ is divided by 8.

13. The number N is represented by a three-digit base ten numeral. N is equal to the cube of its units digit, and is also equal to the square of a two-digit numeral formed by its other two digits. Find N.

14. In triangle ACB, $AC = 7$ and $BC = 5$. Squares $ACXY$ and $BCWZ$ are drawn exterior to the triangle (the triangle has no interior point in common with either square). Find the numerical value of $AB^2 + XW^2$.

15. Find the sum of the seventeenth powers of the 17 roots of the equation $x^{17} - 3x + 1 = 0$.

16. A train leaves Main Street for Times Square every 15 minutes, starting at 8 A.M. A train also leaves Times Square for Main Street on the same schedule. All trains run at a constant speed along the same tracks, and the trip either way takes 90 minutes. Including the trains it meets at each terminal, how many trains will the one leaving Main Street at noon encounter on its trip to Times Square?

17. Altitude $\overline{CD}$ to hypotenuse $\overline{AB}$ of right triangle ABC is a diameter of circle O. This circle intersects $\overline{AC}$ in E and $\overline{BC}$ in F. If $AC = 9$ and $BC = 12$, find EF.

18. Find the degree-measure of all angles x such that $0° \le x \le 180°$ and $\cos^6 x - \sin^6 x + \frac{(\sin^2 2x)(\cos 2x)}{4} = 0$.

19. AB and CA are decimal numerals, and A, B, C are distinct digits. If four times AB equals CA, find the ordered triple (A,B,C).

20. Two cardboard rectangles each have dimensions 2 and 8. They are placed on a table so that one pair of diagonally opposite corners coincides, but the other pair of diagonally opposite corners does not coincide. Find the area of the region in which the rectangles overlap.

21. The numbers a and b are both perfect squares, and are both represented by four-digit decimal numerals. The digits of these two numerals are the same but in the reverse order. If the square root of a divides the square root of b, find a.

22. In trapezoid $ABCD$, the ratio of base AB to base CD is 2:3. Diagonals $\overline{AC}$ and $\overline{BD}$ intersect in X, and the line through X parallel to $\overline{AB}$ intersects $\overline{AD}$ at P. Find the ratio of the area of triangle PAX to the area of triangle ABD.

23. From a point P on the hyperbola with equation $x^2 - y^2 = 9 - 4y$, tangents are drawn to the circle centered at the origin with radius 1. What is the minimal possible length of these tangents?

24. Walking along a bus line, a student found that a bus caught up with her (going in the same direction) every 12 minutes and a bus passed her (going in the opposite direction) every 4 minutes. The student travels at a constant rate, and all buses travel at the same constant rate as one another. The buses leave their terminals at equally spaced intervals of time. How long, in minutes, is each such interval?

25. A boy has as many brothers as he has sisters. His sister has twice as many brothers as she has sisters. How many children are in this family?

26. Radii $\overline{OA}$ and $\overline{OB}$ of circle O meet at an angle of 120°. A square has one vertex on $\overline{OA}$, another on $\overline{OB}$, and two more on minor arc $\overset{\frown}{AB}$. If $OA = 13$, the area of the square can be written as $p + q\sqrt{3}$, where p and q are rational. Find the ordered pair (p,q).

27. If $r_1, r_2, \ldots, r_7$ are the roots of $2x^7 - x^6 - 5x^5 + 17x^4 - 419x^2 - 372x - 10 = 0$, find $(1 + r_1)(1 + r_2)\cdots(1 + r_7)$.

28. Find all ordered pairs (x,y) of <u>positive</u> integers satisfying

$$x = \frac{6 - x}{y^2 - x}.$$

29. One hundred pennies are arranged in seven stacks, of which no two stacks contain the same number of pennies. A student counts the number of pennies in each stack and takes 50 pennies in such a way as to disturb the fewest number of stacks. He ends up taking pennies from N stacks. For all such arrangements of pennies, what is the largest possible value of N that will be necessary?

30. A right angle has its vertex at the centroid of an equilateral triangle of side 1 unit. Find the maximum possible area interior to both the angle and the triangle.

Problems—Fall 1984

1. (91%) In square $ABCD$, points P, Q, R, S are chosen on sides $\overline{AB}$, $\overline{BC}$, $\overline{CD}$, $\overline{DA}$, respectively, so that $AP:PB = BQ:QC = CR:RD = DS:SA = 1:3$. Find the ratio of the area of $PQRS$ to that of $ABCD$.

2. (62%) A student guesses at random at three true-false questions. What is the probability that she gets at least two correct answers?

3. (31%) In rectangle $ABCD$, $AB = 6$ and diagonal $BD = 10$. Circle O (with center O) is inscribed in triangle CBD, and circle P (with center P) is inscribed in triangle BAD. Find OP.

4. (05%) The cube of a certain integer has a decimal representation consisting of ten digits, of which the two leftmost, as well as the rightmost, are the digit 7. Find the integer whose cube has this form.

5. (24%) Find all ordered pairs (x,y) of real numbers such that $3^{x^2-2xy} = 1$ and $2 \log_3 x = \log_3 (y + 3)$.

6. (25%) Find the numerical value of $\cos 15°(\sin 75° + \cos 45°) + \sin 15°(\cos 75° - \sin 45°)$.

7. (89%) For all real nonzero numbers, $f(x) = 1 - \frac{1}{x}$ and $g(x) = 1 - x$.

   If $h(x) = f[g(x)]$ for what value of x does $h(x) = 8$?

8. (52%) Find the numerical value of $\dfrac{\cos 15° + \sin 15°}{\cos 15° - \sin 15°}$.

9. (28%) Side $\overline{BC}$ of triangle ABC is extended through C to X so that $BC = CX$. Similarly, side $\overline{CA}$ is extended through A to Y so that $CA = AY$, and side $\overline{AB}$ is extended through B to Z so that $AB = BZ$. Find the ratio of the area of triangle XYZ to that of triangle ABC.

10. (21%) Find the value of c for which the roots of $x^3 - 6x^2 - 24x + c = 0$ form an arithmetic progression.

11. (43%) Three ferryboats start at a terminal at noon and go to different destinations. Ferryboat A reaches its destination after 20 minutes, boat B after 15 minutes, and boat C after 32 minutes. Upon reaching their destinations, the boats return to the terminal, then make another trip, and so on. The trip back to the terminal in each case is the same length, and takes the same time, as the trip out. What is the least number of hours after which the three ferries will again dock at the terminal simultaneously?

12. (22%) In right triangle ABC, leg $AC = \sin\theta$ and leg $BC = \cos\theta$. Find the length of the longer leg if the length of the median to the hypotenuse $\overline{AB}$ is $\tan\theta$.

13. (56%) Points M, N, and P are the respective midpoints of sides $\overline{AB}$, $\overline{BC}$, and $\overline{CA}$ of triangle ABC. A point X is chosen outside of the plane of triangle ABC. Points D, E, F are chosen such that M, N, and P are respective midpoints of $\overline{XD}$, $\overline{XE}$, and $\overline{XF}$. Find the ratio of the area of triangle DEF to that of triangle ABC.

14. (30%) For all real numbers x, the function $f(x)$ satisfies $2f(x) + f(1 - x) = x^2$. Find $f(5)$.

15. (31%) In triangle ABC, $AB = 5$ and $AC = 8$. Point P is on $\overline{BC}$ and $BP:PC = 3:5$. Find the ratio of the radius of the circle through A, B and P to the radius of the circle through A, C, and P.

16. (09%) If x and y are real numbers, with $x > y$ and $xy = 1$, find the minimum possible value for $\dfrac{x^2 + y^2}{x - y}$.

17. (43%) If $[x]$ denotes the "greatest integer" function, find the largest prime number p such that $\left[\dfrac{n^2}{3}\right] = p$ for some integer n.

18. (02%) Square $ABCD$ has area 1 square unit. Point P is 5 units from its center. Set S is the set of points that can be obtained by rotating point P 90° counterclockwise about some point on or inside the square. Find the area of set S.

19. (17%) Circle O passes through vertex D of square $ABCD$, and is tangent to sides $\overline{AB}$ and $\overline{BC}$. If $AB = 1$, the radius of circle O can be expressed as $p + q\sqrt{2}$. Find the ordered pair of rational numbers (p,q).

20. (35%) Find all ordered pairs (x,y) of real numbers for which $x^2 + xy + x = 14$ and $y^2 + xy + y = 28$.

21. (17%) In a rectangular coordinate system, a tangent from the point $(24,7)$ to the circle whose equation is $x^2 + y^2 = 400$ has point of tangency (a,b) where $b > 0$. Find a.

22. (14%) Angle ABC is a right angle and $CB = 1$. D is a point on ray $\overrightarrow{BC}$ such that $DB = 3$ and E is the point on ray $\overrightarrow{BA}$ such that m angle DEC is maximum. Find the distance BE.

23. (27%) An ordinary pack of playing cards is shuffled, and two cards are dealt face up. Find the probability that at least one of these is a spade.

24. (5%) In convex quadrilateral $PQRS$, diagonals $\overline{PR}$ and $\overline{QS}$ intersect at T, with $PT:TR = 5:4$ and $QT:TS = 2:5$. Point X is chosen between T and S so that $QT = TX$, and $\overline{RX}$ is extended its own length to Y. If point Y is outside the quadrilateral, find the ratio of the area of triangle PSY to that of triangle QRT.

25. (62%) Point P is chosen along leg $\overline{BC}$ of right triangle ABC so that $BP = PA$. If leg $BC = 10$ and leg $AC = 4$, find BP.

26. (65%) Five identical black socks and five identical brown socks are in a drawer. Two socks are picked at random. Find the probability that the two socks picked will match.

27. (76%) The roots of $f(x) = 0$ are 2, 3, 7, 5, and 9.
The roots of $g(x) = 0$ are 3, 5, 7, 8, and -1.
Find all solutions of the equation $\frac{f(x)}{g(x)} = 0$.

28. (44%) A set of distinct, nonzero real numbers is placed along the circumference of a circle. Each of the numbers is equal to the product of the two numbers adjacent to it. What is the least possible number of numbers in the set.

29. (26%) In equilateral triangle ABC of edge length 1, D is on $\overline{BC}$ so that angle $DAC = 45°$. Find the area of triangle DAC.

30. (15%) A regular 11-gon is inscribed in a circle. How many triangles are there whose three vertices are all vertices of the 11-gon and whose interiors contain the center of the circle?

ANSWERS

Spring 1975

1. 419
2. $\frac{1}{100}$, 10, $\frac{\sqrt{10}}{10}$, or equivalent
3. 63
4. $32\frac{8}{11}$
5. $\frac{60}{7}$
6. 0
7. 18, 45, 90, 99
8. (2,8)
9. m
10. (3,7)
11. $\frac{288}{5}$
12. $-1 + \sqrt{2}$
13. -1
14. (1, 5, 5)
15. $\frac{8}{7}$
16. $a = b + 4$
17. 5
18. $c = -a - b$
19. $\{x \mid x \le -3\}$
20. $\{x \mid x \ge 3 \text{ or } 0 < x < 1\}$
21. 6
22. 0, $\frac{2}{3}$, $\frac{7}{2}$
23. $\frac{3}{4}$
24. $\frac{3 \pm \sqrt{5}}{2}$
25. 135, 137
26. d, b, a, c
27. $\frac{5}{4}$ or 5:4
28. 2, 22
29. 12
30. $xyz = \frac{a - a^3}{3}$

Fall 1975

1. 8
2. 248
3. (-1,-4), (0,-3), (1,-2), (2,-1), (3,0), (4,1)
4. 100
5. 15
6. -3
7. 300
8. $-\frac{3}{14}$
9. 17
10. $(5,0)$, $\left(-\frac{5}{2}, \frac{3\sqrt{7}}{2}\right)$, $\left(-\frac{5}{2}, -\frac{3\sqrt{7}}{2}\right)$
11. 900
12. 3:2 or 3/2
13. $m = 29 - 2d$
14. 2, 10
15. 9
16. 1.333
17. 20
18. 7
19. 0, 2, 5, 11
20. $(4\sqrt{3}, 4)$
21. (8,-2)
22. $\frac{3}{4}$ or 3:4
23. 58
24. $\frac{3}{5}$
25. $\frac{1}{9}$, $\frac{1}{16}$
26. $x^2 + y^2 = \frac{a^3 + 2b}{3a}$
27. $\left(2, \frac{1}{9}\right)$, (2,9)
28. 343
29. 12
30. 6

Spring 1976

1. $\frac{1}{3}$ 2. $\frac{\sqrt{2}}{2}$ 3. 4:1 or 4 4. 3, 24

5. 13 6. -24 7. 10 8. 54

9. 20 10. 10201 or 101^2 11. 18

12. 19 13. 12 14. (5,6) 15. 9:2 or $\frac{9}{2}$

16. $\frac{6}{5}$ 17. 365 18. (25,36) 19. 12, 24, 36, 48

20. 10 21. $\left(\frac{2}{7}, 1\right)$ 22. (4,2,2), (2,2,4)

23. 112 24. $b_n = n^2 + 3n - 1$ 25. (1,41,2)

26. 8 27. $\sqrt{97}$ 28. 2, 10, 14 29. $\sqrt{5}$:2 or equivalent

30. $\frac{3}{4}, -\frac{3}{4}, 1, -1, \frac{4}{3}, -\frac{4}{3}$ 31. $\frac{7}{25}$

32. $\left(6, \frac{8}{3}, \frac{3}{2}\right), \left(-6, -\frac{8}{3}, -\frac{3}{2}\right)$

33. $°F = \frac{12}{5}°C - 20$ OR $°F = \frac{12}{5}°C - 20$ or equivalent equation required

34. 10 35. (1,-4), (9,12) 36. (3,5)

Fall 1976

1. k^{32} 2. 0, -1, $\frac{-1 \pm i\sqrt{3}}{2}$ 3. (1,1)

4. $\frac{84}{13}$ 5. 1:2

6. (2,2,2), (-3,-3,-3), (5,1,1), (1,5,1), (1,1,5) 7. 12

8. (8,6,5) 9. 17 10. 3 11. 10^{10}

12. (1,1) 13. 1:2 14. $-\frac{4}{3}$ 15. $\frac{21}{8}$

16. 4 17. 35 18. -2, 3, $\frac{-1 \pm i\sqrt{3}}{2}$

19. 9 20. $\frac{5}{12}$ 21. 1:5 22. $\frac{4}{3}$

23. 12   24. 16   25. $\frac{16}{65}$ 26. 252 or $\binom{10}{5}$

27. -1   28. 20   29. 98   30. 7, 21, 35

Spring 1977

1. 5:3:2   2. $\frac{44}{199}$ 3. 1   4. $0, -\frac{4}{3}$

5. (37,12)   6. 2:1 or 2   7. $\left(\frac{-1+\sqrt{5}}{2}, \frac{3-\sqrt{5}}{2}\right)$

8. $\frac{2}{3}, -\frac{2}{3}$ 9. $3a + 2b$ 10. $\left(\frac{3}{2}, 3\right)$ 11. 21

12. $24(2+\sqrt{3})$ 13. $\frac{1}{6}$ 14. 60

15. $0, -\frac{1}{4}$ 16. any one of: (550,275), (605,110), (451,418)

17. 4   18. 1:2 or $\frac{1}{2}$ 19. $\frac{625}{16}$ 20. (4,7), (28,29)

21. (-2,-3), (3,2)   22. 9:2 or $\frac{9}{2}$ 23. $\frac{77}{85}$

24. 2   25. 11111   26. $\frac{18}{35}$ 27. (40,40)

28. 2   29. -1   30. $\frac{\pi}{4}, \frac{\pi}{2}, \frac{3\pi}{4}$ NOT 45°, 90°, 135°

Fall 1977

1. (1,0), (-1,0), (1,2), (-1,-2)   2. $\sqrt{10}$

3. $\frac{44}{5}$ 4. 50   5. (5,4)   6. $\frac{5}{12}$

7. 4, -4   8. (9, 40, 41)   9. $0, \pm\frac{\sqrt{2}}{2}$

10. $\left(\frac{1}{7}, \frac{2}{7}\right)$, (-2,-4), $\left(1, \frac{1}{2}\right)$, $\left(-\frac{8}{7}, -\frac{4}{7}\right)$, $\left(\frac{1}{3}, \frac{1}{3}\right)$, $\left(-\frac{4}{3}, -\frac{4}{3}\right)$

11. $1 \le x < \sqrt{2}$ or equivalent   12. 20, 24, 25, $\frac{117}{5}$, $\frac{336}{25}$

13. -1   14. $3a:2b:6c$ 15. $\frac{1}{336}$ or equivalent

16. 75   17. $\sqrt{3}$ 18. 83,886,082 or $20 \cdot 2^{22} + 2$

19. 2   20. 5, 7

21. $c = ab$ or equivalent EQUATION solved for c 22. 5, 845

23. $\frac{3}{4}$ 24. $(2^2)(3^{1000})$ 25. 12

26. 72 27. 100, 289 28. $\left(-1, \frac{1}{16}\right)$ 29. (3,4)

30. 1, -1

Spring 1978

1. $\frac{1}{12}$ 2. $\sqrt{5}$ 3. 4/5 4. (6, 5, 4)

5. 13 6. $100^{99} - 99$ 7. 81 8. (25, 39)

9. $\left(-\frac{1}{2}, 1\right)$, (0,1) 10. 257 11. 400π

12. $\sqrt{3}$ 13. $\frac{337}{625}$ 14. $P(x) = 3x + 2$ or equivalent EQUATION

15. $157\frac{1}{2}$ 16. 1680 17. 8 18. $\frac{10}{13}$

19. $1 + \sqrt{3}$ 20. $f(n) = n^2 + n - 1$ or equivalent EQUATION

21. 11 or 8 + 2 + 1 22. $\frac{1}{2}\pi$ or 90° 23. 2/9

24. 180 25. $\frac{1}{2}\sqrt{3}$ 26. (8,3), (8,5), (7,7), (5,8), (3,8)

27. $\frac{5}{16}$ 28. $\left(\frac{4}{3}, 0, -\frac{1}{3}, 0\right)$ 29. -196

30. $\sqrt{580}$ or $2\sqrt{145}$ (or equivalent)

Fall 1978

1. 25 2. 0 3. 974 4. 7

5. 2^{3^5} 6. $\frac{1}{512}$ or equivalent 7. 53 or 53 years

8. $\frac{1}{2}(3^{1978} - 1)$ 9. 2, $\frac{1}{2}$, 3, $\frac{1}{3}$

10. $\frac{3}{4}$ 11. $\frac{2a}{b}$ or equivalent 12. 20

13. 125 14. 5 15. 30 16. 90 or 90°

17. 13 18. 8 19. (1,8), (-3,0), (5,2)

20. 1800 or $1800 21. 7 22. 3

23. 41 24. 0, 1, 2, 3

25. $\left(\frac{11}{2}, 6\right)$ or equivalent 26. $-\frac{1}{4}$ 27. -1

28. 7 29. 1, 2, 3 30. 0

Spring 1979

1. 1 2. $3\sqrt{3} + \frac{3\sqrt{21}}{7}$ 3. 10 4. 80 or 80°

5. -2 6. $894.73 7. 18 8. 45 or 45°

9. 20 10. $2^{15} + 1, 2^{21} + 1, 2^{35} + 1, 2^{105} + 1$

11. 4 12. 45 or equivalent 13. (576, 64)

14. $\sqrt{2} - 1$ 15. $\sqrt{3}$ 16. 1 17. $\frac{1}{2}$, 1, 2

18. sec 6° or sec(π/30) or exact equivalent 19. B

20. (-1, -1, 1) 21. $\frac{11}{16}$ 22. 36

23. 3^{2n} 24. 15 25. nuts 26. 10

27. z 28. 8 29. 118864

30. .20736 or $(2^3 \cdot 3^4)/5^5$ or 648/3125 or equivalent

Fall 1979

1. 54 2. 405 3. $a + b - 2$ or equivalent

4. $\left(-\frac{3}{4}, 1\right)$ or equivalent ordered pair 5. $\frac{4}{5}$

6. $-\frac{1}{4}$ 7. 20 8. $\frac{1}{6}$ 9. -2

10. $\frac{1}{2}$ 11. 2 12. $c = \frac{3ab - a^3}{2}$ or equivalent equation

13. 57 14. 1 15. $\frac{y^2}{2}$ or $\frac{1}{2}y^2$

16. 2 and -3 (both required) 17. 48 or 60 (both required)

18. x 19. 31 20. $\frac{1}{2}$, 2 21. 11

22. $\frac{1}{2}$ 23. (1,3,2), (-1,3,-2), (-1,-3,2), and (1,-3,-2)

24. $\frac{1}{4}$ 25. 16 26. (13/2, 5/2)

27. $\frac{91}{100}$ 28. 3840 29. 22 30. (4,-1)

Spring 1980

1. 72 2. $(x^2 + 1)(2x^2 + x + 2)$ 3. 4

4. 4 5. 80° 6. $\frac{4}{3}$ 7. 30

8. $\frac{3}{2}$ 9. 792 10. $x^4 + 4x^2 - 21$

11. $\left(\frac{1}{5}, \frac{2}{5}\right)$ 12. 6/5 13. 4

14. $\frac{1}{2} + \frac{1}{2}i$, $\frac{1}{2} - \frac{1}{2}i$ or equivalents; $-\frac{1}{2} + \frac{1}{2}i$, $-\frac{1}{2} - \frac{1}{2}i$ (all four required)

15. 5/2 16. 2/3 17. 0°, 180° 18. 15

19. 2 20. 3 21. 10 22. $x^2 + 1$

23. (-4,-3) (ordered pair required) 24. $\frac{50}{51}$

25. 6 26. (16,15), (9,2) (both required)

27. 8 28. (5,1) 29. $6 - 3\sqrt{2}$ 30. 6

Fall 1980

1. $m + ks - 12s$ 2. (1,1,-8) 3. 5

4. 112 5. 17 6. $\frac{-8}{9}$ 7. $\left(\frac{1}{3}, 1\right)$

8. (4,-3) 9. $n = q + 4 + 4/q$ (equation required)

10. 9 11. (100,100) 12. $3\sqrt{2}$ 13. 5/6

14. 17 15. 423 16. $\frac{1}{2}$ 17. $\frac{\sqrt{2}}{\sqrt{3}}$ or $\frac{\sqrt{6}}{3}$

18. 10 19. $\frac{-1 + \sqrt{5}}{2}$ 20. 7 21. 96

22. $\frac{5}{2}$ or 5:2 or equivalent 23. 3 24. 8/165

25. 13/2 26. (0,-6,-12) 27. $\frac{3\pi}{10}$ 28. 2

29. $\frac{-1 \pm i\sqrt{3}}{2}$ 30. 1/8

Spring 1981

1. $\frac{r+1}{r-1}$
2. $\frac{3-\sqrt{5}}{2}$
3. (1,-1), (-1,-1) (both required)
4. $\frac{a}{1-a}$
5. $\sqrt{5}$
6. $\frac{11}{16}$
7. 11
8. $\frac{b}{1-2b^2}$
9. $C = \frac{AB}{1+A+B}$ (equation required)
10. 1, 3, 9
11. $\left(\frac{\sqrt{3}}{2}, \frac{1}{2}\right)$
12. 5
13. $c = \sqrt{a^2 + 4b}$ (equation required)
14. 500/11 or 45 $\frac{5}{11}$ or equivalent
15. 5
16. 11
17. 20
18. 50/3
19. 81
20. (3, 5)
21. 15
22. 4
23. 3:1 or 3
24. 9
25. $\frac{1}{2-r}$
26. $\sqrt{17} + 2\sqrt{2}$
27. 0
28. (2,6)
29. 4/5
30. 6

Fall 1981

1. 8 or 8¢
2. 2
3. 72°
4. $x^2 + 2x + 4$ and $x^2 - 2x + 4$
5. 5
6. $f(x) = x + 1$ (equation required)
7. $3\sqrt{3}$
8. $\frac{1}{2}$
9. 103
10. $\frac{1}{9}$ (gram)
11. 2
12. 40
13. 4
14. $\frac{1}{7}$
15. (1,2)
16. 7
17. $\frac{27}{11}$
18. 601
19. 20
20. $\frac{1}{6^4}$ or $\frac{1}{1296}$
21. FOUR or 4
22. $\frac{56}{33}$
23. 15, 255 (both required)
24. 80
25. 4 or 4 grams
26. 4:25 or $\frac{4}{25}$
27. 66
28. 11
29. 30°
30. $\sqrt{39}$

Spring 1982

1. 0
2. 5200
3. 7
4. $\frac{13}{25}$
5. 8
6. 2
7. 6
8. 2796
9. 72
10. $P(x) = x^2 + 8$
11. $\frac{1}{4}$
12. 2/3
13. (8,12)
14. $\frac{1 + \sqrt{5}}{2}$
15. 12
16. $2x^2 + 2x$
17. $y = \frac{a^2 - 2}{2}$ (equation required)
18. $\frac{15}{4}$
19. 3
20. 9
21. $x = \frac{p^2}{100}$ (equation required)
22. 2113
23. $\frac{1 + \sqrt{5}}{2}$
24. 45
25. 61
26. 40
27. 6
28. $\frac{9}{4}$
29. $k = 17$
30. 108°

Fall 1982

1. 3, $\frac{1}{3}$
2. 1 (or "one")
3. 178
4. 2
5. $8\sqrt{2}$ or equivalent
6. $3^{10} - 3 \cdot 2^{10} + 3$ or 55980
7. 140
8. $\frac{8}{3}$, $-\frac{17}{3}$
9. $2\sqrt{5}$
10. $\frac{2}{1 + \sqrt{2}}$ or $2\sqrt{2} - 2$ or equivalent
11. 1
12. 2
13. 25
14. 256, $\sqrt{2}$ or equivalents (both required)
15. 3
16. 12
17. (11,3)
18. $\sqrt{3}$
19. $\sqrt{10}$, $\frac{\sqrt{10}}{3}$
20. -66
21. 9
22. 5
23. $\frac{9}{4}$
24. 45°
25. $\frac{3}{8}$
26. 3
27. 2
28. 16
29. 16
30. 0, $\frac{\pi}{2}$, $\frac{\pi}{8}$, $\frac{5\pi}{8}$

Spring 1983

1. 1111
2. 5:8 or equivalent
3. $\frac{19}{6}$ or $-\frac{19}{6}$
4. 15
5. $(\sqrt[3]{9}, \sqrt[3]{3})$ or equivalent
6. 6
7. $\frac{1}{3}$
8. 4
9. 3
10. $\frac{17}{13}$
11. 15, 45, 18
12. 25
13. 39
14. $\frac{1}{3}$
15. 23
16. 140
17. 10
18. 21
19. 105
20. 90° or $\frac{\pi}{2}$
21. 10
22. 7
23. 60
24. $-\frac{8}{3}$
25. 6
26. 1
27. (1,5,8) and (4,5,7) (both required)
28. $-\frac{3}{2}$
29. 30
30. $\sqrt{\frac{15}{2}}$

Fall 1983

1. -1
2. 3
3. 3
4. $\frac{\pi}{2}$
5. 4
6. 648
7. 15
8. $\frac{13}{6}$
9. $6\sqrt{2}$
10. 9
11. $\sqrt{95}$
12. 16
13. 90°
14. 9
15. 24
16. $4\sqrt{2}$ or equivalent
17. $m = n^3$ (equation required)
18. $4\sqrt{11}$ or equivalent
19. 16
20. 5
21. 0, -1 (both required)
22. $2\sqrt{6}$
23. 90 or 90°
24. 3, 4 (both required)
25. 7
26. $4\sqrt{3}$
27. $\frac{12}{\sqrt[3]{2}}$
28. (1, .2, 2.1)
29. $\frac{1}{3}$
30. $\frac{1}{4}$

Answers

Spring 1984

1. 1250
2. $80°$
3. $7\sqrt{2}$
4. $1 - a$
5. $a_n = n^2 + n - 1$ (equation required)
6. $\frac{1}{7}$
7. 512
8. 4 or 4 minutes
9. 3
10. (9,-3)
11. $\frac{\sqrt{7}}{4}$
12. 5
13. 729
14. 148
15. -17
16. 13
17. $\frac{36}{5}$
18. 45, 135
19. (2,3,9)
20. $\frac{17}{2}$
21. 1089
22. 6:25
23. $\sqrt{6}$
24. 6 minutes
25. 7
26. (156,-39)
27. 19
28. (2,2), (3,2)
29. 3
30. $\frac{1}{3(1 + \sqrt{3})}$ or $\frac{\sqrt{3} - 1}{6}$ or equivalent

Fall 1984

1. 5:8 or equivalent
2. $\frac{1}{2}$
3. $2\sqrt{5}$
4. 1983
5. (2,1)
6. $\frac{3}{2}$
7. $x = \frac{8}{7}$
8. $\sqrt{3}$
9. 7 or 7:1
10. $c = 64$
11. 16
12. $\frac{2\sqrt{5}}{5}$
13. 1
14. $\frac{34}{3}$
15. 5:8
16. $2\sqrt{2}$
17. 5
18. 2
19. (2,-1)
20. (2,4), (-7/3,-14/3) (both required)
21. 12
22. $\sqrt{3}$
23. $\frac{15}{34}$
24. 15:8
25. $\frac{29}{5}$ or equivalent
26. $\frac{4}{9}$
27. 2,9 (both required)
28. 6
29. $\frac{3 - \sqrt{3}}{4}$
30. 55

SOLUTIONS

Solutions—Spring 1975

1. $n = 3a + 2 = 4b + 3 = 5c + 4 = 6d + 5 = 7e + 6$.
Then $n + 1$ is divisible by 3, 4, 5, 6, 7 (since, for example, $n + 1 = 3a + 3 = 3(a + 1)$). The smallest such positive $n + 1$ is $\text{lcm}(3,4,5,6,7) = 2^2 \cdot 3 \cdot 5 \cdot 7 = 420$. Therefore, the smallest such positive n is 419.

2. First note that x must be positive. Then, using base ten logarithms, let $a = 2 + \log x$, $b = -1 + \log x$, $c = 1 + \log x^2 = 1 + 2 \log x$. Observe that $a + b = c$, so this problem is equivalent to $a^3 + b^3 = c^3 = (a + b)^3$.
Therefore, (formula 1.22) $a = 0$, $b = 0$, or $c = 0$; in other words, $2 + \log x = 0$, $-1 + \log x = 0$, or $1 + 2 \log x = 0$, so $\log x = -2, 1, -\frac{1}{2}$, and $x = \dfrac{1}{100}, 10, \dfrac{1}{\sqrt{10}}$.

3. Method 1: Since point D is equidistant from lines AB and AC (see diagram), AD must be the angle bisector of angle BAC. But then, since $DE || AB$ and $DF || AC$, line AD must bisect angle EDF as well. In a similar manner we may prove that the triangles ABC, DEF have the same angle bisector (lines), and hence a common incenter. And, since a tangent to a circle is perpendicular to a radius at its point of contact, the inradius of triangle ABC must be 2 more than that of triangle DEF.
We can use formula 3.62 to find the inradius of triangle ABC. The semiperimeter is 21. The area, by Hero's formula (3.61) or otherwise, is 84. Hence the inradius of triangle ABC is 4 and that of triangle DEF is $4 - 2 = 2$.
Now the triangles ABC, DEF are similar, and the ratio of similitude is the ratio of their inradii, or $4:2 = 2$. Hence the ratio of their areas is $4:1$, and the required area is $3/4$ that of triangle ABC, or 63.

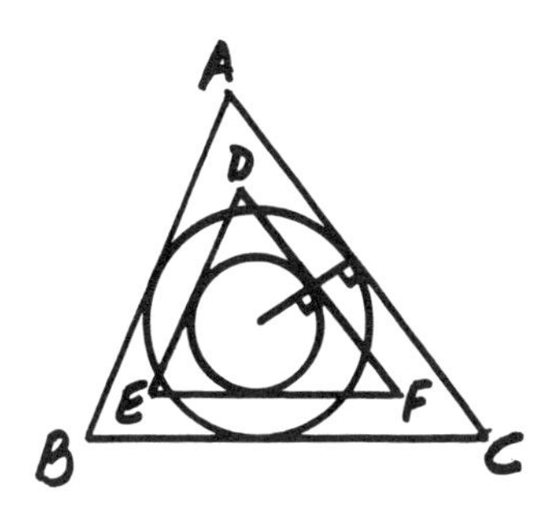

Method 1

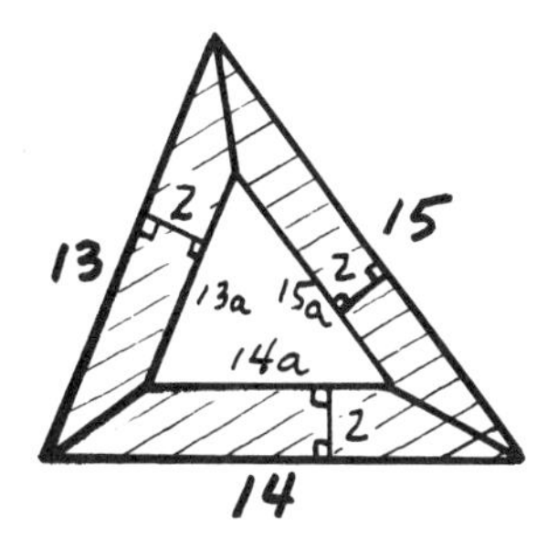

Method 2

3. (Continued)

Method 2: As before, the area of triangle I is 84. If the ratio of similitude of triangles I and II is a, $A_{\text{triangle II}} = 84a^2$. We can analyze the shaded area both as $A_{\text{triangle II}} - A_{\text{triangle I}}$ and as the sum of the areas of three trapezoids, each with $h = 2$. Then $A_{\text{triangle II}} - A_{\text{triangle I}} = A_{\text{3 trapezoids}}$ gives

$84 - 84a^2 = \frac{1}{2} \cdot 2(13 + 13a) + \frac{1}{2} \cdot 2(14 + 14a) + \frac{1}{2} \cdot 2(15 + 15a)$,

$84(1 - a^2) = 42(1 + a)$, so $2(1 - a) = 1$, so $a = \frac{1}{2}$ and the shaded area is equal to $84 - 84a^2 = 84 - 21 = 63$.

The area of triangle I is easily calculated if we draw the altitude to the side with measure 14. If we solve for this altitude using the Pythagorean theorem, we find its length to be 12. Hence triangle I can be thought of as made from two right triangles with integral sides, which are "stuck together," along a common side. A triangle such as triangle I is called Heronian.

4. Let x be the number of minute units the hour hand advances in the desired time interval. The minute hand advances $30 + x$ units in that time. Since the minute hand moves 12 times as fast as the hour hand in any time interval, $30 + x = 12x$, $11x = 30$,

$x = \frac{30}{11} = 2\frac{8}{11}$, so $30 + x = 32\frac{8}{11}$ minutes.

5. Draw radii $\overline{OE}$, $\overline{OF}$ perpendicular to sides $\overline{AB}$, $\overline{BC}$, respectively. Then quadrilateral $BEOF$ is a square. Furthermore, triangle $CFO \sim$ triangle CBA so $\frac{15 - r}{r} = \frac{15}{20}$ and $300 - 20r = 15r$, $35r = 300$,

$r = \frac{60}{7}$.

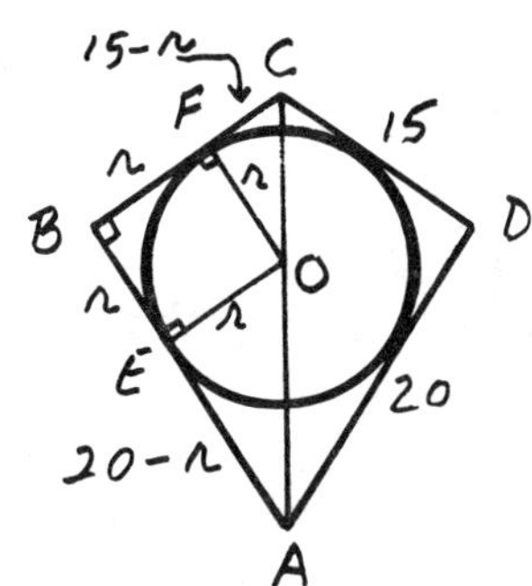

6. Note first that $f(x)$ and $g(x)$ are real if and only if $|x| \geq 2$. Rearranging terms in the original equation, we have:

$\sqrt{x^2 - 4} - x = 2\sqrt{x^2 - 1}$. Squaring both sides, $2x^2 - 4 - 2x\sqrt{x^2 - 4} = 4x^2 - 4$ so $-x\sqrt{x^2 - 4} = x^2$. Squaring again, $x^4 - 4x^2 = x^4$, so $4x^2 = 0$ and $x = 0$. But, since $|x| \geq 2$, $x = 0$ is an extraneous root, so there are no real solutions.

7. The sum of the digits in any multiple of 9 is itself a multiple of 9. Hence, $A + B$ must be a multiple of 9. The possibilities for AB are thus 18, 27, 36, 45, 54, 63, 72, 81, 90, 99. By trial and error, the only solutions are 18, 45, 90, 99.

8. The closest point, M, is on line $\overleftrightarrow{OP}$. Since $OP = \sqrt{(6-(-2))^2 + (5-11)^2} = \sqrt{8^2+6^2} = 10$, and $r = 5$, $OM = \frac{1}{2}OP$, so M is the midpoint of $\overline{OP}$ and M is the point (2,8).

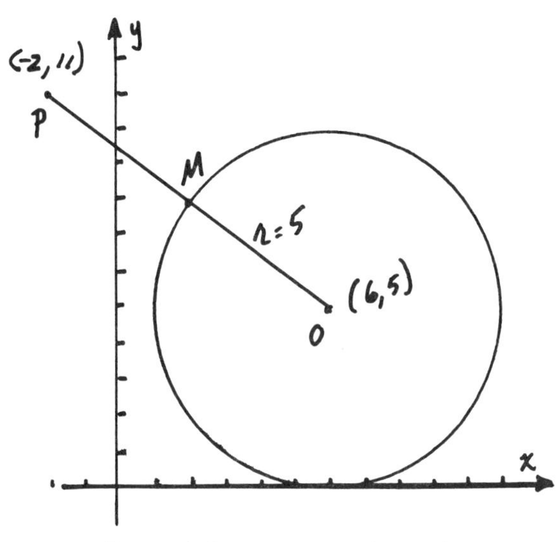

9. The quadrilateral formed by connecting the centers is a square. It follows that $a\sqrt{2} + a = r$, so $a(\sqrt{2}+1) = r$.

$$a = \frac{r}{\sqrt{2}+1} = r(\sqrt{2}-1) = m(\sqrt{2}-1),$$

and $r = m$.

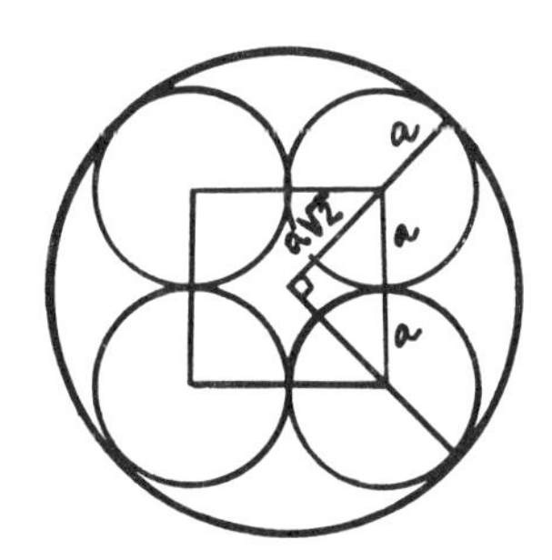

10. $\sqrt{10+\sqrt{84}} = \sqrt{a} + \sqrt{b}$, where a and b are positive integers. Square both sides to get $10 + 2\sqrt{21} = a + b + 2\sqrt{ab}$. Equate rational and irrational parts to get $a + b = 10$ and $ab = 21$. Solve to obtain $(a,b) = (7,3)$ or $(3,7)$. Since $a < b$, $(a,b) = (3,7)$. For more such problems, see Chrystal (pp. 207-210), or Hall and Knight, Elementary Algebra (pp. 281-284).

11. Method 1: $PQRS$ is a square. (This is easy to prove, for example, using similar triangles.) By the Pythagorean theorem in triangle DEA, $DE = \sqrt{16 + 144} = 4\sqrt{10}$. In similar triangles PAE and ADE, the ratio of similitude is $\frac{AE}{DE} = \frac{4}{4\sqrt{10}} = \frac{1}{\sqrt{10}}$. Therefore,

$$PE = \frac{1}{\sqrt{10}}(4) = \frac{2}{5}\sqrt{10},\quad PA = \frac{1}{\sqrt{10}}(12) = \frac{6}{5}\sqrt{10},\quad SD = PA = \frac{6}{5}\sqrt{10}, \text{ and}$$

$$PS = DE - PE - SD = 4\sqrt{10} - \frac{2}{5}\sqrt{10} - \frac{6}{5}\sqrt{10} = \frac{12}{5}\sqrt{10}. \text{ Hence,}$$

$$\text{Area}_{PQRS} = \left(\frac{12}{5}\sqrt{10}\right)^2 = \frac{144}{25}(10) = \frac{288}{5}.$$

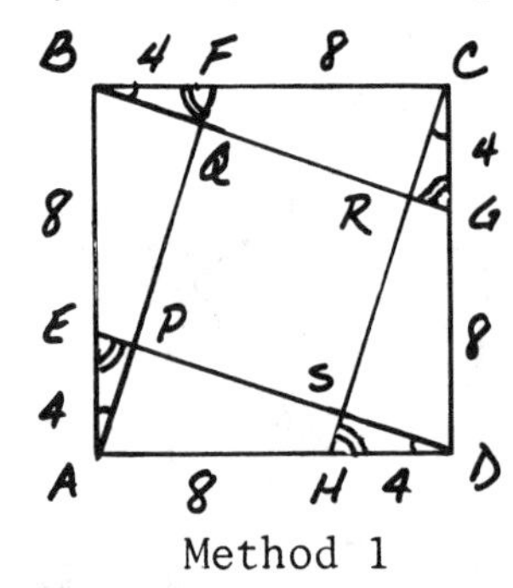

Method 1

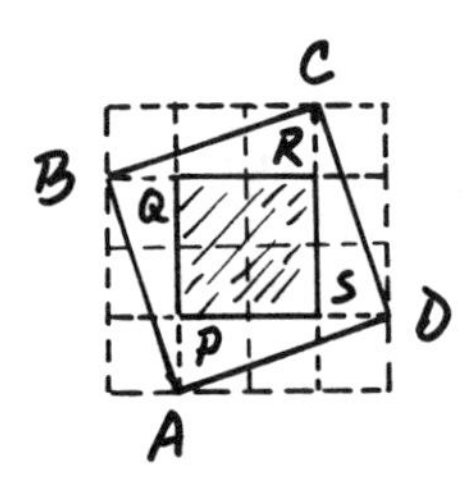

Method 2

Method 2: Area_{PQRS} = 4 boxes. The total unshaded area is 12 boxes, so the unshaded area in $ABCD$ is 6 boxes, and the total area of $ABCD$ is 10 boxes. Thus, $A_{PQRS} = \frac{4}{10}\left(A_{ABCD}\right) = \frac{4}{10}(144) = \frac{288}{5}$.

12. Let $x = \cfrac{1}{2 + \cfrac{1}{2 + \cfrac{1}{2 + \cfrac{1}{2 + \dots}}}}$

Then, assuming convergence, we have $x = \frac{1}{2 + x}$, or $x^2 + 2x = 1$.

If $x^2 + 2x - 1 = 0$, $x = -1 \pm \sqrt{2}$. Since $x > 0$, $x = -1 + \sqrt{2}$ or $\sqrt{2} - 1$. For a discussion of convergence of this type of continued fraction, see Niven and Zuckerman (Chapter 7) or Khinchin or Olds.

13. We have: $\frac{x^3 - x^2 - x + 1}{x^3 - x^2 + x - 1} = \frac{x^2(x - 1) - (x - 1)}{x^2(x - 1) + (x - 1)}$

$$= \frac{(x - 1)(x^2 - 1)}{(x - 1)(x^2 + 1)} = \frac{(x - 1)^2(x + 1)}{(x - 1)(x^2 + 1)} = 0.$$

The only value of x that makes the numerator, but not the denominator, 0 is $x = -1$.

14. Regrouping, we have:

$$\begin{aligned}
&(x + 1)(x + 2)(x + 3)(x + 4) + 1 \\
&= [(x + 2)(x + 3)][(x + 1)(x + 4)] + 1 \\
&= [(x^2 + 5x + 6)(x^2 + 5x + 4)] + 1 \\
&= (y + 6)(y + 4) + 1, \text{ where } y = x^2 + 5x \\
&= y^2 + 10y + 25 \\
&= (y + 5)^2 \\
&= (x^2 + 5x + 5)^2, \text{ so that } (A,B,C) = (1,5,5)
\end{aligned}$$

Note that the product of any four consecutive integers, when increased by one, results in a perfect square.

15. Drawing radius $\overline{OT}$, $\overline{OT} \perp \overline{AT}$, so $AT = 4$.
Using the Pythagorean theorem in triangle OTX, we have
$9 + (4 - x)^2 = (3 + x)^2$, and $x = \frac{8}{7}$.

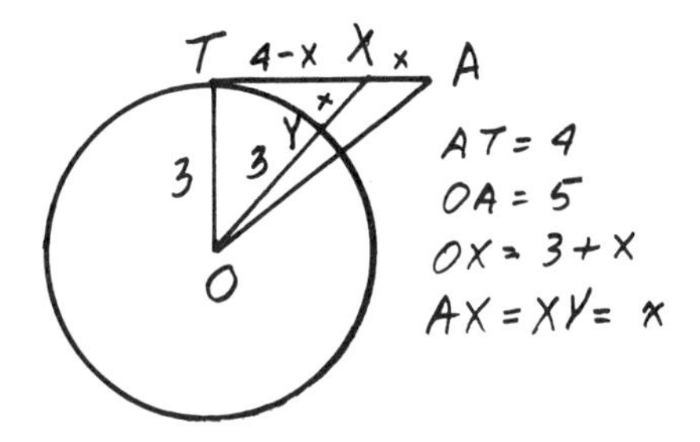

16. Since $a > 0$ and $b > 0$, we know that $\sqrt{ab} \neq 0$ and $a\sqrt{b} = \sqrt{a}\sqrt{ab}$ and $b\sqrt{a} = \sqrt{b}\sqrt{ab}$. The given equation is thus equivalent to

$$\frac{\sqrt{ab}(\sqrt{a} + \sqrt{b})}{\sqrt{ab}(\sqrt{a} - \sqrt{b})} - \frac{\sqrt{ab}(\sqrt{a} - \sqrt{b})}{\sqrt{ab}(\sqrt{a} + \sqrt{b})} = \sqrt{ab} \qquad \text{or}$$

$$\frac{\sqrt{a} + \sqrt{b}}{\sqrt{a} - \sqrt{b}} - \frac{\sqrt{a} - \sqrt{b}}{\sqrt{a} + \sqrt{b}} = \frac{(\sqrt{a} + \sqrt{b})^2 - (\sqrt{a} - \sqrt{b})^2}{a - b} = \frac{4\sqrt{ab}}{a - b} = \sqrt{ab}\ .$$

Since $\sqrt{ab} \neq 0$, dividing by $\sqrt{ab}$ produces $\frac{4}{a - b} = 1$. Therefore, $4 = a - b$ and $a = b + 4$.

17. $\frac{W}{L} = \frac{L}{L + W}$, so $L^2 = WL + W^2$. Therefore, $L^2 - WL - W^2 = 0$ and $L = \frac{W \pm \sqrt{5W^2}}{2}$. But $L > 0$, so $L = \frac{W(\sqrt{5} + 1)}{2}$. We are given that $LW = 50(\sqrt{5} - 1)$, so $L = \frac{50(\sqrt{5} - 1)}{W}$. Equating these two expressions for L yields $\frac{W(\sqrt{5} + 1)}{2} = \frac{50(\sqrt{5} - 1)}{W}$. Therefore,

17. (Continued)

$W^2 = \dfrac{100(\sqrt{5} - 1)}{\sqrt{5} + 1}$. Rationalizing the denominator, $W^2 = 25(\sqrt{5} - 1)^2$, so $W = 5(\sqrt{5} - 1)$ and $k = 5$. This sort of rectangle is a "golden" rectangle. For more on these, see Coxeter, Posamentier, Excursions (Chapter 9), or Huntley.

18. First note that $a \neq 0$, $b \neq 0$, and $a \neq b$. By the factor theorem (1.311), q is a root of both equations, so: (*) $q^2 + aq + bc = q^2 + bq + ac = 0$, $aq - bq = ac - bc$, $(a - b)q = (a - b)c$. Since $a \neq b$, $q = c$. Substitution in (*) yields $c^2 + ac + bc = 0$ or $c(c + a + b) = 0$, whereupon $c = 0$ or $c = -a - b$. Since $c \neq 0$, $c = -a - b$.

19. This problem can be done purely algebraically, but it is easier to remember that absolute value may denote distance on the number line. Thus, the equation $|x| - |x + 3| = 3$ means that the distance from x to 0 is 3 more than the distance from x to -3. This happens whenever x is to the "left" of -3 on the number line, or $x \leq -3$.

20. Note that $x \geq 0$ for $\sqrt{x}$ to be meaningful and $x \neq 0,1$ for the denominators to be nonzero. Rationalizing denominators and adding the resulting fractions yields $\dfrac{2x}{x^2 - x} \leq 1$ or $\dfrac{2x}{x(x - 1)} \leq 1$.

Since $x \neq 0$, $\dfrac{2}{x - 1} \leq 1$. If $x > 1$, $2 \leq x - 1$ or $x \geq 3$.

If $0 < x < 1$, $2 \geq x - 1$ or $x \leq 3$, so $0 < x < 1$. Thus the solution set is $\{x | 0 < x < 1 \text{ or } x \geq 3\}$.

21. Method 1: Use coordinate geometry, let $A(0,0)$, $B(12,0)$, $X(x,0)$. It follows that $Y\left(\frac{x}{2}, \frac{x}{2}\right)$ and $Z\left(6 + \frac{x}{2}, 6 - \frac{x}{2}\right)$ since Y is the midpoint of $\overline{AC}$ and Z is the midpoint of $\overline{XE}$. The coordinates of the midpoint of $\overline{YZ}$ are then $W\left(3 + \frac{x}{2}, 3\right)$. Note that the y-coordinate of W is always 3. Since $0 \leq x \leq 12$, W moves along a horizontal line from (3,3) to (9,3); this is a path of length 6.

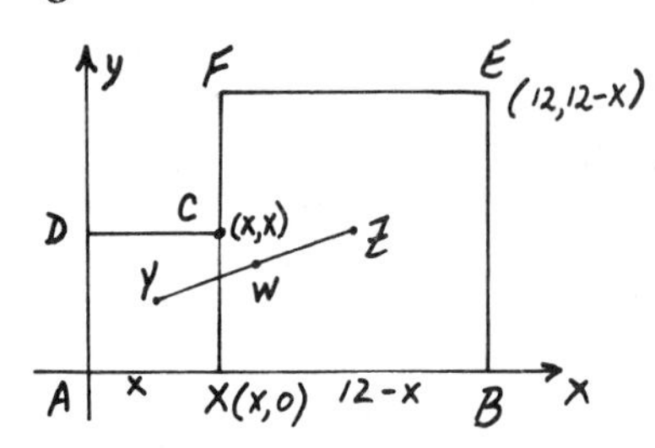

Method 1

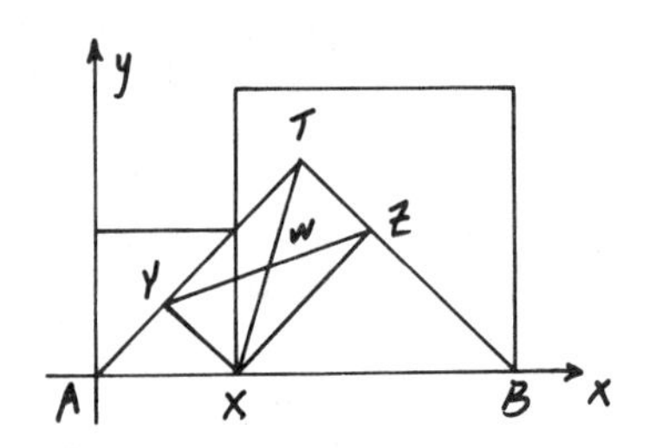

Method 2

21. (Continued)

Method 2: Since triangles AYX, XZB are right-angled, we can extend $\overline{YZ} = \overline{XT}$, and W is their common midpoint. Thus the locus of W is the locus of midpoints of the segments TX as X varies from A to B. This is easily seen to be a line segment parallel to AB and equal to $\frac{1}{2}AB$. One way to prove this final result is by using homothecy. See, for example, Coxeter and Greitzer, or Altshiller-Court.

22. This equation has the form $\dfrac{a + b}{a - b} = \dfrac{a + c}{a - c}$ where $a = 3x^3 - 2x^2$, $b = x + 1$, and $c = 5x - 13$. By 1.12 it is equivalent to $\dfrac{a}{b} = \dfrac{a}{c}$. Hence $a = 0$ or $b = c$. Thus $3x^3 - 2x^2 = x^2(3x - 2) = 0$ or $x + 1 = 5x - 13$. Considering both cases we obtain $x = 0, \dfrac{2}{3}, \dfrac{7}{2}$.

23. Method 1: Consider the special case where triangle ABC is equilateral, so D coincides with X and E with Y. Then triangle $AXY \sim$ triangle ABC and $XY = \frac{1}{2}BC$, $\text{Area}_{\text{triangle } AXY}$ $= \frac{1}{4}\text{Area}_{\text{triangle } ABC}$ and $\text{Area}_{XYBC} = \dfrac{3}{4}\text{Area}_{\text{triangle } ABC}$.

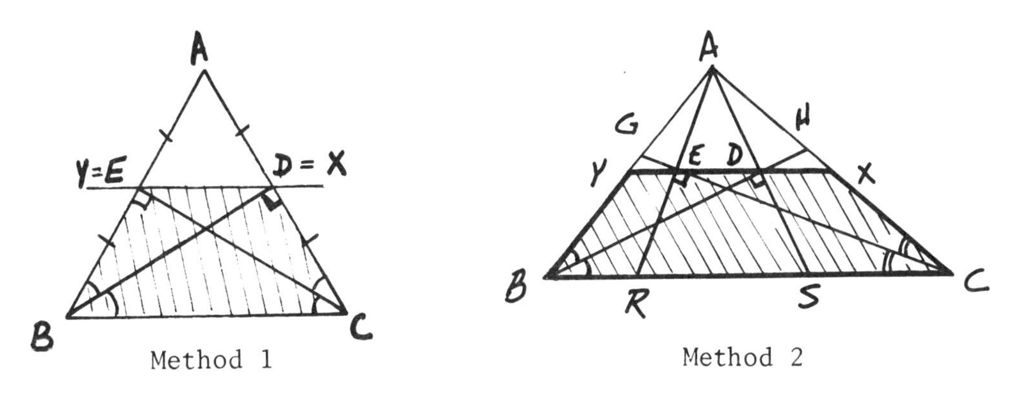

Method 1 Method 2

Method 2: (This solution is general, and points D and E need not be interior to ABC.) Extend $\overline{AE}$ and $\overline{AD}$ to meet $\overline{BC}$ in R and S, respectively. Then triangle $AEC \cong$ triangle REC (by A.S.A.), so $AE = ER$. Similarly, triangle $ADB \cong$ triangle SDB, so $AD = DS$. Hence $\overleftrightarrow{DE} \| \overleftrightarrow{RS}$, making $\overline{DE}$ bisect $\overline{AB}$ and $\overline{AC}$. Therefore, $\text{Area}_{\text{triangle } AXY} = \frac{1}{4}\text{Area}_{\text{triangle } ABC}$, so Area_{XYBC} $= \dfrac{3}{4}\text{Area}_{\text{triangle } ABC}$.

24. Method 1: Adding x^2 to both sides, we get $x^4 - 4x^3 + 6x^2 - 4x + 1$ $= x^2$, or $(x - 1)^4 = x^2$, so $(x - 1)^2 = \pm x$. Solving these two quadratic equations for real x yields only two solutions, namely, $x = \dfrac{3 \pm \sqrt{5}}{2}$.

24. (Continued)

Method 2: We can divide through by x^2, to get $x^2 - 4x + 5 - \frac{4}{x} + \frac{1}{x^2} = 0$. If we let $y = x + \frac{1}{x}$, then $y^2 - 2 = x^2 + \frac{1}{x^2}$, and we find that $y^2 - 2 - 4y + 5 = y^2 - 4y + 3 = 0$. Thus $y = 3$ or $y = 1$. Substituting and solving for x, we find, as before, the real solutions $\frac{3 \pm \sqrt{5}}{2}$.

The substitutions worked because the coefficients of the given equation were symmetric (from the left and right terms in towards the middle). Hence if r is a root, $\frac{1}{r}$ is also a root. The substitution $y = x + \frac{1}{x}$ "ties" the roots together in pairs, halving the degree of the equation. For more of this sort of technique, see Larsen (problem 1.3.2).

25. Method 1: Let $30AB5 = 225n = 3^2 \cdot 5^2 \cdot n$. Since $30AB5$ is a multiple of 9, the sum of the digits, $8 + A + B$, is a multiple of 9. Since each digit is at most 9, it follows that $8 + A + B = 9$ or 18, so $A + B = 1$ or 10. Since $30AB5$ is also an odd multiple of 25, B must be either 2 or 7. Therefore, $A + B = 10$ and either $A = 8$ (when $B = 2$) making $n = \frac{30825}{225} = 137$, or $A = 3$ (when $B = 7$) so $n = \frac{30375}{225} = 135$.

Method 2: Since $30{,}000 < 225n < 31{,}000$, it follows that $134 \leq n \leq 137$. Since $30AB5$ is odd, n must be odd, so the only possible values for n are 135 or 137.

26. $a = \sqrt{.16} = .4$; $b = \sqrt[3]{.0639} < \sqrt[3]{.064} = .4$, so b is slightly less than $.4$; $c = \sqrt[6]{.0041} = \sqrt[6]{4.1 \times 10^{-3}} = \sqrt[6]{4100 \times 10^{-6}} = \sqrt[6]{4100} \times 10^{-1} > \sqrt[6]{4096} \times 10^{-1} = 4 \times 10^{-1} = 0.4$, so c is slightly greater than .4. (Note: The suspicion that 4100 was close to 4^6 motivated computing $4^6 = (4^3)^2 = 4096$.) Lastly, $d = (.2)^2 = .04$. Thus, in increasing order, we have d, b, a, c.

27. Using mass points, assign weights of 3 to C and 4 to A. Then there must be a weight of 2 at B, and hence a weight of 5 at A'. Therefore, $x:y = 5:4$ where $x = AP$ and $y = PA'$. (See Appendix B.)

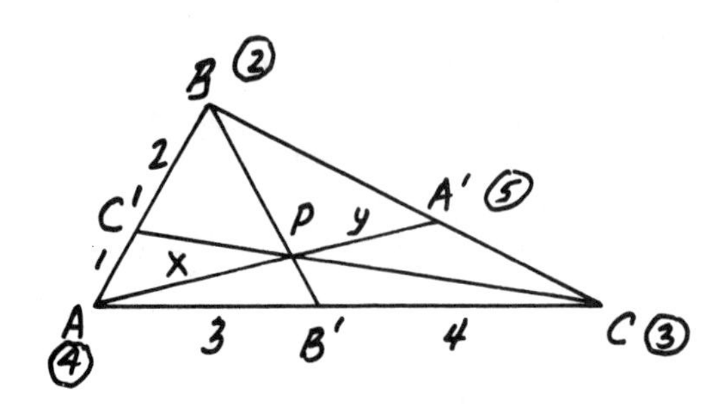

28. Method 1: Let ℓ be the line perpendicular to m: $3x + 4y = 12$ at $Q(0,3)$. The slope of ℓ is $\frac{4}{3}$, so its equation is $y = \frac{4}{3}x + 3$. We are looking for points $P\left(a, \frac{4}{3}a + 3\right)$ that are on ℓ such that $PQ = 2$. By the distance formula, $(PQ)^2 = a^2 + \left(\frac{4}{3}a\right)^2 = 2^2$, so $a = \pm\frac{6}{5}$. When $a = \frac{6}{5}$, $P\left(\frac{6}{5}, \frac{23}{5}\right)$ is on line m_1: $3x + 4y = k_1$, so $k_1 = 22$. When $a = -\frac{6}{5}$, $P\left(-\frac{6}{5}, \frac{7}{5}\right)$ is on m_2: $3x + 4y = k_2$, so $k_2 = 2$.

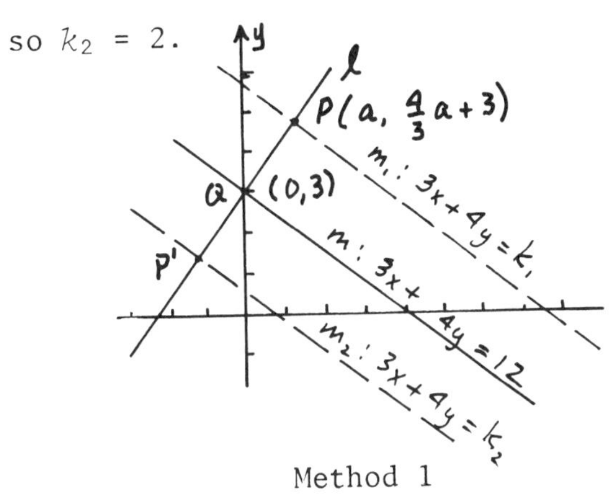

Method 1

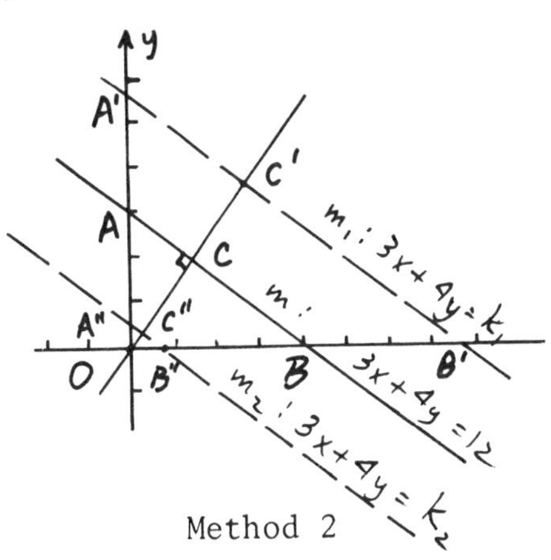

Method 2

Method 2: Draw line ℓ perpendicular to line m and passing through 0. Triangle $OAC \sim$ triangle BAO, which is a 3-4-5 triangle. Therefore $\frac{OA}{OC} = \frac{AB}{OB}$, or $\frac{3}{OC} = \frac{5}{4}$, so $OC = \frac{12}{5}$. We are given that $CC' = 2$, so $OC' = \frac{12}{5} + 2 = \frac{22}{5}$. The triangles $OA'C'$, OAC are similar, so $\frac{OA'}{OC'} = \frac{OA}{OC}$. Hence $\frac{OA'}{\frac{22}{5}} = \frac{3}{\frac{12}{5}}$ and $OA' = \frac{11}{2}$. Thus, $A'\left(0, \frac{11}{2}\right)$ lies on m_1: $3x + 4y = k_1$, so $k_1 = 22$. Similarly, in $OA''C''$, $OC'' = \frac{12}{5} - 2 = \frac{2}{5}$ and $OA'' = \frac{1}{2}$. Thus, $A''(0,\frac{1}{2})$ lies on m_2: $3x + 4y = k_2$, so $k_2 = 2$.

29. Using the special relationships in $30^\circ - 60^\circ - 90^\circ$ triangle MEP and in isosceles right triangles CPE and CMB, we have $CM = \frac{x}{2} + \frac{x\sqrt{3}}{2} = 12$, so $x(1 + \sqrt{3}) = 24$, and $x = \frac{24}{\sqrt{3} + 1}$. Rationalizing the denominator, $x = 12(\sqrt{3} - 1)$ so $k = 12$.

29. (Continued)

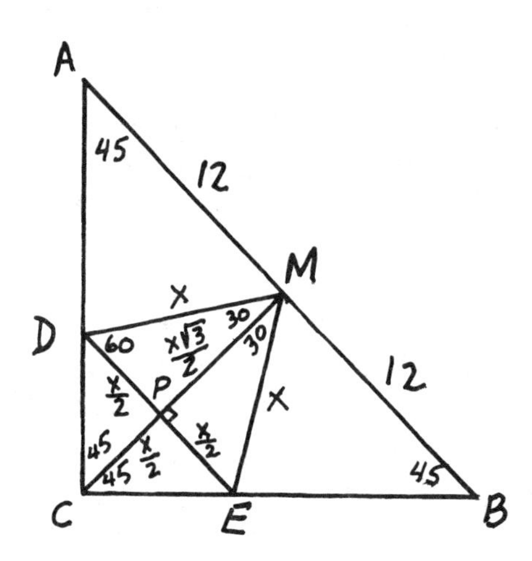

30. Let x, y, and z be the roots of the equation $g^3 - k_1g^2 + k_2g - k_3 = 0$. Then $x^3 - k_1x^2 + k_2x - k_3 = 0$, $y^3 - k_1y^2 + k_2y - k_3 = 0$, $z^3 - k_1z^2 + k_2z - k_3 = 0$. Adding these equations gives $a - k_1a^2 + k_2a - 3k_3 = 0$. By the conditions of the problem, and 1.4, $k_1 = a$, and we need to find k_3. Hence we need only express k_2 in terms of a. To do this, note that $a^2 = (x + y + z)^2 = x^2 + y^2 + z^2 + 2(xy + yz + xz) = a^2 + 2k_2$. Hence $k_2 = 0$ and $k_3 = \frac{a - a^3}{3}$. The trick of forming an equation with roots x, y, and z was motivated by noting that the functions given in the problem are all symmetric in x, y, and z. For many more such problems, see Krechmar (section 5, problems 16-39).

Solutions—Fall 1975

1. If $0 < x < 12.5$, $\left[\frac{x}{12.5}\right] = 0$, so $f(x) = 0$.

If $x \geq 12.5$, $\left[-\frac{12.5}{x}\right] = -1$, so $f(x) = -\left[\frac{x}{12.5}\right]$.

Then, if n is a positive integer and $12.5n \leq x < 12.5(n + 1)$, $f(x) = -n$. When $87.5 \leq x < 100$, $f(x) = -7$, so the range of f contains 8 values, namely, 0, -1, -2, ..., -7. (See 5.21 and 5.22.)

2. Note that all digits are less than 7. We have $81A + 9B + C = 49C + 7B + A$, so $80A + 2B - 48C = 0$ and $B = 8(3C - 5A)$. Since B is a multiple of 8 and $B < 7$, $B = 0$. Therefore, $3C = 5A$. Since $C < 7$, $C = 5$ and $A = 3$. The number is $305_9 = 503_7 = 248$ in base ten notation.

3. A careful graph shows that the six points (-1,-4), (0,-3), (1,-2), (2,-1), (3,0), (4,1) on the line $y = x - 3$ are also inside the circle $x^2 + y^2 = 25$. The points (-2,-5) and (5,2) are outside the circle. To check this algebraically, we can write $x^2 + (x - 3)^2 = 2x^2 - 6x + 9 \leq 25$, or $x^2 - 3x - 8 \leq 0$. Since the coefficient of x^2 is positive, the expression takes on negative values <u>between</u> the two roots (which are real). These roots are $\frac{3}{2} \pm \frac{\sqrt{41}}{2}$, or approximately 4.7 and -1.7. The possible integral values for x are thus -1, 0, 1, 2, 3, 4, and the y-values can be found by substitution.

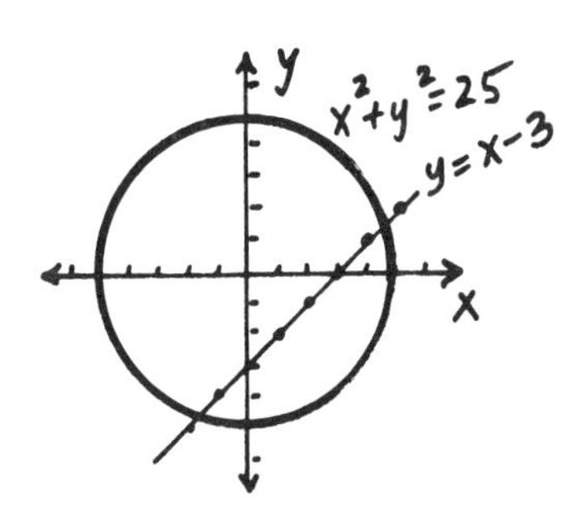

4. We have:

	number of people	cost per person	total cost
originally	n	x	nx
10 drop out	$n - 10$	$x + 1$	$(n - 10)(x + 1)$
15 more drop out	$n - 25$	$x + 3$	$(n - 25)(x + 3)$

The total cost is always the same, so $\begin{cases}(n - 10)(x + 1) = nx \\ (n - 25)(x + 3) = nx\end{cases}$

or $\begin{cases} n - 10x = 10 \\ 3n - 25x = 75\end{cases}$. Solving simultaneously (by eliminating x), we find that $n = 100$.

5. By the angle bisector theorem (3.31), $\frac{x}{8 - x} = \frac{6}{10}$, or $10x = 48 - 6x$, so $x = 3$ and $8 - x = 5$. $\text{Area}_{\text{triangle } ABD} = \frac{1}{2}bh = \frac{1}{2} \cdot 5 \cdot 6 = 15$. Note that if we use the fact that the ratio of the areas of triangles with the same altitude is equal to the ratio of their bases, we need not even solve explicitly for DB:

$\text{Area}_{\text{triangle } ABD} = \frac{DB}{CB} \cdot \text{Area}_{\text{triangle } ACB} = \frac{5}{8} \cdot 24 = 15.$

5. (Continued)

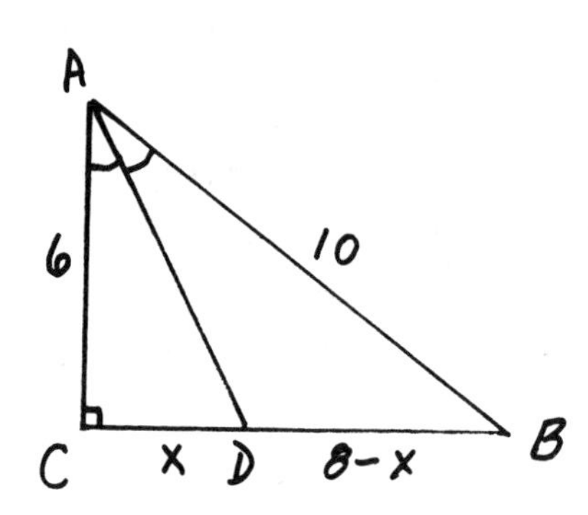

6. We have $x^3 + 2px^2 - px + 10 = 0$. By the rational root theorem (1.7), the only possible rational roots are -10, -5, -2, -1, 1, 2, 5, 10. The only sequences in arithmetic progression are -5, -2, 1 and -1, 2, 5. Since the product of the roots is -10 (see 1.4), the roots must be -1, 2, 5. The sum of the roots is $-2p = 6$, so $p = -3$.

7. The train travels $300 + x$ meters in 20 seconds. Since the light is above the train for 10 seconds, the train travels x meters in 10 seconds, so it would travel $2x$ meters in 20 seconds. Thus, $300 + x = 2x$ and $x = 300$.

8. If a, b, c are the three roots of $x^3 + 3x + 14 = 0$, then (see 1.311), $x^3 + 3x + 14 = (x - a)(x - b)(x - c) = x^3 - (a + b + c)x^2 + (ab + ac + bc)x - abc$. Thus $a + b + c = 0$, $ab + ac + bc = 3$, $abc = -14$ (See 1.4.) Hence $\frac{1}{a} + \frac{1}{b} + \frac{1}{c} = \frac{bc + ac + ab}{abc} = \frac{3}{-14} = -\frac{3}{14}$.

9. Since the trapezoid is a cyclic quadrilateral, it must be isosceles (see 3.92).

Method 1: By Ptolemy's theorem (3.91)

$200 + 89 = x^2$,

so $x = 17$.

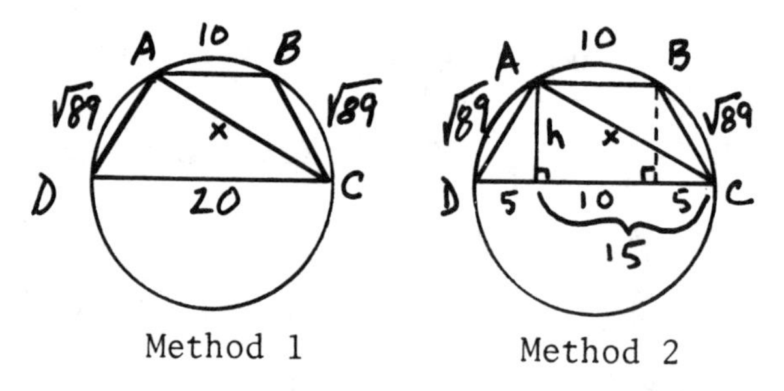

Method 2: Drop an altitude from A. Using the Pythagorean theorem twice,

$$h^2 + 225 = x^2,$$
$$h^2 + 25 = 89.$$

Subtracting, $200 = x^2 - 89$, so $x = 17$.

10. Let $z = a + bi = x + \frac{1}{x}$, where $z \varepsilon C$: $a, b \varepsilon R$.

$z^3 = \left(x + \frac{1}{x}\right)^3 = x^3 + 3x + \frac{3}{x} + \frac{1}{x^3}$

$z^3 = 3\left(x + \frac{1}{x}\right) + x^3 + \frac{1}{x^3}$, so $z^3 = 3z + 110$, $z^3 - 3z - 110 = 0$.

The rational root theorem (1.7) suggests possible rational roots $\pm 1, \pm 2, \pm 5, \pm 10, \pm 11, \pm 22, \pm 55, \pm 110$. Substitution shows that 5 is a root, so $z - 5$ is a factor (see 1.311).

$(z - 5)(z^2 + 5z + 22) = 0$.

$z = 5$ or $z = \frac{-5 \pm 3i\sqrt{7}}{2}$, so $(a,b) = (5,0)$, $\left(-\frac{5}{2}, \frac{3\sqrt{7}}{2}\right)$, $\left(-\frac{5}{2}, -\frac{3\sqrt{7}}{2}\right)$.

11. Using the recursive formula and the given initial values, we find $a_4 = 7$ and $a_5 = 9$, and can conjecture that $a_n = 2n - 1$. This is easily proved by induction: we've already established the first few cases. If the conjecture is true for all integers less than n,

$$a_n = a_{n-1} + a_{n-2} - a_{n-3}$$
$$= (2n - 3) + (2n - 5) - (2n - 7)$$
$$= 2n - 1.$$

Thus, we are looking for the sum of the first 30 odd integers, which is $30^2 = 900$. (Or, use formula 1.512.)

12. We use mass points (see Appendix B), placing weights of 3 at E, 2 at A, 1 at B, and 4 at F. For balance, there must be weights of 2 at C and 3 at D. Therefore, $x{:}y = 3{:}2$.

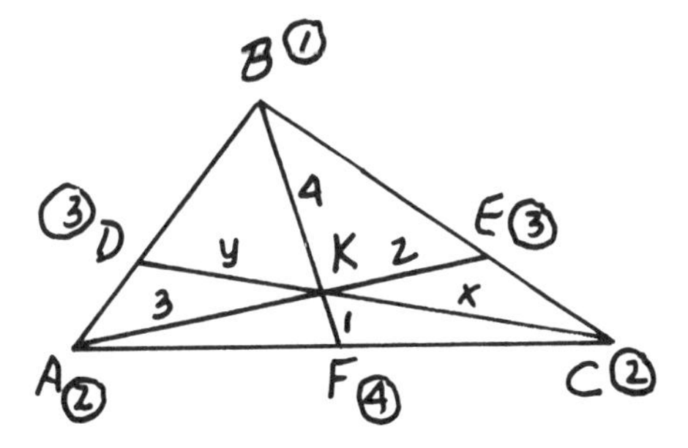

13. When the sum of the ages of father and daughter is 132, the daughter will be $\frac{132}{3} = 44$ and the father will be 88. Thus we will always have $f = d + 44$. Since $f + d + m = 73$, $d + 44 + d + m = 73$, so $m = 29 - 2d$. (Equation required)

14. We use the Pythagorean theorem. In triangle ADG, $AD = BC = \sqrt{145}$. In triangle REC, $16 + (12 - x)^2 = a^2$. In triangle RFB, $25 + x^2 = 145 - a^2$. Adding these equations, $2x^2 - 24x + 195 = 145$.

$$x^2 - 12x + 20 = 0$$
$$(x - 2)(x - 10) = 0$$
so $x = 2, 10$.

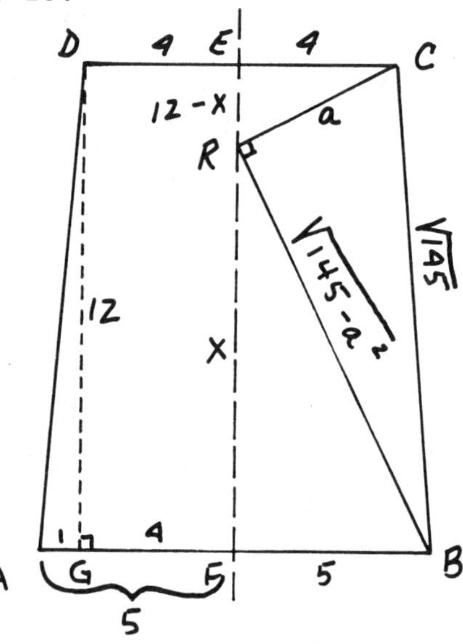

15. By Ptolemy's theorem (3.91), $3a + 6a = ax$, $9a = ax$, so $x = 9$. More generally, if equilateral triangle ABC is inscribed in circle O and P is any point on minor arc AB, $PC = PA + PB$.

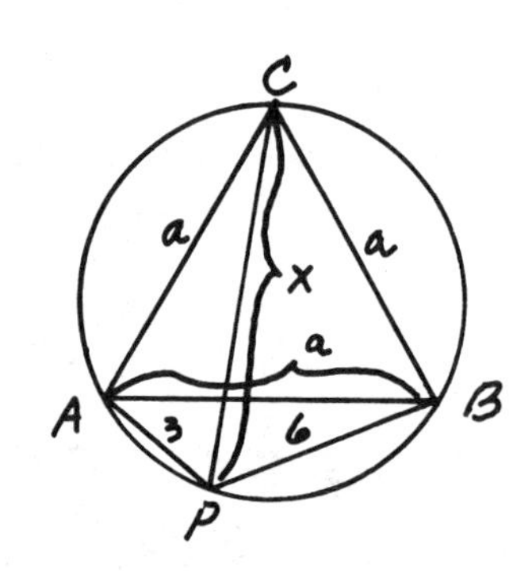

16. Consider the limit $x^{x^{x^{x^{\cdot^{\cdot^{\cdot}}}}}} = a$, where the dots indicate endless exponentiation. This is equivalent to $x^a = a$, whence $x = a^{\frac{1}{a}}$. In particular, when $a = \frac{4}{3}$, $x^{\frac{4}{3}} = \frac{4}{3}$, so $x = \left(\frac{4}{3}\right)^{\frac{3}{4}} = X$. Therefore, the limit of the given sequence is $a = \frac{4}{3}$, so $a_{1,000,000}$ is very close to 1.333. For a formal treatment of such "exponential towers" and the validity of the above sort of argument see Knoebel.

17. In isosceles triangle BCD, the measure of angle BCD = the measure of angle BDC = 70, so the measure of angle $ACD = 90 - 70 = 20$.

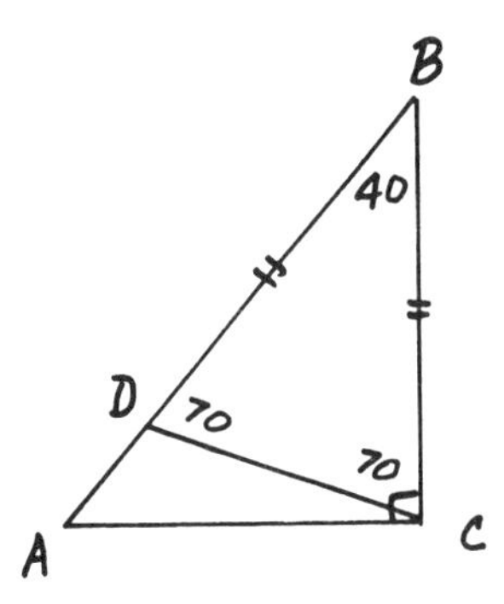

18. Eventually you want to leave your opponent with 1 toothpick. Your strategy is to leave him with $8n + 1$ toothpicks at each turn. (Practically speaking, if he takes x toothpicks, you take $8 - x$ toothpicks.) Initially, you take 7 toothpicks, leaving him with $1000 - 7 = 8(125) - 7 = 8(124) + 1$ toothpicks. For a fuller discussion, see Kasner and Newman (p. 172).

19. $x^3 - 3x^2 - x + 3 = x^2(x - 3) - (x - 3) = (x - 3)(x^2 - 1)$, so

$$\frac{12(x^2 - 4x + 3)}{x^3 - 3x^2 - x + 3} = \frac{12(x - 1)(x - 3)}{(x - 3)(x + 1)(x - 1)} = \frac{12}{x + 1} \text{ if } x \neq 1,3.$$

The expression $\frac{12}{x + 1}$ is a positive integer if $x + 1 = 1, 2, 3, 4, 6, 12$ or $x = 0, 1, 2, 3, 5, 11$. But $x \neq 1, 3$, so $x = 0, 2, 5, 11$ are the only possibilities.

20. The minimum (m) occurs when A is folded to land at the midpoint of BC, so $m = \frac{1}{2} \cdot 8 = 4$. The maximum (M) occurs when A is folded to coincide with B (or C), so $M = 4\sqrt{3}$ and $(M,m) = (4\sqrt{3},4)$.

21. The area of pentagon $ABCDE$ is $5 \cdot 2 + \frac{1}{2} \cdot 6(8 + 2) = 40$. The equation of $\overleftrightarrow{DE}$ is $y = 8 - x$, so the coordinates of P are $(k, 8 - k)$. Then the area of trapezoid $AQPE$ is $\frac{1}{2}k(8 + 8 - k) = 20$, so $k^2 - 16k + 40 = 0$ and $k = 8 \pm 2\sqrt{6}$. Since $k < 8$, $k = 8 - 2\sqrt{6}$ and $(a,b) = (8,-2)$.

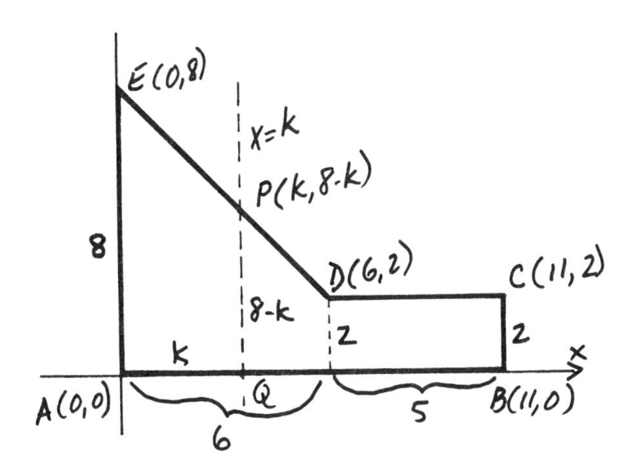

22. Method 1: Let a, b be the rates of A, B, respectively. Distance equals rate multiplied by time, so $PX = a$ miles and $PY = b$ miles. Had they exchanged destinations, we would have $\left(\text{since } T = \frac{D}{R}\right)$, $\frac{b}{a} = \frac{a}{b} + \frac{35}{60}$. Letting $r = \frac{a}{b}$, $\frac{1}{r} = r + \frac{7}{12}$, and $12r^2 + 7r = 12$ $= (3r + 4)(4r - 3) = 0$. Since $r > 0$, we have $r = \frac{3}{4}$.

    Method 2: Let rb, b be the rates of A, B, respectively. The distances are now $PX = rb$, $PY = b$, so, when they exchange destinations, the time equation becomes $\frac{1}{r} = r + \frac{7}{12}$. As above, $r = \frac{3}{4}$.

23. It is not hard to see that, in decimal notation, the integers x and x^5 have the same last digit. Here, if $x^5 = 656356768$, the last digit of x must be 8. Next observe that $10^5 < x^5 < 10^{10}$, so $10 < x < 100$. Furthermore, $x^5 = 6563.56768 \times 10^5$, $5^5 = 3125$, and $6^5 = 7776$, so $50 < x < 60$ and $x = 58$.
    To see that x^5 and x have the same units digit, we can use trial and error. Since we need to keep track only of the units digits, there are only ten cases.

24. $AM = 3\sqrt{3}$. Using mass points (Appendix B), put weights of 1 at A, 2 at C, and 2 at B. This gives a weight of 4 at M, so $AU{:}UM = 4{:}1$, $MU = \frac{1}{5}(AM) = \frac{1}{5}(3\sqrt{3}) = \frac{3}{5}\sqrt{3}$, and $k = \frac{3}{5}$.

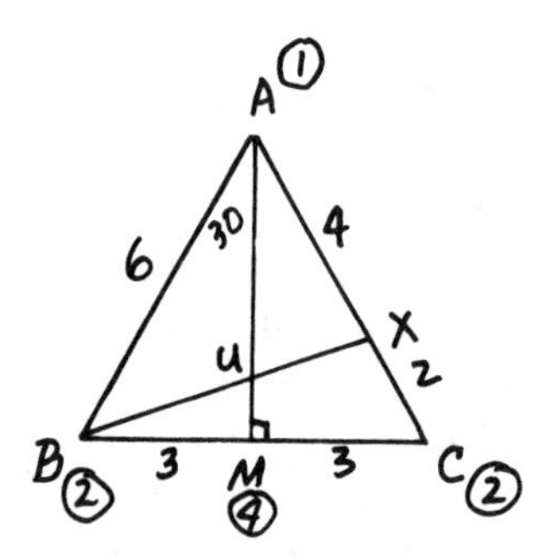

25. If the unit fractions are $\frac{1}{a^2}$ and $\frac{1}{b^2}$, then their sum is $\frac{a^2 + b^2}{a^2b^2}$. The denominator of the sum is always a square. We need the numerator to be a square as well; that is, we need $a^2 + b^2 = c^2$ for some integer c. But this is just the requirement that (a,b,c) be a Pythagorean triple (see 2.1). The largest possible fractions will result from the smallest possible (a,b), which is $(3,4)$. The fractions are $\frac{1}{9}$ and $\frac{1}{16}$.

26. <u>Method 1:</u> $(x + y)^3 = x^3 + y^3 + 3xy(x + y)$,
so $a^3 = b + 3axy$ and $xy = \dfrac{a^3 - b}{3a}$. Now $x^2 + y^2 = (x + y)^2 - 2xy$,
so $x^2 + y^2 = a^2 - \dfrac{2(a^3 - b)}{3a}$ or $x^2 + y^2 = \dfrac{a^3 + 2b}{3a}$.

<u>Method 2:</u> Let $d = x^2 + y^2$. Factoring, $b = x^3 + y^3$
$= (x + y)(x^2 - xy + y^2) = a(d - xy)$, so $\dfrac{b}{a} = d - xy$, so $xy = d - \dfrac{b}{a}$.
Also, $a^2 = (x + y)^2 = d + 2xy$, so $a^2 = d + 2\left(d - \dfrac{b}{a}\right) = 3d - \dfrac{2b}{a}$,
$3d = a^2 + \dfrac{2b}{a} = \dfrac{a^3 + 2b}{a}$, and $d = x^2 + y^2 = \dfrac{a^3 + 2b}{3a}$. (Equation required)

27. Letting $n = \sqrt{x - 2y}$ $(n \geq 0)$, we have $n + \dfrac{1}{n} = \dfrac{10}{3} = 3 + \dfrac{1}{3}$,
so $n = \dfrac{1}{3}$ or $n = 3$. Therefore, $x - 2y = \dfrac{1}{9}$ or $x - 2y = 9$;
so $x = 2y + \dfrac{1}{9}$ or $x = 2y + 9$ and $(a,b) = \left(2, \dfrac{1}{9}\right)$ or $(2,9)$.

28. <u>Method 1:</u> When A is folded to D, triangle AXY lands in position triangle DXY, so triangle $DXY \cong$ triangle AXY. We use the law of cosines three times: In triangle CYD, $a^2 = 12^2 + (15 - a)^2 - 2 \cdot 12(15 - a)(\tfrac{1}{2})$ so $a = \dfrac{21}{2}$. In triangle BXD, $b^2 = 3^2 + (15 - b)^2 - 2 \cdot 3(15 - b)(\tfrac{1}{2})$ so $b = 7$. Then, in triangle AXY,
$x^2 = a^2 + b^2 - 2ab(\tfrac{1}{2})$

$$= \left(\frac{21}{2}\right)^2 + 7^2 - \left(\frac{21}{2}\right) \cdot 7, \text{ so } x^2 = 7^2\left(\frac{9}{4} + 1 - \frac{3}{2}\right) = 7^2\left(\frac{7}{4}\right), \text{ so}$$

$x = \dfrac{7}{2}\sqrt{7} = \tfrac{1}{2}\sqrt{n}$, and $n = (49)(7) = 343$.

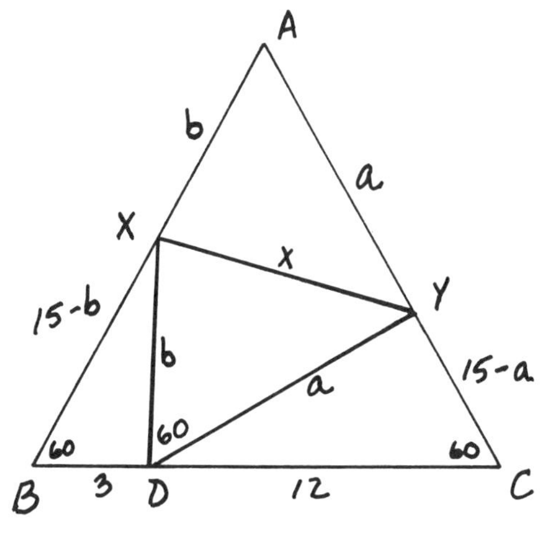

Method 1

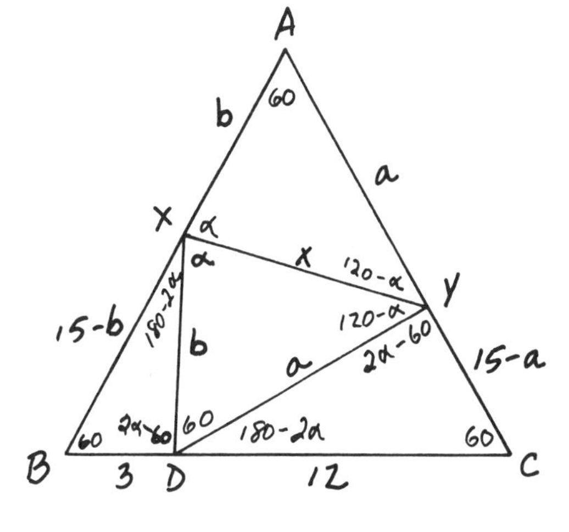

Method 2

28. (Continued)

Method 2: As before, triangle $DXY \cong$ triangle AXY. Analyzing the angles as shown, triangle $BXD \sim$ triangle CDY, so $\frac{b}{15 - b} = \frac{a}{12}$ and $15a - ab = 12b$. Also $\frac{b}{3} = \frac{a}{15 - a}$, so $15b - ab = 3a$. (*) Subtracting, $15a - 15b = 12b - 3a$, so $18a = 27b$ and $a = \frac{3}{2}b$. Substituting in (*), $15b - \frac{3}{2}b^2 = \frac{9}{2}b$, so $\frac{3}{2}b^2 = \frac{21}{2}b$. Since $b \neq 0$, $b = 7$ and $a = \frac{21}{2}$. Again, using the law of cosines in triangle AXY, $x = \frac{7\sqrt{7}}{2}$, so $n = 343$.

29. The shaded area is

$2(\text{Area}_{\text{semicircle on diameter } AD} - \text{Area}_{\text{semicircle on diameter } AC})$

$= 2(\frac{1}{2} \cdot 16\pi - \frac{1}{2} \cdot 4\pi) = 12\pi$, so $k = 12$.

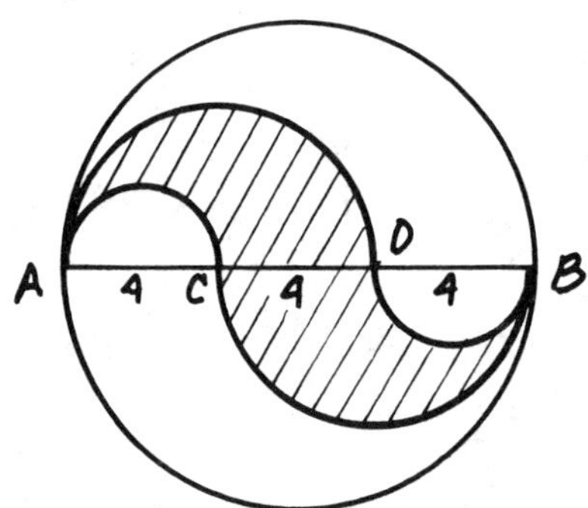

30. $10^4 = 10{,}000$ and $20^4 = 160{,}000$, so $ABCDE$ is the fourth power of a two-digit number whose first digit is 1. Since $A + C + E = B + D$ is a sufficient condition for divisibility by 11, $ABCDE = 11^4 = 14641$, so $C = 6$. Note that 11^4 is easily calculated using the binomial theorem (1.21) on $(10 + 1)^4$.

Solutions—Spring 1976

1. If the nth term is denoted by a_n, we have $\frac{a_n}{a_{n-1}} = \frac{7^{n-1}}{10^n} \cdot \frac{10^{n-1}}{7^{n-2}} = \frac{7}{10}$. Since this ratio is constant, the series is geometric, and the sum to infinity is $\frac{\frac{1}{10}}{\frac{3}{10}} = \frac{1}{3}$. (see 1.523.)

2. We apply two angle bisector theorems:

$\frac{a}{b} = \frac{p}{q}$ and $t^2 = ab - pq$. (See 3.31, 3.32.)

Let $AE = k$, $AC = 1$, $BC = x$. Then $BE = kx$ and $CE = k\sqrt{x}$. Now the second theorem gives $k^2x = x - k^2x$, so $2k^2 = 1$ and $k = \frac{\sqrt{2}}{2}$. Note that the measures of angles C, A, and B are unimportant.

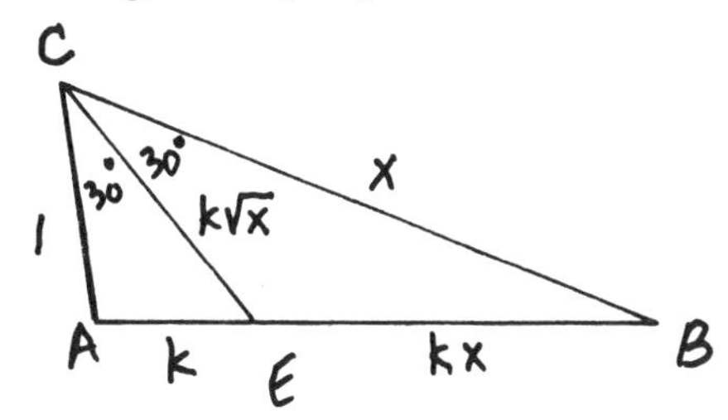

3. Using mass points (see Appendix B), we assign weights 2, 1, 2 to A, B, C, respectively. Then $2A + 1B = 3D$, $2A + 2C = 4F$, and $1B + 4F = 2C + 3D = 5G$. Hence $\frac{BG}{GF} = \frac{4}{1}$.

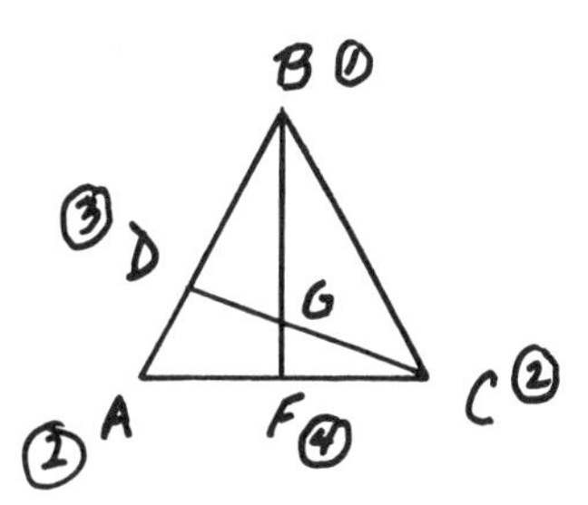

4. $A \cdot 50^3 + B \cdot 50^2 + B \cdot 50 + A = x^3 = (50C + C)^3 = C^3 \cdot 50^3 + 3C^3 \cdot 50^2 + 3C^3 \cdot 50 + C^3$. Now $C^3 < 50$ (otherwise the right side would be a five-digit numeral). Thus $A = C^3$. Then $B \cdot 50^2 + B \cdot 50 = 3C^3 \cdot 50^2 + 3C^3 \cdot 50$ and $3C^3 < 50$. Then $C = 1$ or 2 and $B = 3C^3 = 3$ or 24.

5. Using the law of cosines, $(20 - b)^2 = (b + 1)^2 + b^2 + b(b + 1)$, or $2b^2 + 43b - 399 = 0$, or $(2b + 57)(b - 7) = 0$, and $BC = 13$.

6. Let $A = \frac{x^2 + 1}{x}$. Then $A + \frac{1}{A} = \frac{29}{10}$, so $10A^2 - 29A + 10 = (2A + 5)(5A + 2) = 0$, and $A = \frac{2}{5}$ or $\frac{5}{2}$. If $A = \frac{x^2 + 1}{x} = x + \frac{1}{x} = \frac{5}{2} = 2 + \frac{1}{2}$, then $x = 2$ or $\frac{1}{2}$. If $A = \frac{2}{5}$, x will be complex, since for real x, $\left|x + \frac{1}{x}\right| \geq 2$ (6.1). In fact, $x + \frac{1}{x} = \frac{2}{5}$ leads to $x = \frac{1 \pm \sqrt{-24}}{5}$, and $k = -24$.

7. Method 1: We have $\frac{n}{2}[a + (a + n - 1)] + 100$

$= \frac{n}{2}[(a + n) + (a + n + n - 1)]$ (see 1.512), which simplifies to

$n^2 = 100$, and $n = 10$.

Method 2: Since each integer in the second sum is n more than the corresponding integer in the first sum, and since there are n integers in each sum, $\text{Sum}_1 + 100 = \text{Sum}_1 + n^2$, from which $n = 10$.

8. A rhombus is symmetric in its diagonals. Hence, lines AC and BD are lines of symmetry for the entire figure. Thus the opposite sides of quadrilateral Q are parallel, making it a parallelogram. Also, the diagonals of a rhombus are perpendicular, so that $AC \perp BD$. Since the pairs of sides of Q are parallel to AC and BD, the adjacent sides of Q are perpendicular, making it a rectangle. Now the four squares drawn are congruent, so $AH = AI = BI = BJ$ (see diagram). Also, angle HAI = angle HAD + angle DAB + angle BAI $= 45° + 30° + 45° = 120°$. Similarly, it is easy to see that angle $IBJ = 120°$. Thus triangle $HAI \cong$ triangle IBJ, and Q is in fact a square. It is not hard to see that $HA = 3\sqrt{2}$, and the law of cosines then shows that $HI = \sqrt{54}$, so the area of Q is 54.

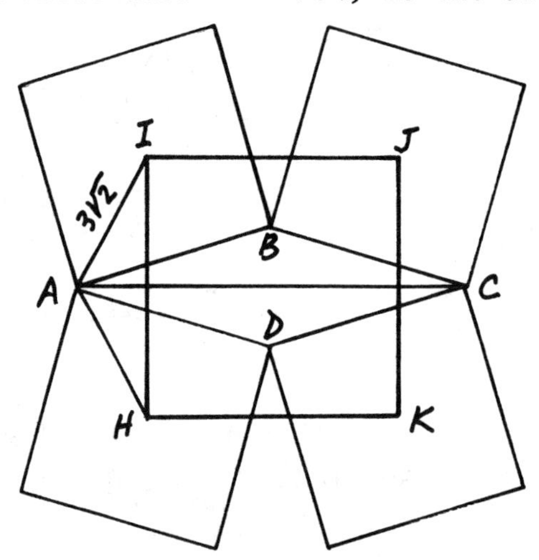

9. Since $625 = 25^2$, we first need to find all right triangles whose hypotenuse is 25. There are two, 7,24,25 and 15,20,25. This produces two solutions in the first octant and four solutions in the first quadrant: (7,24), (24,7), (15,20), (20,15). Altering of signs yields 16 solutions in Quadrant I - Quadrant IV. These, plus the four solutions on the axes, yield a total of 20. One may also proceed by using the relationship: $25^2 = [k(m^2 + n^2)]^2$ (see 2.1). Then $(k,m,n) = (1,4,3)$ or $(5,2,1)$ or $(1,5,0)$ generate the 20 solutions.

10. If we notice that the given number is slightly larger than 100^3, it is not hard to see that it is, in fact, 101^3. Since 101 is prime, the greatest proper divisor is $101^2 = 10201$.

11. The intersections are (5,4), (4,5), (-5,-4), and (-4,-5), forming a rectangle whose dimensions are $\sqrt{2}$ and $\sqrt{162}$ and whose area is 18.

12. If he bought x oranges at 3 for 16¢, he paid a total of $\left(x \cdot \frac{16}{3} + 2x \cdot \frac{21}{4}\right)$¢. To make 20% profit, he must make $\frac{120}{100}$ of this; that is to be equivalent to selling $3x$ oranges at $\frac{k}{3}$ ¢ each. Thus $\frac{120}{100}\left(\frac{16x}{3} + \frac{42x}{4}\right) = 3x \cdot \frac{k}{3}$, leading to $k = 19$.

13. Let the number of adults and the number of minors be represented by A and M, respectively. Then $\frac{A + 24}{M} = \frac{2}{1}$, so $2M - 24 = A$. Also, $\frac{A}{M - x} = \frac{2M - 24}{M - x} = \frac{2}{1}$, leading to $x = 12$.

14. Method 1: The center of circle O lies on angle bisector $AD = 5\sqrt{3}$. In 30-60-90 triangle AOG (see diagram), $OG = 3$, so $AO = 6$. Thus the circle lies inside triangle ABC, and $OD = AD - AO = 5\sqrt{3} - 6$, so $(a,b) = (5,6)$.

Method 2: Once we see that 0 is inside the triangle, we can use 3.7. This theorem shows that $OG + OH + OD = 5\sqrt{3}$, so that $OD = 5\sqrt{3} - 6$.

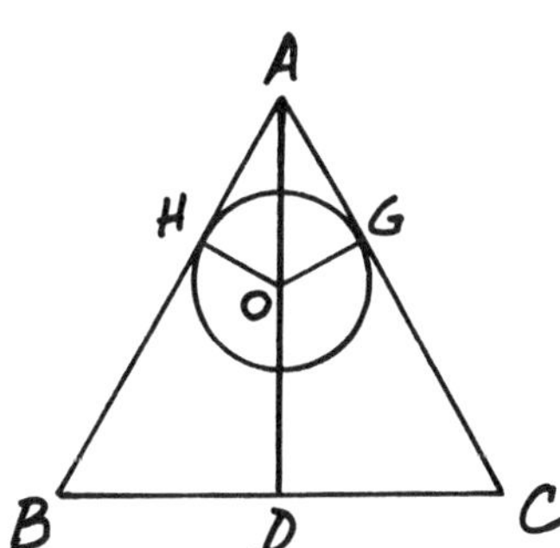

15. Using mass points, we assign weights of 2 and 3 to B and C, respectively. Then $2B + 3C = 5D$; now assign a weight of 6 to A to have the proper ratio of AE to ED. Then $6A + 3C = 9F$, so $\frac{BE}{EF} = \frac{9}{2}$.

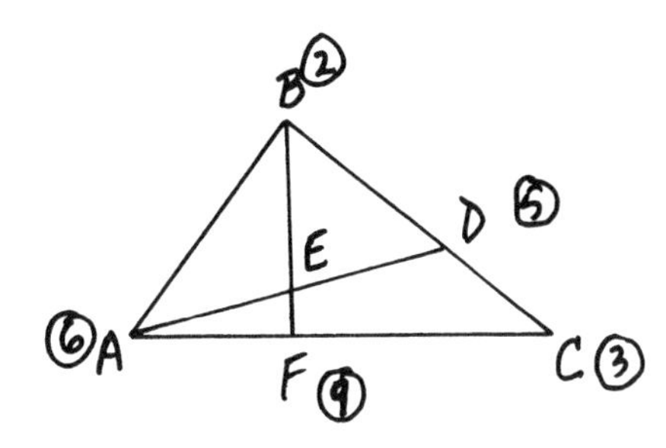

16. We have $1 + \frac{1}{2} - \frac{1}{4} - \frac{1}{8} + \frac{1}{16} + \frac{1}{32} - \frac{1}{64} - \frac{1}{128} + \ldots = S$.

Combining the terms in groups of two,

$S = \frac{3}{2} - \frac{3}{8} + \frac{3}{32} - \frac{3}{128} + \ldots$ and $S = \dfrac{\frac{3}{2}}{1 - \left(-\frac{1}{4}\right)} = \dfrac{\frac{3}{2}}{\frac{5}{4}} = \frac{6}{5}$. (See 1.523.)

17. We have $a^2 = c^2 - b^2 = b^2 + 2b + 1 - b^2 = 2b + 1$.

Let $a = A^3$. Then $A^6 = 2b + 1$, which is an odd integer, so A is also an odd integer. For the least possible c, we need the least possible value of a. Taking $A = 1$ leads to $b = 0$. Taking $A = 3$ gives $A^6 = 729$, $b = 364$, and $c = 365$.

18. Since triangle $BAE \cong$ triangle $FDE \cong$ triangle FCB, triangle BEF is equilateral. By the law of cosines, $EB^2 = 100 + 48\sqrt{3}$, and the area of triangle $EBF = \frac{\sqrt{3}(100 + 48\sqrt{3})}{4} = 25\sqrt{3} + 36$, so $(a,b) = (25,36)$.

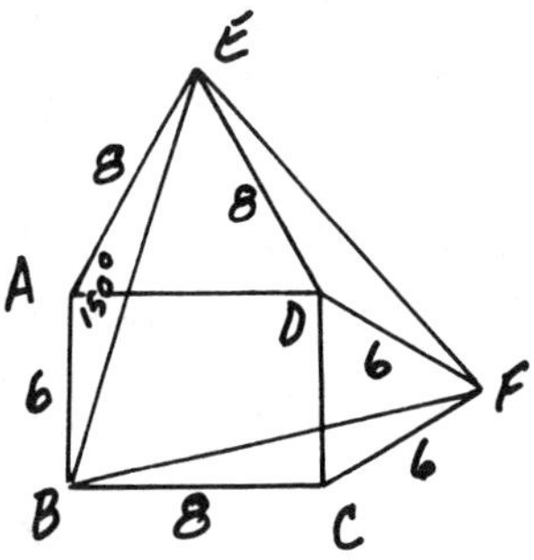

19. We have $3(t + u) + 10t + u = 10u + t$, which simiplifies to $u = 2t$. Solutions are 12, 24, 36, 48.

20. Method 1: Extend median BP to point D so that $MP = PD$. Then $AMCD$ will be a parallelogram (since AC and MD bisect each other). Now triangle AMD is a 6-8-10 triangle, and angle MAD is a right angle, so $AMCD$ is a rectangle and $AC = MD = 10$.

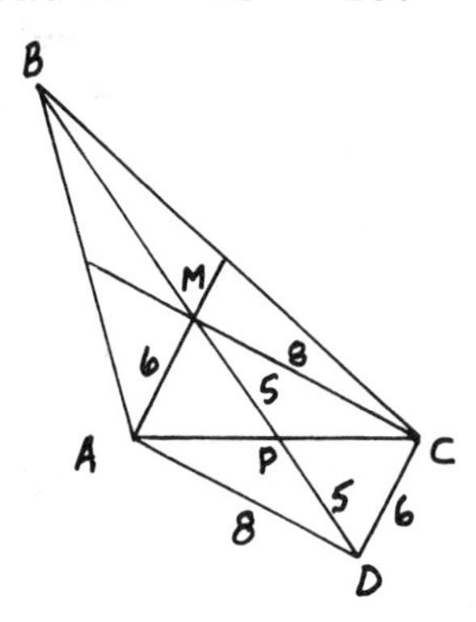

Method 2: Using formula (3.2) in triangle AMC, we have $25 = \frac{1}{2}(36 + 64) - \frac{1}{4}c^2$, so $c = 10$.

21. Using similar triangles:

$\frac{4 - y}{y + 3} = \frac{3}{4} \rightarrow y = 1,$

$\frac{x + 1}{2 - x} = \frac{3}{4} \rightarrow x = \frac{2}{7}.$

Thus, $(x,y) = \left(\frac{2}{7}, 1\right).$

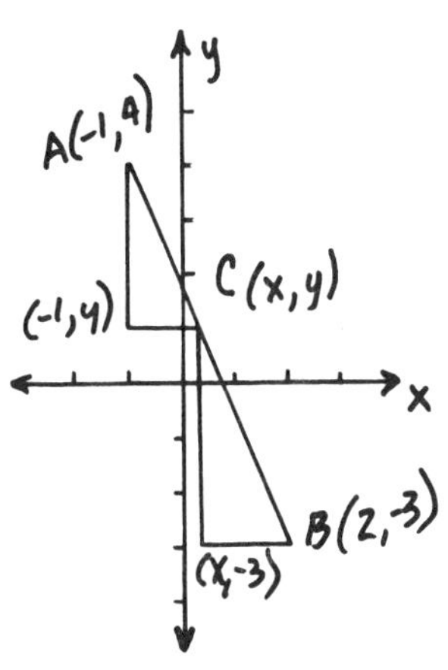

22. Subtracting the third equation from the second gives $y = 2$. Adding the same equations gives $x + z = 6$. Thus $\sqrt{4} = \sqrt{x} + \sqrt{z} - \sqrt{2}$, or $2 + \sqrt{2} = \sqrt{x} + \sqrt{z}$. Squaring and simplifying, $\sqrt{xz} = 2\sqrt{2}$, from which $xz = 8$. But $x + z = 6$. Solving these two equations simultaneously, $(x,y,z) = (4, 2, 2)$ or $(2, 2, 4)$.

23. If the legs are a and c, we have $a^2 + c^2 = (c + 1)^2 = c^2 + 2c + 1$. Simplifying shows $a^2 = 2c + 1$. Thus we are looking for an odd square (a^2) that is greater than 201. The least such square is 225, so $a^2 = 225$ and $c = 112$.

24. Method 1: Since the successive differences of the sequence b form an arithmetic progression, the second order of differences (i.e., the differences between the differences) is constant, making b_n quadratic, Substituting several values into $b_n = px^2 + qx + r$ and solving the resulting three linear equations, we find $b_n = n^2 + 3n - 1$. See Hall and Knight, Higher Algebra (p. 322) or Spiegel, for a fuller discussion of these finite difference techniques.

Method 2: (A telescoping sum): It is easy to see that $a_n = 6 + 2(n - 1) = 2n + 4$.

Then $b_n - b_{n-1} = a_{n-1} = 2(n - 1) + 4$

$$b_{n-1} - b_{n-2} = a_{n-2} = 2(n - 2) + 4$$

$$\vdots$$

$$b_2 - b_1 = a_1 = 2(1) + 4$$

Adding, $b_n - b_1 = 2\left(\frac{(n - 1)n}{2}\right) + 4(n - 1)$ (See 1.53.)

$$= n^2 + 3n - 4.$$

Since $b_1 = 3$, we again find $b_n = n^2 + 3n - 1$.

25. Let $x = \sqrt{10 + \sqrt{10 + \sqrt{10 \ldots}}}$. Then $x^2 = 10 + \sqrt{10 + \sqrt{10 + \ldots}}$ $= 10 + x$.
Solving, and rejecting the negative root, we find $x = \dfrac{1 + \sqrt{41}}{2}$,
and $(a,b,c) = (1,41,2)$. For a discussion of the rather advanced topic of convergence of this sort of expression, see Rudin (p. 78, problem 3) or Larsen (problem 7.6.8).

26. When the graph of $|x| + |y| = 2$ is centered about $(0,1)$, we have the graph of $|x| + |y - 1| = 2$, a square whose diagonals are 4 units long. The area $= \frac{1}{2} \cdot 4 \cdot 4 = 8$. See translation of axes in any book on analytic geometry.

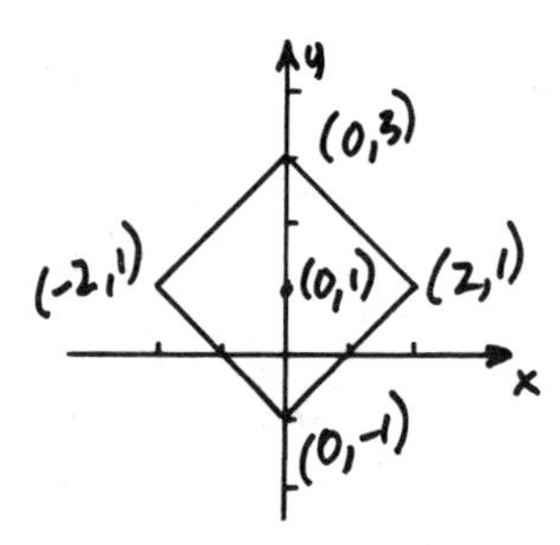

27. In right triangle AHM, $HM = 3$. Since triangle ABM is isosceles, we have $HB = HM = 3$; and since AM is a median, $BM = MC = 6$. Hence from right triangle AHC, $AC^2 = 4^2 + 9^2 = 97$, so $AC = \sqrt{97}$.

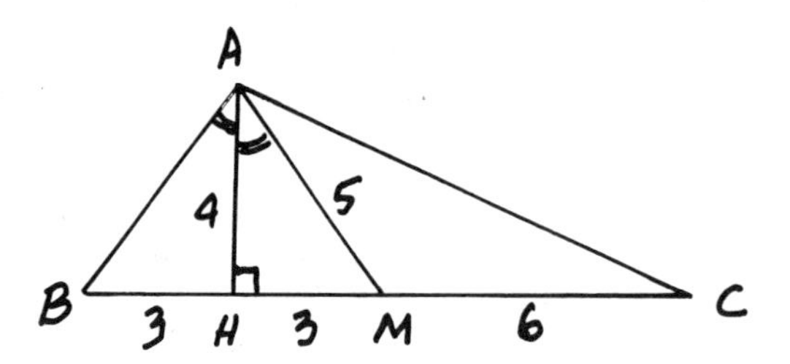

28. Let the integers by x, y, and $x + 12$. Then $(x + 12)^2 - y^2 = y^2 - x^2$, and $x^2 + 12x + 72 = y^2$, or $(x + 6)^2 + 6^2 = y^2$. Since the only right triangle with a leg of 6 is the 6-8-10 right triangle, x, y, and z are, respectively, 2, 10, 14.

29. Let $EB = x$. Then $CB = AD = A'D = AE = 2x$, and from right triangle ECB, $EC = x\sqrt{5}$. Hence $EC:CB = \sqrt{5}:2$.

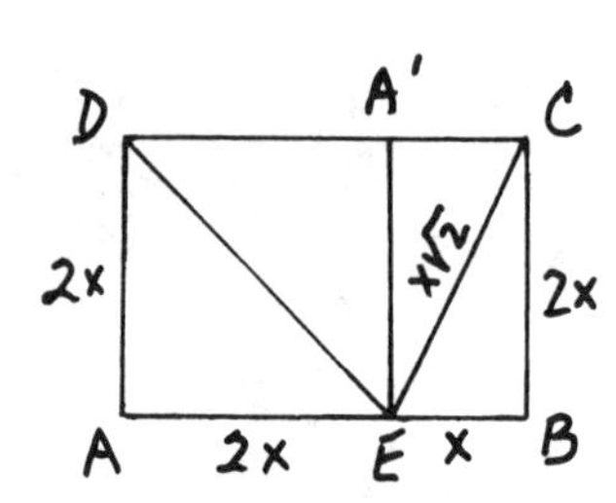

30. Let $a = 16x^2 - 9$, $b = 9x^2 - 16$. Then $a + b = 25x^2 - 25$, and we can use (1.22). The possibilities are $16x^2 - 9 = 0$, $9x^2 - 16 = 0$, or $25x^2 - 25 = 0$, giving rise to the six solutions: $\frac{3}{4}, -\frac{3}{4}, \frac{4}{3}, -\frac{4}{3}, 1, -1$.

31. Squaring produces $\sin^2 x + 2 \sin x \cos x + \cos^2 x = 1 + \sin 2x = \frac{1}{25}$, so $\sin 2x = -\frac{24}{25}$. Recognizing the ratios from a 7-24-25 right triangle (or otherwise), we find $\cos 2x = \frac{7}{25}$ (a positive value, since $2x$ is in quadrant IV).

32. Subtracting the first equation from the second gives $xy - yz = 12$; adding this to the third equation gives $xy = 16$. From this, $yz = 4$ and $xz = 9$. Multiplying these last three equations gives $(xyz)^2 = 16 \cdot 4 \cdot 9$, so $xyz = \pm 4 \cdot 2 \cdot 3 = \pm 24$. If we now divide this equation by xy, xz, and yz successively, we find $z = \pm\frac{24}{16} = \pm\frac{3}{2}$, $x = \pm 6$, and $y = \pm\frac{8}{3}$. Noting also that x and y must have like signs, as must y and z, the only acceptable triples are $(x,y,z) = \left(6, \frac{8}{3}, \frac{3}{2}\right)$ and $\left(-6, -\frac{8}{3}, -\frac{3}{2}\right)$.

33. We wish to find a linear equation whose graph passes through the points (°C,°F) = (25,40) and (125,280). The slope of the line determined by the two points is $\frac{12}{5}$, so $\frac{°F - 40}{°C - 25} = \frac{12}{5}$, and $°F = \frac{12}{5}°C - 20$.

34. In general, if two triangles are situated as in diagram 1, then $\text{area}(PBQ):\text{area}(ABC) = \left(\frac{1}{2} pq \sin B\right):\left(\frac{1}{2} ac \sin B\right) = \frac{pq}{ac}$. In this problem (see diagram 2), the area of triangle PQR can be obtained by subtracting the areas of the three outer triangles from that of triangle ABC:

$$\frac{kx}{35} = x - \left(\frac{3}{5} \cdot \frac{2}{6} \cdot x + \frac{2}{5} \cdot \frac{4}{7} \cdot x + \frac{4}{6} \cdot \frac{3}{7} \cdot x\right) = \frac{10x}{35}, \text{ so } k = 10.$$

For more such problems, see "the Menelaus triangle," in Kay (p. 208).

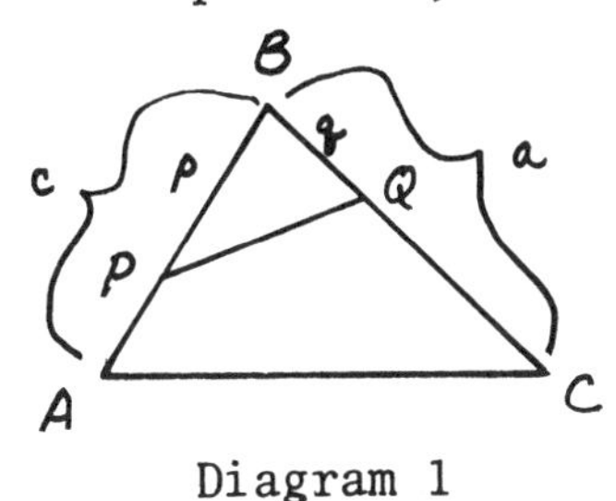

Diagram 1

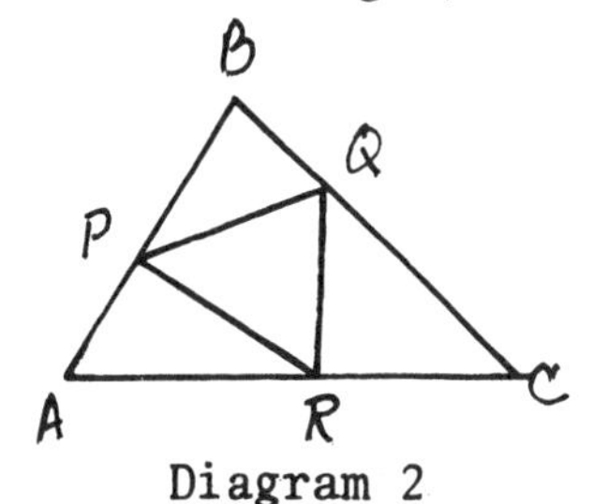

Diagram 2

35. Since $y - x = x - 6$, we have $y = 2x - 6$. And since $\frac{16}{y} = \frac{y}{x}$, we have $16x = y^2 = 4x^2 - 24x + 36$. Thus $4x^2 - 40x + 36 = 0$, so $x = 1$ or 9. Then $(x,y) = (1,-4)$ or $(9,12)$.

36. Substituting the points into the equation produces:

$$\begin{aligned} -11 &= a - b + c \qquad [A] \\ 1 &= a + b + c \qquad [B] \\ 4 &= 4a + 2b + c \qquad [C] \end{aligned}$$

Equations [A] and [B] imply that $b = 6$. Substituting in [B] and [C] leads to $a = -1$, $c = -4$. Now the vertex of the parabola $y = -x^2 + 6x - 4$ has abscissa $-\frac{b}{2a} = 3$, so its ordinate is $y = -9 + 18 - 4 = 5$. Hence $(x,y) = (3,5)$.

Solutions—Fall 1976

1. Let $\frac{4}{y} = x$, or $y = \frac{4}{x}$. Then $8y = \frac{32}{x}$ and $\left[f\left(1 + \frac{4}{y}\right)\right]^{8y}$

$= [f(1 + x)]^{32/x} = k^{32}$.

2. $x^5 + 5x^4 + 10x^3 + 10x^2 + 5x + 1 = x^5 + 1$, or $5x(x^3 + 2x^2 + 2x + 1) = 0$. The factor theorem (1.311) shows that $x + 1$ is a factor of the polynomial. Dividing, we get $5x(x + 1)(x^2 + x + 1) = 0$ and $x = 0, -1 \frac{-1 \pm i\sqrt{3}}{2}$.

3. $\tan\left(\frac{\pi}{4} + x\right) = \frac{1 + \tan x}{1 - \tan x} = \frac{\cos x + \sin x}{\cos x - \sin x}$ (See 4.341).

$A \sec 2x + B \tan 2x = \frac{A + B \sin 2x}{\cos 2x} = \frac{A + B \sin 2x}{\cos^2 x - \sin^2 x}$

Comparing the denominators suggests that we convert the first into the second by multiplying by $\frac{\cos x + \sin x}{\cos x + \sin x}$ getting

$\frac{\cos^2 x + 2 \sin x \cos x + \sin^2 x}{\cos^2 x - \sin^2 x} = \frac{1 + \sin 2x}{\cos^2 x - \sin^2 x}$. Thus $(A,B) = (1,1)$.

4. Since triangle $RBS \sim$ triangle CBA, $\frac{12 - x}{x} = \frac{12}{14}$, and $x = \frac{84}{13}$.

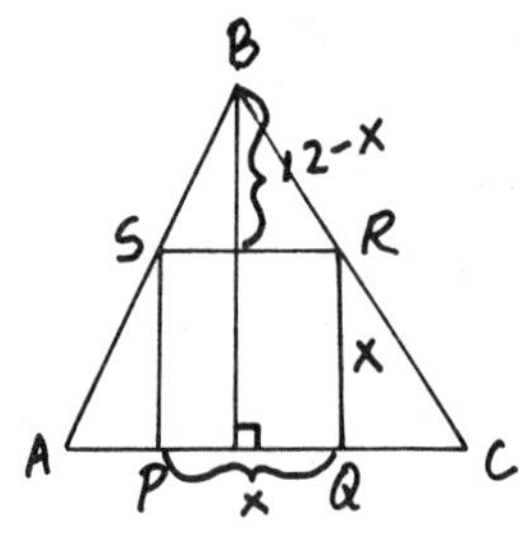

5. Using mass points, we assign weights 1, 2, 1 to A, B, C, respectively. Then $2B + 1C = 3D$, $1A + 1C = 2E$, and $2B + 2E = 3D + 1A = 4F$. Hence $BD:DC = 1:2$.

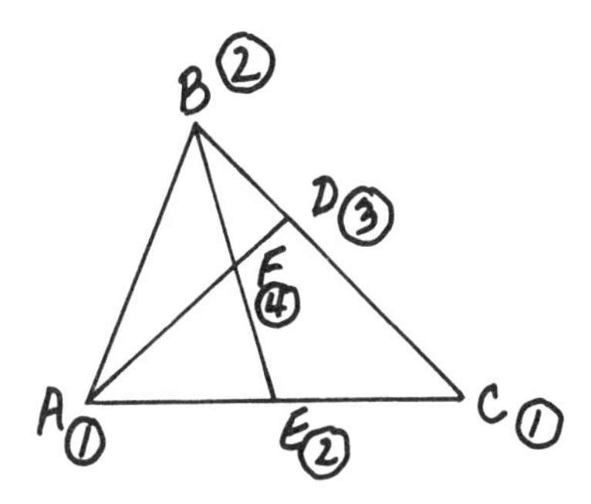

6. Subtracting the second equation from the first, $(x - y) - z(x - y) = (x - y)(1 - z) = 0$. Thus $z = 1$ or $x = y$. If $z = 1$ we have $x + y = 6$ and $1 + xy = 6$, and $(x,y) = (5,1)$ or $(1,5)$. This produces $(5,1,1)$ and $(1,5,1)$. If $x = y$ we have $x + xz = 6$ and $z + x^2 = 6$. Subtracting, we find $x^2 - x - xz + z = 0$ or $x(x - 1) - z(x - 1) = (x - z)(x - 1) = 0$. Thus $x = 1$ or $x = z$. If $x = 1$, we have $y = 1$ and $z = 5$ producing $(1,1,5)$. If $x = z$, we have $x = y = z$, so $x + x^2 = 6$ and $x = 2$ or -3 producing $(2,2,2)$ and $(-3,-3,-3)$.

7. Method 1: Using Hero's formula (see 3.61) with $s = 8$ gives area $K = 4\sqrt{(8 - x)(x - 2)}$. Since the factors of the radicand have a constant sum, their product is maximal when they are equal (see 6.3); this gives $x = 5$ and $K = 12$.

Method 2: The locus of points such that the sum of the distances to two fixed points is a constant (e.g., 10) is an ellipse with the fixed points as foci. Clearly the triangle has maximum area when the third vertex is at the end of the minor axis, giving a triangle of sides 5,5,6. Again, we find $K = 12$.

8. $2^{3x} + 2^{4y} = 2^{3x}(1 + 2^{4y-3x}) = 2^{5z}$. Thus the factor in parentheses must be a power of 2, which is only possible if $4y = 3x$. Then $2^{3x}(1 + 1) = 2^{5z}$ leads to $3x + 1 = 5z$. Trying multiples of 4 for x (since $3x = 4y$) quickly produces $(8,6,5)$ as the required solution. For a more general treatment, see Liff.

9. Note that $4624 = 17^2 \cdot 2^4$ so $x^3 - x^2 = x^2(x - 1) = 17^2 \cdot 16$, from which $x = 17$.

10. We can solve a slightly more general problem just as easily. Let R and r represent the radius-lengths of the large and small circles, respectively. Clearly, angle $AOX = 60°$, so OAX is a 30-60-90 triangle.

Then $\frac{OA}{AX} = \frac{2}{\sqrt{3}} = \frac{R - r}{r}$.

10. (Continued)

Solving, we find that $r = \dfrac{3R}{3 + 2\sqrt{3}}$.

Here, $r = 3$. For generalizations of this problem of the "kissing circles," see Coxeter (pp. 13-16).

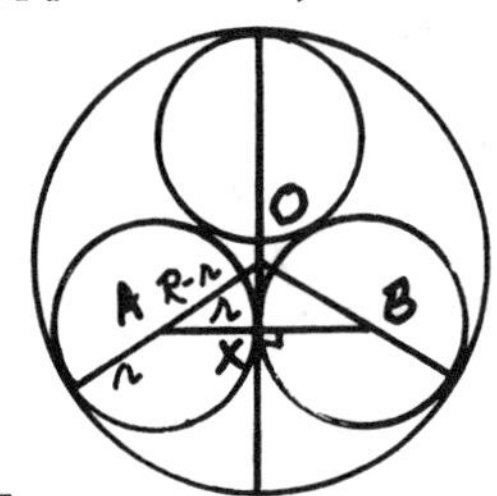

11. Since $\log_{10} n = L \leftrightarrow 10^L = n$, we have $\log[\log(\log x)] = 10^0 = 1$. Then, $\log(\log x) = 10^1 = 10$. Then, $\log x = 10^{10}$. Finally, $x = 10^{10^{10}}$, and $k = 10^{10}$.

12. The left side of the given expression simplifies to

$\dfrac{(\cot x + \tan x)(\cot x - \tan x)}{(\cot x + \tan x)(\cot x + \tan x)}$, which, after cancelling the common factor, converting to $\sin x$ and $\cos x$, and simplifying the resulting complex fraction, becomes $\cos^2 x - \sin^2 x$. When this is added to $2 \sin^2 x$, we get the expression $\cos^2 x + \sin^2 x$. Since (see 4.11) the given expression is identically equal to 1, we have $(m,M) = (1,1)$.

13. Using mass points, we assign weights of 4, 6, 9 to A, B, C, respectively. Then $4A + 6B = 10D$, $6B + 9C = 15E$, and $15E - 10D = 5F$. Hence $DE:EF = 1:2$.

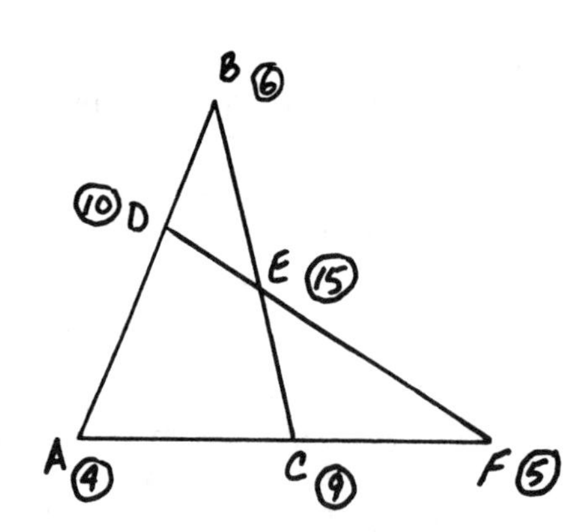

14. Squaring both sides, we get $\sin^2 x + \cos^2 x + 2 \sin x \cos x = \frac{1}{25}$. Thus, $2 \sin x \cos x = -\frac{24}{25}$, or $\sin 2x = -\frac{24}{25}$, and $\cos 2x = \pm \frac{7}{25}$ (see 4.21 and 4.11). Since $\tan x = \frac{\sin 2x}{1 + \cos 2x}$ (see 4.23), we have

$$\tan x = \frac{-\frac{24}{25}}{1 \pm \frac{7}{25}} = -\frac{24}{32} \text{ or } -\frac{24}{18} = -\frac{3}{4} \text{ or } -\frac{4}{3}.$$

But $-\frac{3}{4}$ doesn't check, as this implies (in Quadrant II) that $\sin x = \frac{3}{5}$ and $\cos x = -\frac{4}{5}$, so $\sin x + \cos x = -\frac{1}{5}$. This extraneous root came from the original squaring we did.

15. Method 1: The given series is equal to

$$\left(\frac{2}{1} + \frac{1}{3}\right) + \left(\frac{2}{9} + \frac{1}{27}\right) + \left(\frac{2}{81} + \frac{1}{243}\right) + \cdots$$

$$= \frac{7}{3} + \frac{7}{27} + \frac{7}{243} + \cdots$$

$$= \frac{\frac{7}{3}}{1 - \frac{1}{9}} = \frac{21}{8}. \quad (1.523)$$

Method 2: The given series can also be written as

$$\left(\frac{2}{1} + \frac{2}{9} + \frac{2}{81} + \cdots\right) + \left(\frac{1}{3} + \frac{1}{27} + \frac{1}{243} + \cdots\right) = \frac{18}{8} + \frac{3}{8} = \frac{21}{8}$$

$$\text{or } \left(1 + \frac{1}{3} + \frac{1}{9} + \cdots\right) + \left(1 + \frac{1}{9} + \frac{1}{81} + \cdots\right) = \frac{3}{2} + \frac{9}{8} = \frac{21}{8}. \quad (1.523)$$

For particulars about rearranging terms of an infinite series (without affecting convergence), see any advanced calculus text.

16. At the end, the amount of wine in one jug must always equal the amount of water in the other. Hence, both jugs must be $\frac{1}{2}$ wine, $\frac{1}{2}$ water. We examine first the jug that started with all wine. After the first transfer, it has ($x - 2$ wine):(x total); the second transfer removes wine in that ratio (and adds wine in the ratio $2:x$ from the other jug). Thus the amount of wine there at the end is $(x - 2) - \left(\frac{x - 2}{x}\right) \cdot 2 + \left(\frac{2}{x}\right) \cdot 2$, which must also equal $\frac{x}{2}$. Simplifying produces $x = 4$.

17. Let M be the midpoint of BC. For all real numbers k, $0 < k \leq 1$, and for each triangle ADP (with vertex P on $\overline{MB}$), altitude $7 + k$, there is another triangle ADP' (with vertex P' on $\overline{MC}$), with altitude $7 - k$. Hence the average area $= \dfrac{5(7 + k) + 5(7 - k)}{2} = 35$.

The above solution is the one given in the original contest, and seems intuitively clear. However, depending on how one thinks of the concept of "averaging" an uncountable set of data, one can arrive at a variety of answers to this type of problem. Such continuous probability distributions are discussed in any advanced book on probability.

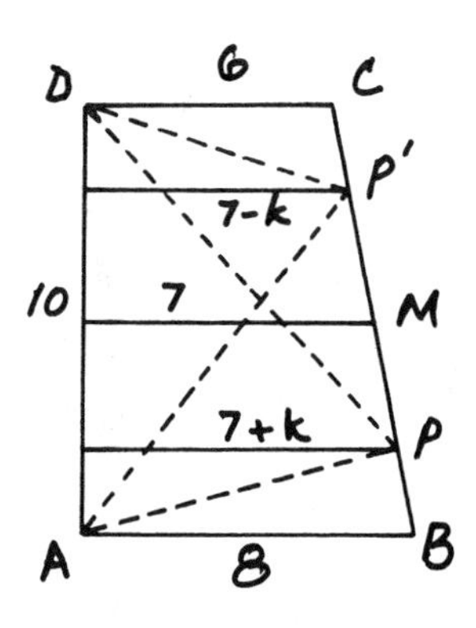

18. $(x)(x - 1)(x^2 + x + 1) - 6(x^2 + x + 1) = 0 = (x^2 + x + 1)(x^2 - x - 6)$ $= (x^2 + x + 1)(x - 3)(x + 2)$. The solutions are -2, 3, $\dfrac{-1 \pm i\sqrt{3}}{2}$.

19. Note that when dealing with base b, a numeral such as 123.456... (for example) becomes in base ten, $1(b)^2 + 2(b)^1 + 3(b)^0 + 4(b)^{-1}$ $+ 5(b)^{-2} + 6b^{-3} + \ldots$. The "decimal" part is therefore $\dfrac{4}{b} + \dfrac{5}{b^2}$ $+ \dfrac{6}{b^3} + \ldots$. Now (see 1.523) $.\overline{17} = \dfrac{1}{x} + \dfrac{7}{x^2} + \dfrac{1}{x^3} + \dfrac{7}{x^4} + \ldots$

$$= \frac{x + 7}{x^2} + \frac{x + 7}{x^4} + \ldots = (x + 7)\left(\frac{1}{x^2} + \frac{1}{x^4} + \ldots\right) = (x + 7)\left(\frac{\frac{1}{x^2}}{1 - \frac{1}{x^2}}\right)$$

$= (x + 7)\left(\dfrac{1}{x^2 - 1}\right)$. Then $\dfrac{1}{5} = \dfrac{x + 7}{x^2 - 1}$, leading to $x = -4$ or $+9$.

There can be no digit 7 in base (-4), so $x = 9$ only.

20. Relating perimeter, diagonal, length, and width gives $\dfrac{p}{2} = d + \dfrac{\ell}{3}$, or $w + \ell = \sqrt{w^2 + \ell^2} + \dfrac{\ell}{3}$, so $w + \dfrac{2\ell}{3} = \sqrt{w^2 + \ell^2}$. Squaring each side, $w^2 + \dfrac{4}{3}w\ell + \dfrac{4}{9}\ell^2 = w^2 + \ell^2$ and $\dfrac{4}{3}w\ell = \dfrac{5}{9}\ell^2$. Thus $\dfrac{w}{\ell}$ $= \tan\theta = \dfrac{5}{12}$.

21. Let b and c represent the respective rates of bicycle and car, and say they pass each other after h hours. B has gone a distance bh and C has gone ch. Obviously, B must now go a distance ch and C must go bh to complete the trips. Since time $= \frac{\text{distance}}{\text{rate}}$, we have $\frac{ch}{b} = 25\left(\frac{bh}{c}\right)$, or $\frac{c}{b} = 25\left(\frac{b}{c}\right)$. This leads to $\frac{b}{c} = \frac{1}{5}$.

22. Let the side-length of the square be s and the side-length of the regular hexagon be h. Then $s^2 = 6h^2\sqrt{3}/4$ and $s^2/h^2 = \sqrt{27/4}$. Thus, $s/h = \sqrt[4]{27/4}$. Since $4s/6h = 2s/3h = \sqrt[4]{(16 \cdot 27)/(81 \cdot 4)} = \sqrt[4]{4/3}$, we have $k = 4/3$.

23. Let the first two primes be p and $p + 2$. Of every three consecutive integers, exactly one is divisible by 3. Neither p nor $p + 2$ is divisible by 3 (since each is prime and greater than 3), so 3 is a divisor of $p + 1$. But $(p) + (p + 2) = 2(p + 1)$. Since p is odd, $p + 1$ is both even and divisible by 3. Hence $p + 1$ is divisible by 6. Thus the expression $2(p + 1)$ is divisible by 12. Since $5 + 7 = 12$, the answer is exactly 12, and not more.

24. Let the center be O, and $AT = x$. Then $AH = x\sqrt{2}$. By the law of cosines in triangle AHO, $(x\sqrt{2})^2 = 32 - 16\sqrt{2}$, or $2x^2 = 32 - 16\sqrt{2}$. The area of the quadrilateral is the sum of the areas of the 12 triangles pictured. Thus, the area $= 2x^2 + 8\left(\frac{16\sqrt{2}}{4}\right) = 16(2 + \sqrt{2})$ and $k = 16$.

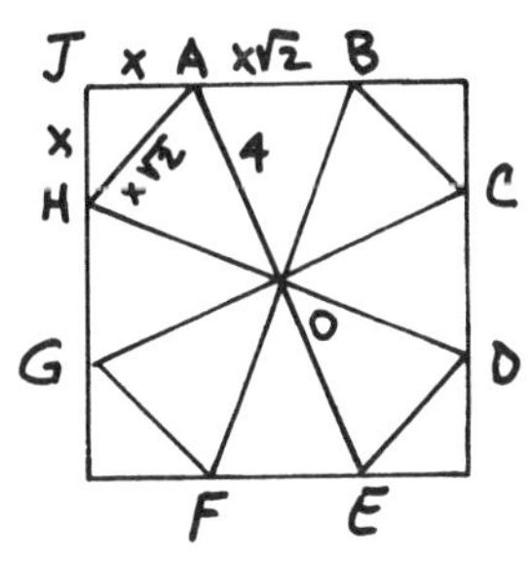

25. $\text{Cos}(x + y) = \frac{3}{5} \cdot \frac{12}{13} - \frac{4}{5} \cdot \frac{5}{13} = \frac{16}{65}$. (See 4.313.)

26. The following solution is adapted from Niven, Mathematics of Choice (pp. 27,28). Let E denote the act of driving one block east, and N denote the act of driving one block north. Thus, to each route, there corresponds a string of five E's and five N's arranged in a row. The number of ways one can fill 10 spaces with 5 E's (the remainder being N's) is given by the binomial coefficient $\binom{10}{5} = \frac{10!}{5!5!} = 252$.

27. Method 1: Since $b = a + c$, we have $ax^2 + (a + c)x + c$ $= (ax + c)(x + 1) = 0$. The larger root is -1.

Method 2: Using the quadratic formula, the two roots are seen to be $\dfrac{-(a+c) \pm \sqrt{(a+c)^2 - 4ac}}{2a} = \dfrac{-(a+c) \pm \sqrt{(a-c)^2}}{2a}$ $= \dfrac{-(a+c) \pm |a - c|}{2a}$ and the larger root is -1. Note that $|a - c| = c - a$ since $c > a$.

28. The two vertices must be endpoints of a major diagonal. Of the two possibilities that exist when two adjacent faces are unfolded, the first diagram provides the shortest path, and the length of the hypotenuse is 20.

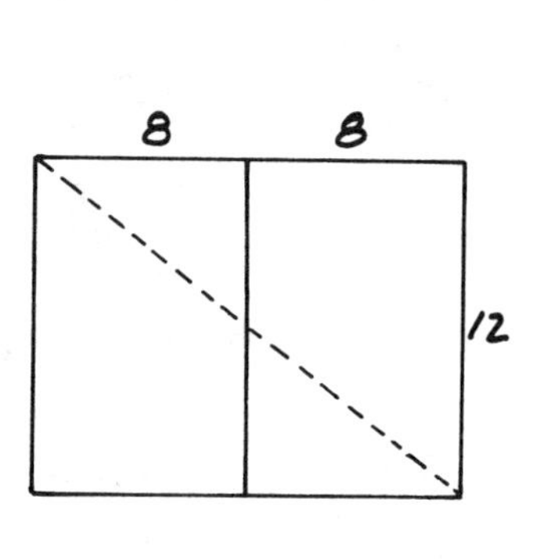

Figure 1

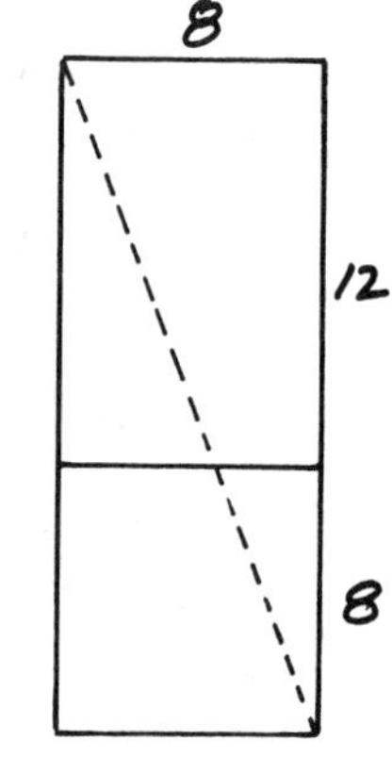

Figure 2

29. Since $\dfrac{\sqrt{2} - \frac{7}{5}}{\sqrt{2}} = \dfrac{5\sqrt{2} - 7}{5\sqrt{2}} = 1 - \dfrac{7\sqrt{2}}{10} = 1 - \dfrac{\sqrt{98}}{10}$, we have $k = 98$.

30. Since $c^2 = a^2 + (49 - c)^2$, we have $a = 7\sqrt{2c - 49}$. Since $2c - 49$ must be an odd perfect square less than 49 (since the hypotenuse < 49), the values of a are 7, 21, 35.
A solution using (2.1) is also possible, although rather more involved.

Solutions—Spring 1977

1. In each of the three methods below, we first find that the ratio $(x + y):z$ equals 4:1 and the ratio $x:(y + z) = 1:1$. Solving for x and y in terms of z then shows that $x:y:z = 5:3:2$.

Method 1: Using mass points, assign weight 1 to point A and weight 2 to each of points B and C. Then $2B + 2C = 4M$, $2B + 1A = 3D$, and the center of mass of the system is at G. Hence $1A + 4M = 5G$, so $(x + y):z = 4:1$. To determine $x:(y + z)$, we assign weight 1 to each of points B and C and weight 2 to point A, and proceed in the same way.

Method 2: Use Menelaus' theorem (see 3.83). We first use the theorem in triangle ABM, with transversal CD. We find that $(1/2)((x + y)/z)(1/2) = 1$, so $(x + y):z = 4$. We then use the theorem on triangle ABM again, but this time with transversal CE. We find that $(2/1)(x/(y + z))(1/2) = 1$, so $x:(y + z) = 1$.

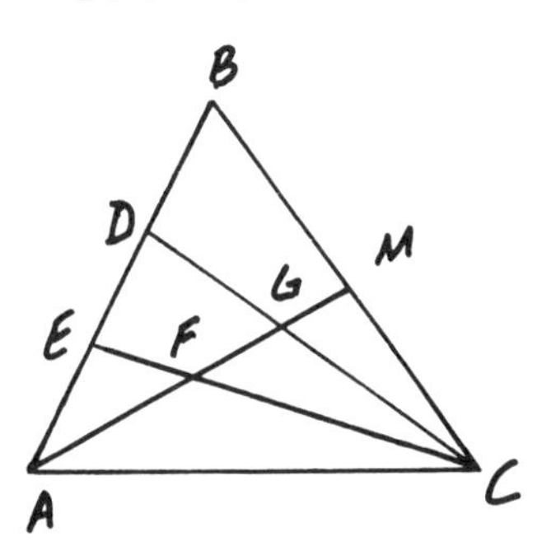

Method 3: Use areas. We use the fact that the ratio of the areas of triangles with equal bases (altitudes) is the same as the ratio of their altitudes (bases). We will use absolute value to denote area.

First we draw line BG. Since $BM = MC$ and triangles BGM, CMG have the same altitude from G, we find that $|BGM| = |CMG|$. Since triangles BAM, MAC have the same altitude from A, we find also that $|BAM| = |MAC|$. Subtracting, we have $|GAB| = |GAC|$. Since $AD:DB = 2:1$ and triangles AGD, BGD have the same altitude from G, we have $|AGD|:|BGD| = 2:1$. Let $|BGD| = 4k$. Then $|AGD| = 8k$ and $|GAB| = 12k$, so $|CAD| = |GAC| + |AGD| = |GAB| + |AGD| = 12k + 8k = 20k$. Now triangles CAD, CBD have the same altitude from C, so $|CAD|:|CBD| = AD:DB = 2:1$, or $|CAD| = (2/3)(|CAB|)$. Hence $|CAB| = 30k$. But clearly $|CAB| = |GAB| + |GAC| + |BMG| + |CMG| = 24k + 2|BMG|$, so that $|BMG| = 3k$. Lastly, we find that $|GAB|:|GMB| = 12k:3k = 4:1$. But triangles GAB, BMG have the same altitude from B, so this ratio is also the ratio of their bases, or $(x + y):z = 4:1$.

To calculate $x:(y + z)$ we draw segment BF and proceed similarly. For more on this method see Kay (pp. 197-207).

2. There are 199 integers in the given set. Let $M(n)$ denote the number of integers in the set that are divisible by n, and let $M(7 \text{ or } 11)$ denote the number of integers that are multiples of 7 or 11. Then:

$$M(7 \text{ or } 11) = M(7) + M(11) - M(77). \quad \text{(See 7.2.)}$$

Using $[x]$ to represent the greatest integer function, we have:

$$M(7) = [299/7] - [100/7] = 42 - 14 = 28$$
$$M(11) = [299/11] - [100/11] = 27 - 9 = 18$$
$$M(77) = [299/77] - [100/77] = 3 - 1 = 2,$$

and $M(7 \text{ or } 11) = 44$. Hence the desired probability is 44/199. For more on the inclusion-exclusion principle, see Niven, Mathematics of Choice (Chapter 5).

3. Letting $y = 0$ gives $f(x \cdot 0) = f(x) \cdot f(0)$, or $f(0) = f(x) \cdot f(0)$. Since $f(0) \neq 0$, we can divide both sides by $f(0)$, obtaining $f(x) = 1$ for all real values of x. Hence $f(1977) = 1$.

4. Squaring both sides, we obtain $\sin^2 \frac{x}{2} = 2 - 2\sqrt{1 - \sin^2 x}$.

Noting that $1 - \sin^2 x = \cos^2 x$, we have $\sin^2 \frac{x}{2} = 2 - 2\sqrt{\cos^2 x}$ $= 2 - 2|\cos x|$. Using 4.24 we have $(1 - \cos x)/2 = 2 - 2|\cos x|$. We consider two cases: Case I: $0 \leq x \leq \pi/2$. In this interval, $\cos x \geq 0$, so $|\cos x| = \cos x$, and we have $(1 - \cos x)/2$ $= 2 - 2 \cos x$. Solving yields $\cos x = 1$, so that $\tan x = 0$. Case II: $\pi/2 < x \leq \pi$. Here $\cos x < 0$, so $|\cos x| = -\cos x$, and we have $(1 - \cos x)/2 = 2 + 2 \cos x$. Solving yields $\cos x = -\frac{3}{5}$, so that $\tan x = -\frac{4}{3}$. The required solutions are therefore 0 and $-\frac{4}{3}$.

5. Let $a = \sqrt{x + y}$ and $b = \sqrt{x - y}$ (where $a \geq 0$ and $b \geq 0$). Since $a^2 = x + y$ and $b^2 = x - y$, we may rewrite the given equations as:

$$a^2 + a = 56$$
$$b^2 + b = 30.$$

Then $a^2 + a - 56 = (a - 7)(a + 8) = 0$, or $a = 7$ (rejecting $a = -8$). Similarly, we find that $b = 5$. We now know that $x + y = 49$ and $x - y = 25$. Solving this system, we get $(x,y) = (37,12)$.

6. From the binomial theorem (see 1.21), we have $(x + y)^3 = x^3 + 3x^2y + 3xy^2 + y^3$. Hence the given information shows that $x^3 + y^3 = 3x^2y + 3xy^2 = 3xy(x + y)$. Since $x + y \neq 0$, we can divide both sides by $x + y$ to get $3xy = x^2 - xy + y^2$ (see 1.321), or $2xy = x^2 - 2xy + y^2 = (x - y)^2$. Hence $(x - y)^2 : xy = 2:1$.

7. We have the equation $x - y = x^2 - y^2$. If $x - y \neq 0$, this leads to $x + y = 1$. Substituting in the equation $xy = x - y$ gives $x - x^2 = 2x - 1$. This leads to $(x,y) = \left(\frac{-1 + \sqrt{5}}{2}, \frac{3 - \sqrt{5}}{2}\right)$ and another pair of values that are not both positive. We can be sure, in dividing by $x - y$, that we are not dividing by 0. For if $x - y = 0$, then the problem asserts that $xy = 0$ as well, which would mean that either $x = 0$ or $y = 0$, contrary to the conditions of the problem.

8. This problem is an exercise in factoring. Let us write s for $\sin\theta$ and c for $\cos\theta$. Then:

$$\begin{aligned} s^6 + c^6 &= (s^2)^3 + (c^2)^3 \\ &= (s^2 + c^2)(s^4 - s^2c^2 + c^4) \quad (1.321) \\ &= 1 \cdot (s^4 - s^2c^2 + c^4) \\ &= (s^2 + c^2)^2 - 3s^2c^2 = 1 - 3s^2c^2. \end{aligned}$$

Hence $1 - 3s^2c^2 = \frac{2}{3}$ and $s^2c^2 = \frac{1}{9}$. But this implies that $4\sin^2\theta\cos^2\theta = \frac{4}{9}$ and $\sin 2\theta = \pm\frac{2}{3}$.

9. Since $f(xy) = f(x) + f(y)$ for all $x, y \geq 0$,

$$\begin{aligned} f(72) &= f(9 \cdot 8) = f(9) + f(8) \\ &= f(3 \cdot 3) + f(2 \cdot 4) \\ &= 2f(3) + 3f(2) \\ &= 2b + 3a. \end{aligned}$$

It is not hard to guess that f must be a logarithmic function. For a nice development of these functions from the functional equation $f(xy) = f(x) + f(y)$, see Apostol, (Vol. I, pp. 174-181).

10. We have $(x + \sqrt{y})^2 = x^2 + y + 2x\sqrt{y} = \frac{21}{4} + 3\sqrt{3}$.

Comparing rational and irrational parts, we see that

$x^2 + y = \frac{21}{4}$ and $2x\sqrt{y} = 3\sqrt{3}$, or $4x^2y = 27$.

Eliminating y by substitution gives $4x^4 - 21x^2 + 27 = (4x^2 - 9)(x^2 - 3) = 0$.

10. (Continued)

The only positive, rational value for x is $\frac{3}{2}$, so $(x,y) = \left(\frac{3}{2}, 3\right)$.

Note that the equation $2x\sqrt{y} = 3\sqrt{3}$ guarantees that x must be positive as well as rational since if $x < 0$, $2x\sqrt{y} < 0$ as well. In general, we can solve this type of problem by noting that $(\sqrt{a} + \sqrt{b})^2 = (a + b) + 2\sqrt{ab}$. In the present problem we can express the radicand in the form $(a + b) + 2\sqrt{ab}$:

$$\sqrt{\frac{21}{4} + 3\sqrt{3}} = \sqrt{\frac{21 + 12\sqrt{3}}{4}} = \sqrt{\frac{21 + 2\sqrt{108}}{4}} = \frac{\sqrt{21 + 2\sqrt{108}}}{2}.$$

Letting $a + b = 21$ and $ab = 108$, we quickly see that $a = 9$ and $b = 12$ is a solution. Therefore, $\frac{\sqrt{21 + 2\sqrt{108}}}{2} = \frac{\sqrt{9} + \sqrt{12}}{2} = \frac{3}{2} + \sqrt{3}$.

For references, see problem 10, Spring 1975.

11. In base x, the numeral $7y3$ has the value $7x^2 + yx + 3$. Similarly, $3y7$ has the value $3x^2 + yx + 7$. Hence we have $7x^2 + yx + 3 = 2(3x^2 + yx + 7)$, or $x^2 - xy - 11 = 0$. Since both x and y are positive integers, we can write $x^2 - xy = x(x - y) = 11$. Since 11 is prime, $x = 11$ or $x = 1$. If $x = 11$, then $y = 10$. Similarly, $x = 1$ implies that $y = -10$, which is impossible. Therefore, $x + y = 21$.

12. Let s_1, s_2, s_3, ... be the lengths of the sides of the successive hexagons. For each s_k, we find (see diagram) that $s_{k+1}:s_k = \sqrt{3}:2$. Hence $6s_{k+1}:6s_k = \sqrt{3}:2$, and the perimeters form a geometric progression whose first term is 12 and whose common ratio is $\sqrt{3}:2$. Their infinite sum is thus $\frac{12}{1 - \frac{\sqrt{3}}{2}} = 24(2 + \sqrt{3})$. (See 1.523.)

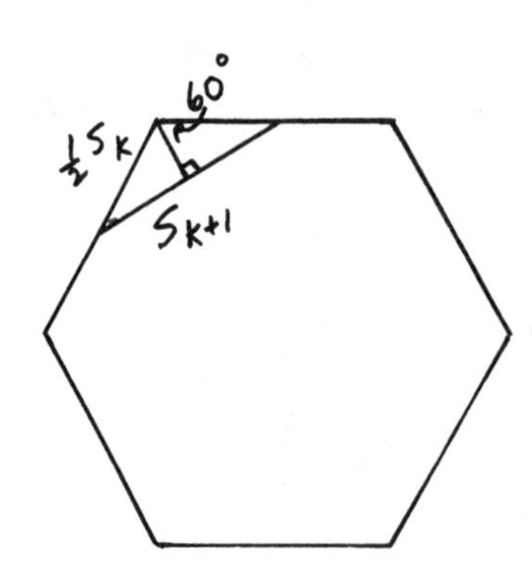

13. Using 4.4, it follows that $\dfrac{2x + 3x}{1 - (2x)(3x)} = 1$ or $6x^2 + 5x - 1 = (6x - 1)(x + 1) = 0$. The root $x = -1$ is extraneous, as the problem demands principal values for Arctan, so $x = \dfrac{1}{6}$.

14. In the figure, note that triangle $CBD \sim$ triangle EBF and triangle $CDF \sim$ triangle ABF.

Therefore, $\dfrac{24}{x} = \dfrac{a}{a + b}$ and $\dfrac{24}{40} = \dfrac{b}{a + b}$. Adding, we get $\dfrac{24}{x} + \dfrac{24}{40} = 1$, which leads to $x = 60$. (See also 3.81.)

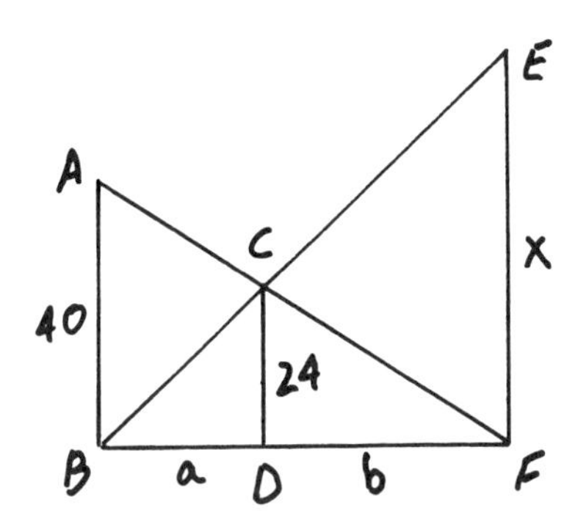

15. The conditions of the problem are equivalent to $9x^2 + 1 = (9x + 1)^2$. Solving for x produces the two solutions $x = 0$ or $x = -\dfrac{1}{4}$.

16. The original expression may be regrouped as $(a^2b^2 + c^2d^2) + (a^2d^2 + b^2c^2)$, which may, in turn, be written as $(ab + cd)^2 + (ad - bc)^2$ or as $(ab - cd)^2 + (ad + bc)^2$. Since $a = 44$, $b = 10$, $c = 33$, and $d = 5$, we obtain the two solutions $x = 605$, $y = 110$ or $x = 550$, $y = 275$.

A third, unrelated solution, may be obtained by noting that $x^2 + y^2 = 378{,}125 = 5^5 \cdot 11^2 = (38^2 + 41^2)(0^2 + 11^2)$. Using the identity $(a^2 + b^2)(c^2 + d^2) = (ac - bd)^2 + (ad + bc)^2$ and letting $a = 38$, $b = 41$, $c = 11$, and $d = 0$, we get $x^2 + y^2 = (38 \cdot 11 - 41 \cdot 0)^2 + (38 \cdot 0 + 41 \cdot 11)^2 = 418^2 + 451^2$. Since $x > y$ we have $x = 451$ and $y = 418$. For a further discussion of these and related identities, see Beiler (pp. 141-145).

17. By expanding and simplifying we obtain

$$\left(x^2 + 2 + \frac{1}{x^2}\right) + \left(y^2 + 2 + \frac{1}{y^2}\right) + \left(x^2y^2 + 2 + \frac{1}{x^2y^2}\right)$$

$$- \left(x^2y^2 + 1 + x^2 + \frac{1}{y^2} + y^2 + \frac{1}{x^2} + 1 + \frac{1}{x^2y^2}\right) = 4.$$

18. We may rewrite the given system as
$$3x - 4y = 6 - p$$
$$6x + 4y = 15$$
for some positive number $p = 6 - (3x - 4y)$.
Adding the equations yields $9x = 21 - p$ or $x = \frac{21 - p}{9}$ and hence $y = \frac{3 + 2p}{12}$. By the hypothesis of the problem, $x = \frac{7 - r}{3}$ and $y = \frac{1 + s}{4}$. We therefore rewrite our solutions as $x = \frac{7 - \frac{p}{3}}{3}$, $y = \frac{1 + \frac{2p}{3}}{4}$ to match the given form. Hence $r = \frac{p}{3}$ and $s = \frac{2p}{3}$ and we have $r{:}s = 1{:}2$.

19. Simplifying the original equation, we obtain $\log_x \frac{25}{4} = \log_x \sqrt{x}$. Hence $\frac{25}{4} = \sqrt{x}$ or $x = \frac{625}{16}$. For another approach to this problem, note that $\log_x \sqrt{x} = \frac{1}{2}$ for all $x > 0$. Hence $\log_x \frac{25}{4} = \frac{1}{2}$, $x^{\frac{1}{2}} = \frac{25}{4}$, and again $x = \frac{625}{16}$.

20. <u>Method 1</u>: Completing the square gives

$$\left(x^2 + x + \frac{1}{4}\right) + 29 = y^2 + \frac{1}{4}, \text{ or} \left(x + \frac{1}{2}\right)^2 - y^2 = \frac{-115}{4},$$

or $4y^2 - (2x + 1)^2 = 115$. Factoring the difference of two squares on the left produces $(2y + 2x + 1)(2y - 2x - 1) = 115$. Since x and y are positive integers, each of the factors on the left are also integers and hence must be factors of 115. This leads to eight possibilities:

$$\begin{matrix} 2y + 2x + 1 = 115 \\ 2y - 2x - 1 = 1 \end{matrix} \text{ or } \begin{matrix} 23 \\ 5 \end{matrix} \text{ or } \begin{matrix} 5 \\ 23 \end{matrix} \text{ or } \begin{matrix} 1 \\ 115 \end{matrix}.$$

We may ignore the last two cases, as the first factor must be larger than the second. Adding, we have:

$$4y = 116 \text{ or } 28$$
$$y = 29 \text{ or } 7$$

20. (Continued)

and by substitution,

$$x = 28 \text{ or } 4.$$

Hence $(x,y) = (28,29)$ or $(4,7)$.

Method 2: Writing the equation as $x^2 + x + 29 - y^2 = 0$ and solving for x using the quadratic formula, we obtain $x = \dfrac{-1 \pm \sqrt{1 - 4(29 - y^2)}}{2}$ or $x = \dfrac{-1 \pm \sqrt{4y^2 - 115}}{2}$. For x to be an integer, it is necessary that $4y^2 - 115$ be a perfect square. This means that $4y^2 - 115 = k^2$ or $4y^2 - k^2 = 115$. Therefore, $(2y - k)(2y + k) = 115$. Following the plan of the first method illustrated we find that $y = 29$ and $k = 7$, or $y = 7$ and $k = 9$. Solving for x gives the same pairs found above.

21. We want to relate the given equations to the binomial expansion of $(x - y)^3$. Multiplying the second equation by 3 yields $3(x^2y - xy^2) = 18$. Since $x^3 - y^3 = 19$ we have $(x - y)^3 = x^3 - y^3 - 3(x^2y - xy^2) = 19 - 18 = 1$. Hence $x - y = 1$ (rejecting possible complex values). Also from the second equation, we have $xy(x - y) = 6$ and since $x - y = 1$, we find $xy = 6$ or $y = \dfrac{6}{x}$. Substituting in the equation $x - y = 1$ and simplifying we obtain $x^2 - x - 6 = 0$, with solutions $x = -2$ or $x = 3$. The corresponding values of y are $y = -3$ or $y = 2$. Hence $(x,y) = (-2,-3)$ or $(3,2)$.

22. Using 3.5 shows that $AZ = AY = 2$, $ZC = CX = 1$, and $XB = BY = 3$ (see figure 1). We may now apply the method of mass points with $\overline{BE} = p$ and $\overline{EZ} = q$. Assign weight 3 to A, 2 to B, and 6 to C (see figure 2). Therefore, $\overline{BE}:\overline{EZ} = p:q = 9:2$.

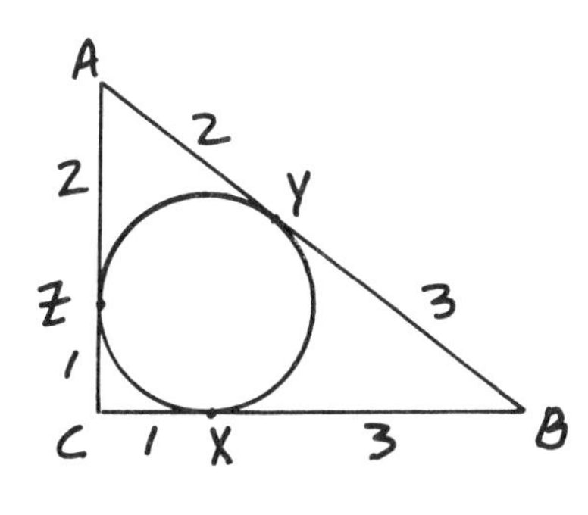

Figure 1

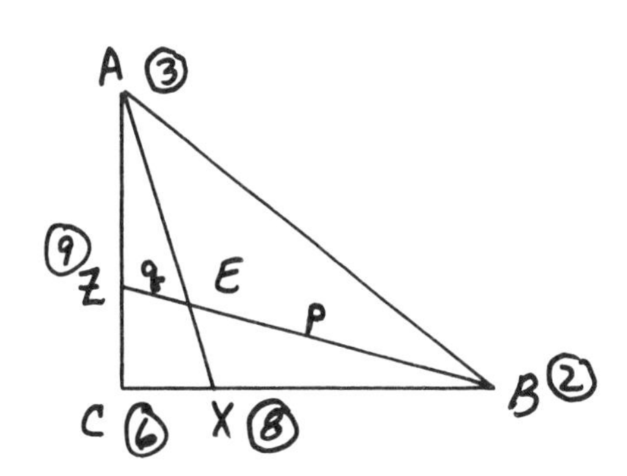

Figure 2

23. Let Arcsin $\frac{3}{5} = A$ and Arccos $\frac{15}{17} = B$.

Then: $\sin x = \sin\left(\text{Arcsin}\,\frac{3}{5} + \text{Arccos}\,\frac{15}{17}\right)$

$= \sin(A + B)$

$= \sin A \cos B + \sin B \cos A$

$= \frac{3}{5} \cdot \frac{15}{17} + \frac{8}{17} \cdot \frac{4}{5}$ (see 4.311)

$= \frac{77}{85}$.

24. Method 1: Squaring both sides yields $k = 2x - 2\sqrt{x^2 - (2x - 1)}$ $= 2x - 2\sqrt{x^2 - 2x + 1} = 2x - 2\sqrt{(x - 1)^2} = 2x - 2|x - 1|$.
Again, $x > 1$ implies that $x - 1 > 0$, hence $|x - 1| = x - 1$.
Therefore, $k = 2x - 2(x - 1) = 2$.

Method 2: Since the given equation is an identity for all $x > 1$, it must certainly hold true for any specific value of x that is greater than 1. Therefore, we can choose a "convenient" value of x such as $x = 5$ and $\sqrt{k} = \sqrt{5 + 3} - \sqrt{5 - 3} = \sqrt{8} - \sqrt{2} = 2\sqrt{2} - \sqrt{2} = \sqrt{2}$.
Hence $k = 2$. The value $x = 5$ was chosen so that $2x - 1$ would be a perfect square.

25. Since N is a numeral in base x, it is helpful to express N as a polynomial in x. By long division, $N = x^4 + x^3 + x^2 + x + 1$. Therefore, $N = 11111$. See 1.33, with $y = 1$.

26. Let us first solve a more general type of problem. Suppose we wish to count the number of distinct paths from point A to point B (on the grid shown below), allowing only moves in a "northerly" or "easterly" direction. One such path is darkened in as an example. We may describe the darkened path by saying, "Go 1 block east, 2 blocks north, 1 block east," etc.

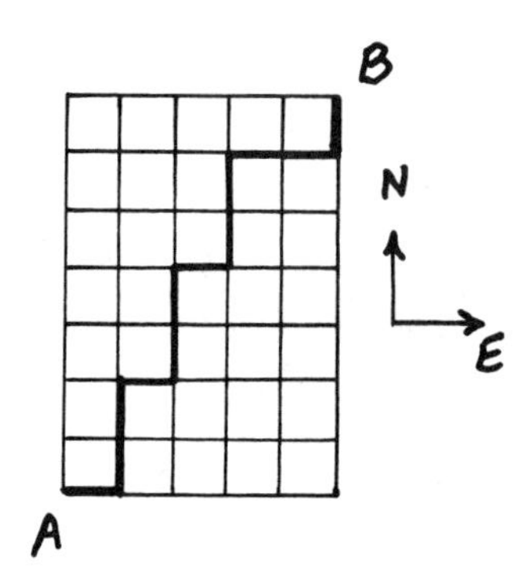

26. (Continued)

If we symbolize 1 block east by the letter E and 1 block north by the letter N, the darkened path may be written as ENNENNENNEEN. Now any path from A to B must go 5 blocks east and 7 blocks north. Hence any such path corresponds to a "word" containing five E's and seven N's. Also, any such "word" corresponds to a unique path. The problem is therefore fully solved by counting the number of distinct permutations of the letters of the "word" EEEEENNNNNNN. Hence the number of distinct paths is $\frac{12!}{5!7!}$. For a thorough treatment of this problem and many other interesting problems of this type, see Niven, Mathematics of Choice.
Returning to the present problem, we first calculate the total number of paths from (0,0) to (4,4). The above discussion shows this to be $\frac{8!}{4!4!} = 70$. Similarly, there are $\frac{4!}{2!2!} = 6$ paths from (0,0) to (2,2) and $\frac{4!}{2!2!} = 6$ paths from (2,2) to (4,4). Thus there are $6 \cdot 6 = 36$ possible paths from (0,0) to (4,4) that pass through (2,2). Hence the desired probability is $\frac{36}{70}$ or $\frac{18}{35}$.

27. From the given information, we have (1) $\frac{3a}{2} + \frac{3b}{2} = 120$ and (2) $2a + b = 120$. Solving this system yields $(a,b) = (40,40)$.

28. The key to the solution of this problem is a well-labeled diagram. Since $\overline{BQ}$ and $\overline{CQ}$ are angle bisectors, and $\overline{BCF} || \overline{DEQ}$, we have angle DBQ = angle CBQ = angle BQD, and angle ECQ = angle FCQ = angle CQE. Therefore, triangle DBQ and triangle ECQ are isosceles triangles. Hence $DQ = 8$ and $EQ = 6$, so $ED = 2$.

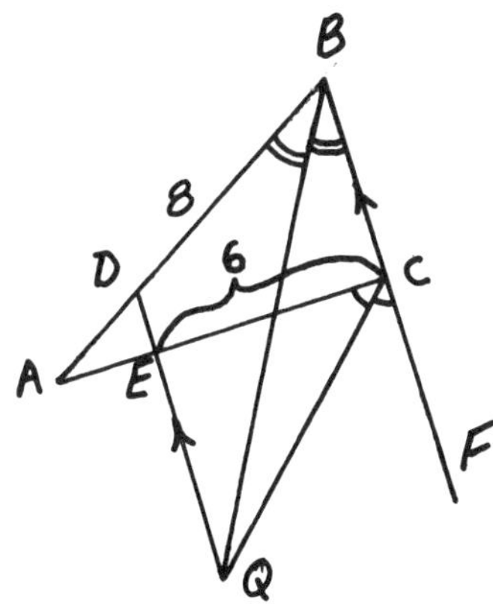

29. We use the fact that if $[x] = n$ and x is not an integer, then $[-x] = -n - 1$ and $([x] + [-x])^5 = (n - n - 1)^5 = (-1)^5 = -1$. See 5.22.

30. Method 1: Since $\sin A + \sin B = 2 \sin \frac{A + B}{2} \cos \frac{A - B}{2}$ (see 4.321), we have $\sin 5x + \sin 3x = 2 \sin 4x \cos x$. Therefore, $2 \sin 4x \cos x = 0$, which implies that $\sin 4x = 0$ or $\cos x = 0$. Since $0 < x < \pi$, we have the three solutions $x = \frac{\pi}{4}$ or $\frac{\pi}{2}$ or $\frac{3\pi}{4}$.

Method 2: We write $\sin 5x$ as $\sin(4x + x)$ and $\sin 3x$ as $\sin(4x - x)$. Using formulas 4.311 and 4.312, we get $\sin 5x = \sin(4x + x)$ $= \sin 4x \cos x + \sin x \cos 4x$ and $\sin 3x = \sin(4x - x)$ $= \sin 4x \cos x - \sin x \cos 4x$. Adding these two identities yields $\sin 5x + \sin 3x = 2 \sin 4x \cos x$. Now proceed as outlined in Method 1.

Solutions—Fall 1977

1. Adding the two equations gives $5x^2 = 5$, so $x = \pm 1$. Substituting these values in either equation gives solutions $(\pm 1,0)$, $(1,2)$, $(-1,-2)$.

2. Using the law of cosines in triangle BEF, and noting that cos angle EBF = sin angle $ABE = \frac{4}{5}$, we have: $x^2 = 9 + 25 - 2(3)(5)\left(\frac{4}{5}\right) = 10$, so $x = \sqrt{10}$. We can avoid using the law of cosines if we draw a square with adjacent sides $\overline{AB}$ and $\overline{BF}$, calling its fourth vertex H (see diagram). If we drop perpendiculars from point E to sides $\overline{AB}$, $\overline{BF}$, the Pythagorean theorem alone will give the required result.

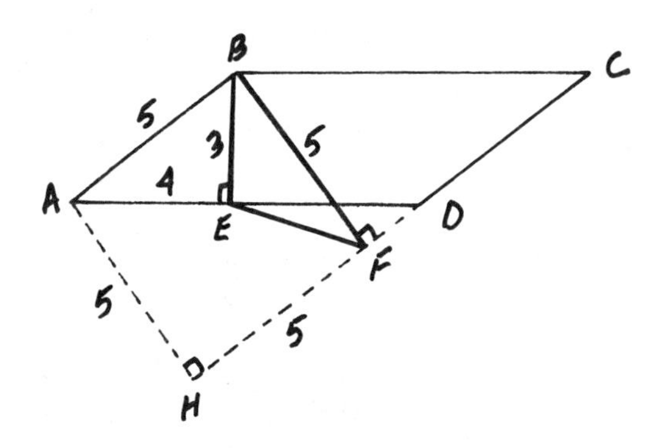

3. Method 1: In right triangle ABD, $BD = 24$. In right triangle ACD, $AC = 15$. Then, using Ptolemy's theorem (3.91), $BC \cdot 25 + 7 \cdot 20 = 15 \cdot 24$ and $BC = \frac{44}{5}$.

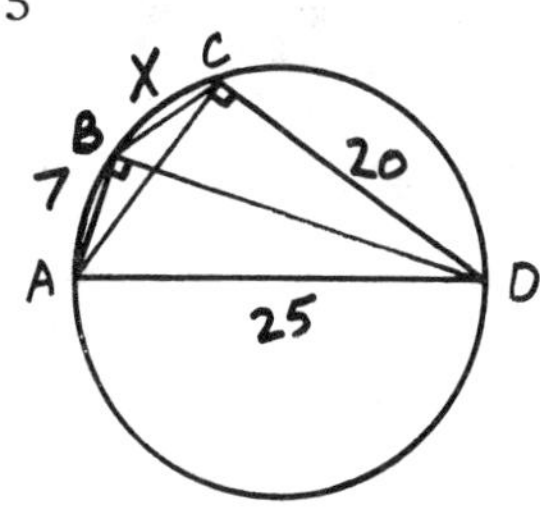

3. (Continued)

<u>Method 2:</u> We have, from right triangles ABD and ACD,

$\cos$ angle $ADB = \frac{24}{25}$ $\qquad$ $\cos$ angle $ADC = \frac{20}{25}$

$\sin$ angle $ADB = \frac{7}{25}$ $\qquad$ $\sin$ angle $ADC = \frac{15}{25}$

Hence $\cos$ angle $BDC = \cos(\text{angle } ADC - \text{angle } ADB)$

$$= \frac{20}{25} \cdot \frac{24}{25} + \frac{15}{25} \cdot \frac{7}{25} = \frac{117}{125} .$$

Now we can use the law of cosines in triangle BDC to find x.

4. Consider A's motion. He first travels $\frac{1}{2}$ hour at x km per hour.

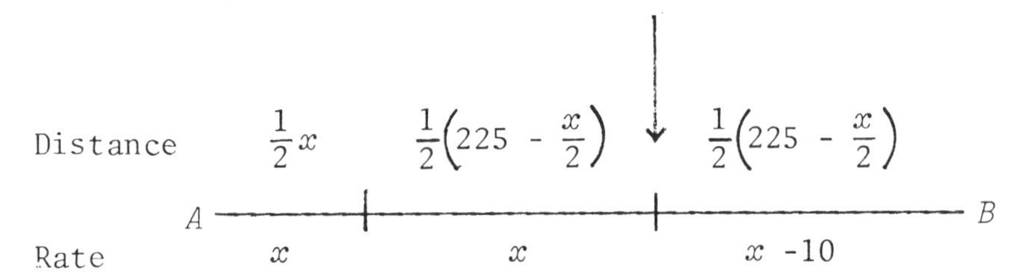

Since A took 5 hours to complete the entire trip and the problem implies that $x > 0$, $\frac{1}{2} + \frac{\frac{1}{2}(225 - \frac{1}{2}x)}{x} + \frac{\frac{1}{2}(225 - \frac{1}{2}x)}{x - 10} = 5$. Then $2x^2 - 109x + 450 = (2x - 9)(x - 50) = 0$, so $x = 50$.

5. Since each of the given polynomials has a lead coefficient of 1, if either of them factors into a quadratic and a linear polynomial, then the lead coefficient of each must be 1. Let the required quadratic factor be $x^2 + px + q$. Since any common factor of two polynomials is a factor of their difference, we must have $x^2 + px + q = \frac{1}{N}[(a - b)x^2 + 3x + 3]$ for some integer N. Since p and q are integers, N must divide 3, so $N = 1$ or $N = 3$. If $N = 1$, $a - b = 1$ as well. Suppose $x^3 + bx^2 + 6x + 3 = (x^2 + 3x + 3)(x + r)$ for some integer r. Clearly $r = 1$, and multiplication produces $b = 4$, so $a = 5$. We then check that $x^3 + 5x^2 + 9x + 6$ is divisible by $x^2 + 3x + 3$, which it is. If $N = 3$, the common quadratic factor would be $x^2 + x + 1$, and we would need $x^3 + bx^2 + 6x + 3 = (x^2 + x + 1)(x + s)$ for some integral

5. (Continued)

s. Clearly $s = 3$, but then $(x^2 + x + 1)(x + 3)$ does not produce a $6x$ term. Thus only $x^2 + 3x + 3$ works, and $(a,b) = (5,4)$.
For a more detailed look at common factors of polynomials, see Borofsky (or any thorough book on the theory of equations).

6. The probability that they roll equal numbers is $\frac{1}{6}$. Hence the probability that they roll unequal numbers is $\frac{5}{6}$. Since neither player has an advantage over the other, the probability of either rolling higher than the other is $\frac{1}{2} \cdot \frac{5}{6} = \frac{5}{12}$.

7. Let a and ar be the first two terms of the progression. The given condition can be written as

$$a(1 + r + r^2 + r^3 + r^4) = 16\left(\frac{r^4 + r^3 + r^2 + r + 1}{ar^4}\right).$$

For real r, it is not hard to see that the numerator of the last expression is not zero. Hence we must have $a = \frac{16}{ar^4}$, or $ar^2 = \pm 4$.
To see that the numerator can never be zero, note that

$$(r - 1)(r^4 + r^3 + r^2 + r + 1) = r^5 - 1. \quad \text{(See 1.33.)}$$

Thus the values that make the numerator vanish are the four complex fifth roots of unity.

8. Since $\frac{(c - a)^2}{b^2} = \frac{(c - a)^2}{c^2 - a^2} = \frac{c - a}{c + a} = \frac{16}{25}$, we have (1.12)

$\frac{c}{a} = \frac{41}{9}$. Let $c = 41k$ and $a = 9k$. Then $b = 40k$. Since k is a common factor of a, b, and c, we conclude that $k = 1$. Therefore $(a,b,c) = (9,40,41)$.

9. By (4.323) $\cos 3x + \text{cox}\ x = 2 \cos 2x \cos x = 0$, so $\cos x = 0$ or $\cos 2x = 0$. The last equation leads to $\cos x = \pm \frac{\sqrt{2}}{2}$.

10. We have (1.22) $a^3 + b^3 = (a + b)^3 \leftrightarrow a = 0$ or $b = 0$ or $a = -b$.
Letting $a = 2x - y$, $b = x - 2y$, we have $y = 2x$ or $x = 2y$ or $x = y$. Letting $\sqrt{\frac{x + 1}{y}} = c$, the second equation can be written as $c + \frac{1}{c} = 2 + \frac{1}{2}$, so $c = 2$ or $\frac{1}{2}$ and $x + 1 = 4y$ or $x + 1 = \frac{y}{4}$.

10. (Continued)

There are six ways to combine the conditions into pairs of simultaneous equations. Each set of equations leads to one of the six solutions:

$$\left(\frac{1}{7}, \frac{2}{7}\right), (-2,-4), \left(1, \frac{1}{2}\right), \left(\frac{-8}{7}, \frac{-4}{7}\right), \left(\frac{1}{3}, \frac{1}{3}\right), \left(\frac{-4}{3}, \frac{-4}{3}\right).$$

11. Since $1 \leq x < 2$, we can write $x = 1 + k$, where $0 \leq k < 1$. Then $[x]^2 = 1$. Since $[x^2] = [(1 + k)^2] = 1$, we must have $1 \leq (1 + k)^2 < 2$, or (since $k \geq 0$), $1 \leq 1 + k < \sqrt{2}$, and $1 \leq x < \sqrt{2}$.

12. The quadrilateral made up of the two triangles (see figure) always has opposite angles supplementary. Hence it is always cyclic (3.92), and we can use Ptolemy's theorem in each case (3.91). For example, in the first case pictured, $25x = 7 \cdot 20 + 15 \cdot 24$ and $x = 20$. The other cases give $\frac{117}{5}$, 25, 24, $\frac{336}{25}$, and 25 again. There are altogether five possible distances.

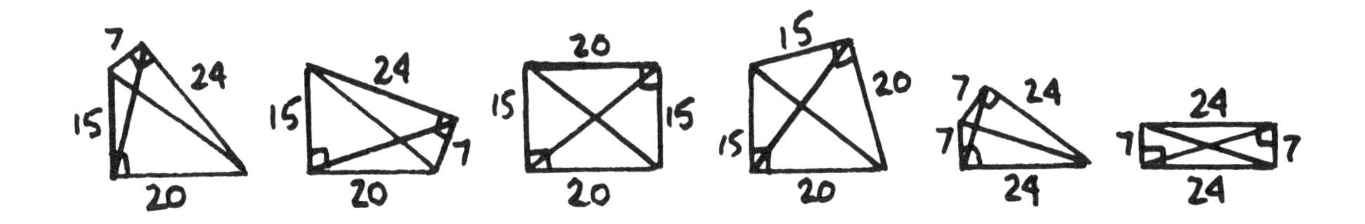

13. If the roots of the equation are r, r, and s, then (1.4) $2r + s = 1$ and $r^2s = -1$. Eliminating s between these produces $2r^3 - r^2 - 1 = (r - 1)(2r^2 + r + 1) = 0$. Since r is real, we have $r = 1$. Substituting this value for x in the original equation gives $m = -1$.

14. We expand, regroup, and simplify:

$(4x^2b^2 - 12abxy + 9a^2y^2) + (4x^2c^2 - 4acxz + a^2z^2)$
$+ (9y^2c^2 - 6bcyz + b^2z^2) = (2xb - 3ay)^2 + (2xc - az)^2$
$+ (3yc - bz)^2 = 0.$

Since each term must equal 0, $\frac{x}{y} = \frac{3a}{2b}$, $\frac{x}{z} = \frac{a}{2c} = \frac{3a}{6c}$, $\frac{y}{z} = \frac{b}{3c} = \frac{2b}{6c}$,

and $x:y:z = 3a:2b:6c$.

15. An integer that is both a perfect square and a perfect cube is a "perfect sixth power." There are four such integers less than or equal to $64^2 = 4^6$. The probability is thus $\binom{4}{2} \div \binom{64}{2} = \frac{1}{336}$.

16. We will use absolute value to denote area. In parallelogram $ABCD$ (see diagram), let $BM = MC = DN = NA$. Since the centroid of a triangle divides each median in the ratio 1:2, we have $MG:GE = 1:2$ and $HG:GH' = MG:(GE + EN) = 1:(2 + 3) = 1:5$ and $GH' = \left(\frac{5}{6}\right)HH'$.

Now $|ABCD| = AD \cdot HH'$ while $|AGD| = \frac{1}{2} AD \cdot \left(\frac{5}{6}\right)HH' = \left(\frac{5}{12}\right)|ABCD|$.

Similarly, $|BGC| = \left(\frac{1}{12}\right)|ABCD|$. Finally, a line through G perpendicular to $\overline{AB}$ and $\overline{CD}$ would be bisected by the parallel lines AB, NM, and DC, so $|ABG| = |CGD| = \frac{1}{4}|ABCD|$. Hence $|AGD|$ is the largest possible, and $|AGD| = \frac{5}{12} \cdot 180 = 75$.

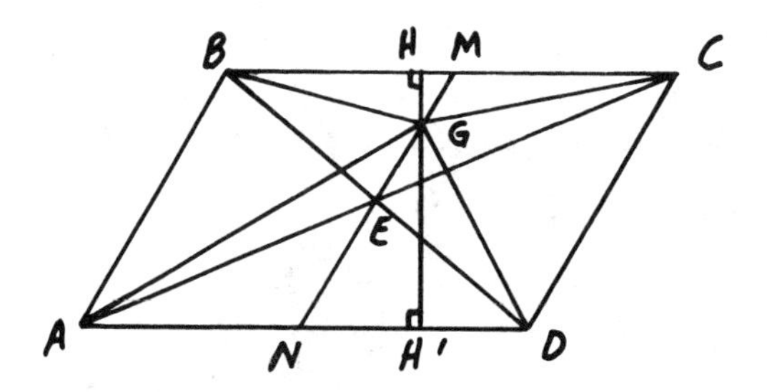

17. Using 4.321 and 4.323, the given expression may be written as

$$\frac{2 \sin 60° \cos 20°}{2 \cos 60° \cos 20°} = \tan 60° = \sqrt{3}.$$

18. Let $S = 1 \cdot 2 + 2 \cdot 2^2 + 3 \cdot 2^3 + \ldots + n \cdot 2^n$.

Then $2S = 2^2 + 2 \cdot 2^3 + \ldots + (n - 1)2^n + n \cdot 2^{n+1}$ and

$$2S - S = -(2 + 2^2 + 2^3 + \ldots + 2^n) + n \cdot 2^{n+1}$$

$$= \frac{-2(2^n - 1)}{2 - 1} + n \cdot 2^{n+1} = S, \text{ and}$$

$$S = -2^{n+1} + 2 + n \cdot 2^{n+1} = 2^{n+1}(n - 1) + 2.$$

When $n = 21$, $S = 20 \cdot 2^{22} + 2$ or 83,886,082.

19. Since $2^{2\log_2 x} = 2^{\log_2 x^2} = x^2$, we have $2x^2 = 8$ and $x = 2$.

20. Let us write $p(x) = x^2 + x + 1$ and $q(x) = (x + 1)^n - x^n - 1$. A quick way to do this problem is to note that if $p(x)$ divides $q(x)$, then the integer $p(1)$ must divide $q(1)$. Now $p(1) = 3$ and $q(1) = 2^n - 2$. Hence we can try $n = 1,2,\ldots$ to find that $n = 5,7$ are the two smallest values that work.

20. (Continued)

But $n = 5,7$ are only necessary conditions for $p(x)$ to divide $q(x)$. We must check to see that these values do actually work. For a more thorough investigation of the problem, we need to know something about the complex solutions to the equation $x^3 = 1$. For a full discussion of this topic, see Hall and Knight, Higher Algebra, pp. 440-444.

We use the factor theorem (1.311) and its corollary (1.312). The roots of $x^2 + x + 1 = 0$ are the complex cube roots of unity. We denote these by ω and ω^2, and we need both to be roots of the equation $q(x) = 0$. That is, we need:

$$(*) \qquad (\omega + 1)^n - \omega^n - 1 = 0$$
$$(\omega^2 + 1)^n - \omega^{2n} - 1 = 0.$$

If we note that $\omega + 1 = -\omega^2$ and $\omega^2 + 1 = -\omega$, the above conditions become $(-\omega^2)^n - \omega^n - 1 = (-\omega)^n - \omega^{2n} - 1 = 0$. First let us suppose that n is odd. Then we have from either equation, $\omega^{2n} + \omega^n + 1 = 0$, so ω^n satisfies $x^2 + x + 1 = 0$ and is a complex cube root of unity. This happens whenever n is a nonmultiple of 3. If n were even, the first equation in (*) above would say that $(-\omega^2)^n - \omega^n - 1 = 0$, so $\omega^{2n} - \omega^n - 1 = 0$. This says that ω^n is a root of $y^2 - y - 1 = 0$, which is false (since the roots of this equation are real, while odd powers of ω are complex).

21. We can denote the roots that sum to zero as r and $-r$. If the third root is s, then (1.4):

(i) $r + (-r) + s = -a$ or $s = -a$

(ii) $-r^2 s = r^2 a = -c$

(iii) $rs - rs - r^2 = -r^2 = b$.

From (ii) and (iii), $c = ab$.

22. If the center of one circle is (x,y), the radius r, and the point of tangency $(x,0)$, we have $y > 0$ and:

(i) $(x - 1)^2 + (y - 9)^2 = r^2$

(ii) $(x - 8)^2 + (y - 8)^2 = r^2$

(iii) $y = r$ (see diagram)

From (i) and (ii), we have $y = 7x - 23$. Then from (ii) and (iii), we have $(x - 4)(x - 124) = 0$. Hence $x = 4$ or 124, and $y = r = 5$ or 845.

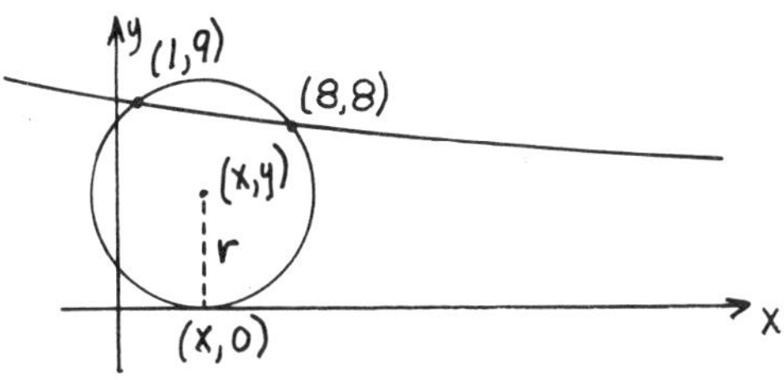

23. Draw altitude SH. Isosceles triangles MRT, TSR are similar, hence angle RST = angle TRM. Now $HR = \frac{1}{2}RT$; and $RE = \frac{2}{3}RM$ $= \frac{2}{3}RT$ (see 3.1), so cos angle RST = cos angle TRM = cos angle HRE $= \left(\frac{1}{2}RT\right) \div \left(\frac{2}{3}RT\right) = \frac{3}{4}$.

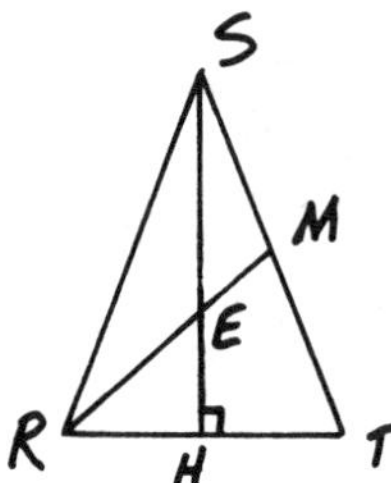

24. To get a picture of how $f(n)$ behaves, we can write: $f(1) = 0$, $f(2) = 1$, $f(3) = 3 + 1$, $f(4) = 3^2 + 3 + 1$, $f(5) = 3^3 + 3^2 + 3 + 1$.

In fact, it is not hard to see that $f(n) = \sum_{i=0}^{n-2} 3^i = \frac{1}{2}(3^{n-1} - 1)$.
(An easy induction will verify this.)

Hence $f(n + 3) - f(n + 1) = \frac{1}{2}[(3^{n+2} - 1) - (3^n - 1)]$

$$= \frac{1}{2}(3^n)(3^2 - 1) = 4 \cdot 3^n.$$

Letting $n = 1000$, we find $f(1003) - f(1001) = 4 \cdot 3^{1000}$ $= 2^2 \cdot 3^{1000}$.

25. By Hero's formula (3.61), the area of the triangle is $\sqrt{45 \cdot 5 \cdot 8 \cdot 32} = \sqrt{9 \cdot 5 \cdot 5 \cdot 4 \cdot 2 \cdot 2 \cdot 16} = 240$.
Since $20h = 240$, $h = 12$.

26. We shall prove that if $a + b + c = 0$, then $a^3 + b^3 + c^3 = 3abc$. We know that $(a^3 + b^3) + c^3 = (a + b)(a^2 - ab + b^2) + c^3$ (see 1.321) and that $a + b = -c$. Then $(a + b)^2 = a^2 + b^2 + 2ab = c^2$. Thus, $a^2 + b^2 - ab = c^2 - 3ab$ and $a^3 + b^3 + c^3$

$$= (a + b)(a^2 + b^2 - ab) + c^3$$
$$= (-c)(c^2 - 3ab) + c^3$$
$$= 3abc.$$

Thus, $abc = \frac{216}{3} = 72$.

For a fascinating application of this result to the solution of cubic equations, see Shklyarsky, Chentsov, and Yaglom, The USSR Olympiad Problem Book (problems 161, 162, and 167).

27. Let $z = x^2$. We have $x^2 - 64 = y^2$, or $x^2 - y^2 = (x + y)(x - y) = 64 = 64 \cdot 1 = 32 \cdot 2 = 16 \cdot 4 = 8 \cdot 8$ (since both $x + y$ and $x - y$ are integers). Since $x + y \geq x - y$, we must solve

$x + y = 64$	$x + y = 32$	$x + y = 16$	$x + y = 8$
$x - y = 1$	$x - y = 2$	$x - y = 4$	$x - y = 8$

Only the second and third give valid solutions, and $x^2 = 100$ or 289.

28. If a quadratic polynomial has rational coefficients, then irrational roots occur in conjugate pairs (see 1.82). Now $\sin^2 15° = \frac{1}{2}(1 - \cos 30°) = \frac{1}{4}(2 - \sqrt{3})$. Hence the other root of the given equation is $\frac{1}{4}(2 + \sqrt{3})$. Then by direct calculation of the sum and product of the roots, $b = -1$ and $c = \frac{1}{16}$.

29. We have:

(i) $$\frac{60}{x} = \frac{60}{y} + 5$$

(ii) $$\frac{60}{x + 1} = \frac{60}{y + 1} + 3$$

From (i), $xy = 12y - 12x$ and from (ii), $xy = 19y - 21x - 1$. Equating the righthand members of the above equations leads to $y = \frac{9x + 1}{7}$. Substituting in (i) then gives $9x^2 - 23x - 12 = 0$, so $(9x + 4)(x - 3) = 0$. Rejecting the negative root, we have $(x,y) = (3,4)$.

30. We have $3^{g(x)\log_3 f(x)} = [3^{\log_3 f(x)}]^{g(x)} = f(x)^{g(x)} = (x + 2)^x$.
Hence we must solve $(x + 2)^x - x + 2$.
In this equation, all quantities are real (this is implied by the inequality in the original question), and the base of the exponent is positive. Hence either the exponent is 1, or the base $(x + 2)$ is 1. Therefore, the only solutions are 1 and -1.

Solutions—Spring 1978

1. There are $\binom{10}{3} = 120$ ways of choosing the three numbers. There are ten choices such that the sum is 15:

10,4,1	9,4,2	8,4,3
10,3,2	8,6,1	7,6,2
9,5,1	8,5,2	7,5,3
		6,5,4

Hence the probability is 10/120 = 1/12.

2. Since the area of the triangle is 6 and its perimeter is 12, we find that the radius of the larger circle is 1 (see 3.62).

Then $PD = PE = 1$ (see diagram) so $AD = AF = 2$. Solving right triangle APD shows that $AP = \sqrt{5}$. Now we use the usual trick for common tangents to two circles: we draw rectangle $QSFH$. Then $QP = QX + XP = 1 + r$ and $PS = PF - SF = PF - QH = 1 - r$. From similar triangles PQS and PAF, we have

$$\frac{QP}{PS} = \frac{AP}{PF} \quad \text{or} \quad \frac{1 + r}{1 - r} = \sqrt{5} .$$

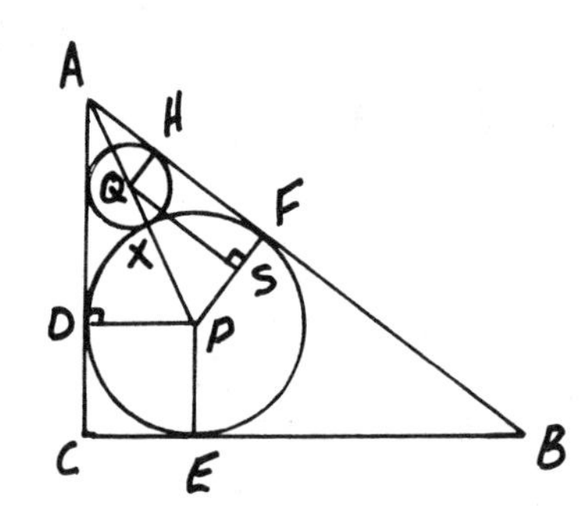

3. Drawing the altitude to the shortest side, we find a 5-12-13 right triangle and a 9-12-15 right triangle, so $\sin \theta = 12/15 = 4/5$. Another method involves calculating the area of the triangle using Hero's formula (see formula 3.61) and equating it with the expression $(\frac{1}{2})(4)(15)(\sin \theta)$(formula 3.64).

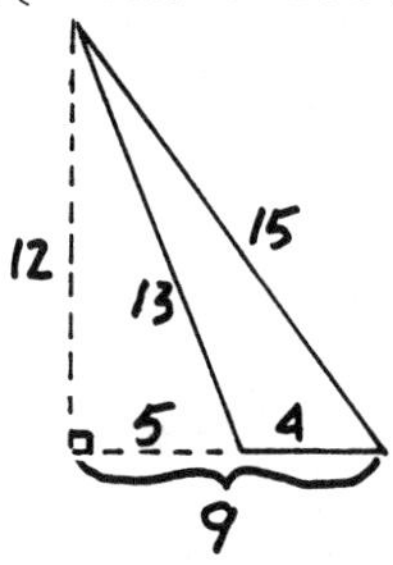

4. We have $(8A + A)^2 = 8^3B + 8^2B + 8C + C$ or $A^2(8 + 1)^2 = 8^2B(8 + 1) + C(8 + 1)$. Dividing through by $(8 + 1)$ gives $9A^2 = 64B + C$, so $A^2 = \dfrac{64B + C}{9} = 7B + \dfrac{B + C}{9}$. But B and C are each less than 8, so if $B + C$ is a multiple of 9 we must have $B + C = 9$. Hence $A^2 = 7B + 1$, or $B = \frac{1}{7}(A + 1)(A - 1)$. For B to be an integer, one of these last factors must be a multiple of 7, and since $A < 8$, we must have $A + 1 = 7$, $A = 6$, $B = 5$, and $C = 4$.

5. The angle subtended at the origin by the line segment is 60° (see diagram). Using the law of cosines, $x^2 = 225 + 49 - 14 \cdot 15 \cos 60°$ $= 169$, so $x = 13$.

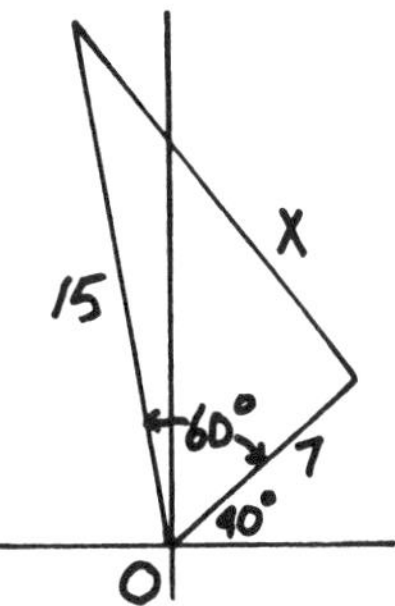

6. We know that the sum of the first n natural numbers is $\frac{n(n+1)}{2}$ (see 1.53). Let the first odd integer be x. Then $x + (x + 2) + (x + 4) + (x + 6) + \ldots + (x + 198) = 100^{100}$, or $100x + 2 + 4 + 6 + \ldots + 198 = 100x + 2(1 + 2 + 3 + \ldots + 99)$

$$= 100x + 99 \cdot 100$$

$$= 100^{100}$$

Then $x + 99 = 100^{99}$, and $x = 100^{99} - 99$.

7. Using the "chain rule" for logarithms (5.11):

$\log_3 x + \log_9 3 \cdot \log_3 x + \log_{81} 3 \cdot \log_3 x = 7$

$\log_3 x + \frac{1}{2}\log_3 x + \frac{1}{4}\log_3 x = 7$

$(\log_3 x)(1 + \frac{1}{2} + \frac{1}{4}) = 7$

$7 \log_3 x = 28$, and $x = 3^4 = 81$.

8. There are n pairs whose sum is $n + 1$, so we need the largest integer n such that $1 + 2 + 3 + \ldots + n \leq 1978$, or $n(n + 1) \leq 2 \cdot 1978$ (see 1.53). But $n^2 < n(n + 1) \leq 2 \cdot 1978 = 3956$, and $62^2 = 3844 < 3956 < 3969 = 63^2$, so $n = 62$. Now $1 + 2 + \ldots + 62 = \frac{62 \cdot 63}{2} = 1953$, so the 1954th pair is (1,63), and this is the first whose sum is 64. We need 24 more pairs to reach the 1978th pair, which is (25,39). This is the arrangement used by Cantor in his proof of the denumerability of the rational numbers. See, for example, Kasner and Newman (pp. 27-64).

9. First we note that if $a^3 + b^3 = c^3$ has a solution in rational numbers, then clearing fractions from the equation will produce a solution in integers. Thus the condition $abc = 0$ must hold also if the equation has rational solutions (see 2.2).
We have $6x(2x + 1) + 1 = 12x^2 + 6x + 1$. Using the hint entails expressing this as the sum of the two cubes. One way to do this is as $8x^3 + 12x^2 + 6x + 1 - 8x^3 = (2x + 1)^3 + (-2x)^3$.

9. (Continued)

Hence we must have $2x + 1 = 0$, $-2x = 0$, or $y = 0$. Only the first two possibilities lead to rational solutions; these are $(-\frac{1}{2},1)$ and $(0,1)$. The theorem given in the hint is one case of the classic "Last Theorem" of Fermat. For a more complete listing of such results, see Edwards.

10. The sides of a "primitive" right triangle--in which the lengths of the sides are relatively prime integers--are given by $a = u^2 - v^2$, $b = 2uv$, $c = u^2 + v^2$, where u,v are relatively prime and of opposite parity (see 2.1). In the present case, $a + c = 2u^2 = x^3$. Now if a prime number p divides a cube, then p^3 divides that cube, so u^2 must be a multiple of 4 and u itself must be even. Let $u = 2r$. Then $x^3 = 2u^2 = 8r^2$, and r^2 must be a cube. Hence r must also be a cube. Letting $r = 1$ yields the the solution given in the problem. Letting $r = 8$ gives $u = 16$. For the least value of c, we can choose $v = 1$ and get $c = 256 + 1 = 257$.

11. Let the length of the diameter be d. Since $DP \cdot PG = MP \cdot PN$, $12 \cdot 12 = 4(d - 4)$, $d = 40$, and the area of the circle is 400π.

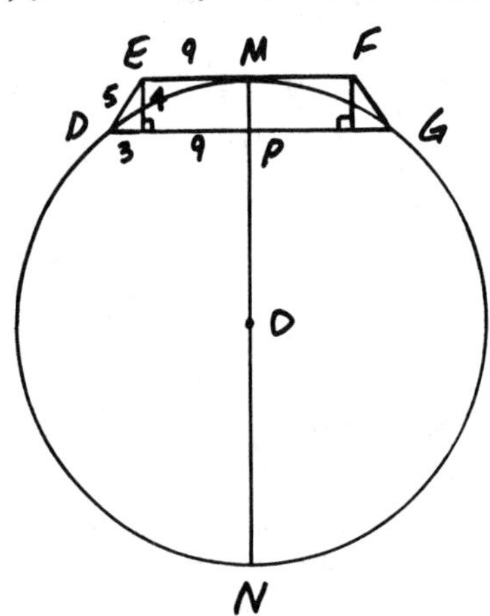

12. We have $\tan(20° + 40°) = \tan 60° = \sqrt{3} = \dfrac{\tan 20° + \tan 40°}{1 - \tan 20° \tan 40°}$.

(See 4.341.) Clearing the fraction, we find that the value sought is $\sqrt{3}$.

13. Since $1 = (\sin^2 x + \cos^2 x)^2 = \sin^4 x + \cos^4 x + 2 \sin^2 x \cos^2 x$, we have $\sin^4 x + \cos^4 x = 1 - 2 \sin^2 x \cos^2 x = 1 - \frac{1}{2}(4 \sin^2 x \cos^2 x)$

$$= 1 - \frac{1}{2}\left(\sin 2x\right)^2$$

$$= 1 - \left(\frac{1}{2}\right)\left(\frac{24}{25}\right)^2 = \frac{337}{625}.$$ (See 4.21.)

14. The wording of the question assumes that $P(x)$ is unique. Since the fraction $\frac{P(x + 2) - I(x)}{(x + 2) - x} = \frac{6}{2} = 3$ is constant, we can take P to be linear with slope 3. Hence $P(x) = 3x + b$. Since $P(0) = 2$, $P(x) = 3x + 2$. That P must be linear is guaranteed by a simple application of the method of finite differences (see problem 24, Spring 1976). Alternatively, if $P(x)$ has degree n, we can write $P(x)$ as $\sum_{i=0}^{n} a_i x^i$. Then $P(x + 2) - P(x)$ $= \sum_{i=0}^{n} a_i[(x + 2)^i - x^i]$. The righthand side of this equation is a polynomial of degree $n - 1$. But the difference on the lefthand side is the constant 3. Hence $n - 1 = 0$, and $n = 1$.

15. We have angle $OVE = 30°$, $VO = 2(OE) = 210$, and $VP = 210 + 105$ $= 315$. Since this is the sum of the lengths of the diameters, the sum of the lengths of the radii is $157\frac{1}{2}$. A more general method, involving summing the geometric progression whose terms are the successive radii, is more cumbersome in this situation.

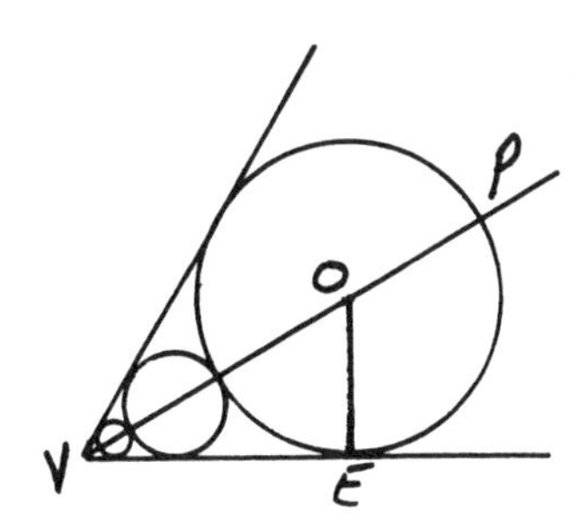

16. A "symmetric" representation of the four terms greatly facilitates the algebraic manipulation. Let the terms be $a - 3d$, $a - d$, $a + d$, and $a + 3d$, with common difference $2d$ and product $(a^2 - d^2)(a^2 - 9d^2)$. Since $2a^2 + 18d^2 = 200$ and $2a^2 + 2d^2 = 136$, $d^2 = 4$ and $a^2 = 64$. Therefore, $(a^2 - d^2)(a^2 - 9d^2) = 60 \cdot 28$ $= 1680$.

17. By the "chain rule" for logarithms (5.11), the given expression is equal to $2^3(\log_a b)(\log_b c)(\log_c a) = 2^3\log_a a = 8$. See why we call it the "chain rule"?

18. We reflect point A in the x-axis to get A' $(0,-2)$, and point D in the line $y = 10$ to get D' $(5,11)$. Now any path from A to B to C to D will be equal in length to the corresponding path from A' to B to C to D', obtained by reflecting $\overline{AB}$ in the x-axis and $\overline{CD}$ in the line $y = 10$. The shortest reflected path clearly lies along line $\overleftrightarrow{A'D'}$, which has slope $\frac{13}{5}$. This line crosses the

18. (Continued)

x-axis at $B\left(\frac{10}{13}, 0\right)$ and the line $y = 10$ at $C\left(\frac{60}{13}, 10\right)$. Note that this problem calls for finding a minimal value, yet is easily solved without using calculus. For more problems like this, see Yaglom, Geometric Transformations I, II, and III.

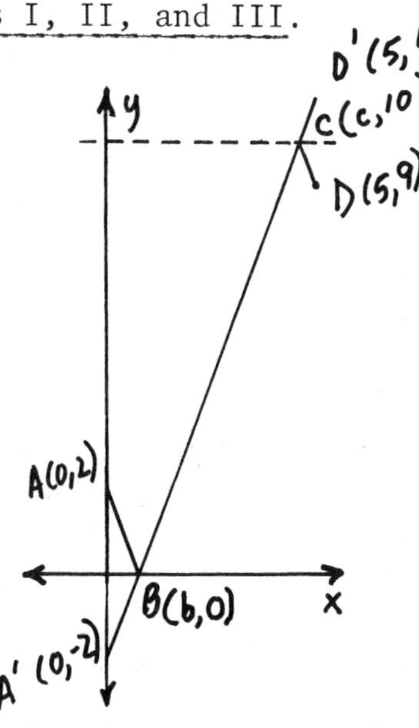

19. If $PX = a$ (see diagram), then $PC = 2a$, $XC = BX = a\sqrt{3}$, and $BC = AC = 2a\sqrt{3}$. Hence $\frac{PC}{PA} = \frac{2a}{2a\sqrt{3} - 2a} = \frac{1}{\sqrt{3} - 1} = \frac{\sqrt{3} + 1}{2}$,

and $k = \sqrt{3} + 1$.

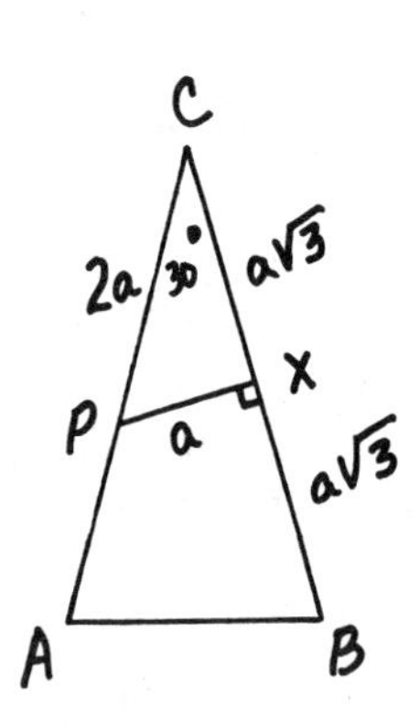

20. We can write:

$$f(2) = f(1) + 2 \cdot 2$$
$$f(3) = f(2) + 2 \cdot 3$$
$$f(4) = f(3) + 2 \cdot 4$$
$$\vdots$$
$$f(n) = f(n - 1) + 2n$$

Adding, we have
$$f(n) = f(1) + 2(2 + 3 + 4 + \ldots + n)$$
$$= 1 + 2(1 + 2 + 3 + \ldots + n) - 2$$
$$= 1 + 2\left(\frac{n(n + 1)}{2}\right) - 2 \quad \text{(see 1.53)},$$

20. (Continued)

and $f(n) = n^2 + n - 1$.
Alternatively, a simple application of the method of finite differences (see problem 24, Spring 1976) shows that $f(n)$ is a polynomial of degree 2. We can then write $f(n) = an^2 + bn + c$, substitute $n = 1, 2, 3$, and solve the resulting system of equations for a, b, c.

21. Since W is odd and $T = 4$, we have $W = 1$.
Since $4(100N + 10E + 1) = 1000D + 100E + 10N + 4$, we have

$$D = \frac{39N - 6E}{100} = \frac{3(13N - 2E)}{100}.$$ Hence 100 divides $13N - 2E$.

But $13N - 2E \leq 13 \cdot 9 - 2 \cdot 1 < 200$, so $13N - 2E = 100$ and $D = 3$.
Finally, we need $13N - 2E = 100$. Clearly, N cannot be too small: in fact, if $N \leq 7$, then $13N - 2E < 91 - 2$. The case $N = 9$ does not lead to an integral value for E. Hence $N = 8$, $E = 2$, and $N + E + W = 8 + 2 + 1 = 11$.

22. We have $\sin^4 x + \cos^4 x = (\sin^2 x + \cos^2 x)^2 - 2\sin^2 x \cos^2 x$
$= 1 - \frac{1}{2}\sin^2 2x$. (See 4.21.)

Now $\sin 2x$ has a basic period of π, so $\sin^2 2x$ has a basic period of $\frac{\pi}{2}$, as does the function under consideration.

23. Since $2t + 5f = 87$, f must be odd, and $1 \leq f \leq 17$. The nine possible solutions for (t,f) are (1,17), (6,15), (11,13), (16,11), (21,9), (26,7), (31,5), (36,3), and (41,1). Of these, only two ordered pairs give prime values for both t and f, so the required probability is 2/9.

24. Summing the infinite geometric progression gives $\dfrac{\frac{1}{50}}{1 - \frac{1}{50}} = \dfrac{1}{49}$ (see 1.523), which equals .0204081632653..., and the required product is 180.

25. Let $BM = MC = m$. Then $BH = AH = m$ and $HC = m\sqrt{3}$. Using mass points, we assign weight $\sqrt{3}$ to A and weight 1 to B and C. The weight at M is then 2. Since $AP:PM = 2:\sqrt{3}$, $k = \sqrt{3}/2$. For a more traditional approach, we can draw $MY \perp BH$. Then angle $BMY = 30°$, so $MY = m\sqrt{3}/2$. From similar triangles APH, MPY, we have $AP:PM = AH:YM = m:m\sqrt{3}/2$, and $k = \sqrt{3}/2$.

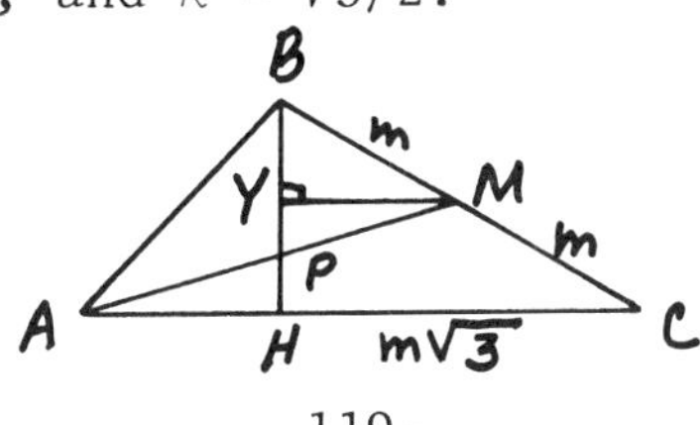

26. We can solve the given equation for x in terms of y:

$$x = \frac{y \pm \sqrt{y^2 - 4(y^2 - 49)}}{2}.$$ Since x is a positive integer,

$196 - 3y^2$ must be a perfect square. This occurs only when $y = 3, 5, 7,$ or 8 and each of these values leads to a corresponding value for x. The solutions are (8,3), (8,5), (7,7), (5,8), and (3,8). Another approach is to write the given condition as $2x^2 + 2y^2 - 2xy = (x - y)^2 + x^2 + y^2 = 98$. Now the problem is to represent 98 as the sum of three squares, and trial and error leads to the solutions quickly.

27. If the number of heads is h and the number of tails is $5 - h$, we must have $h - (5 - h) = 1$, so $h = 3$. There are $\binom{5}{2} = 10$ ways in which the three heads and two tails can be distributed and $2^5 = 32$ possible outcomes altogether, making the required probability $10/32 = 5/16$.

28. We have $\sum_{i=1}^{n} (2i - 1)^2 = \sum_{i=1}^{n} (4i^2 - 4i + 1) = 4\sum_{i=1}^{n} i^2 - 4\sum_{i=1}^{n} i + \sum_{i=1}^{n} 1$

$$= \frac{4n^3}{3} + 2n^2 + \frac{2n}{3} - 2n^2 - 2n + n$$

$$= \frac{4n^3}{3} - \frac{n}{3} \quad \text{(see 1.53)}$$

and $(a,b,c,d) = \left(\frac{4}{3}, 0, -\frac{1}{3}, 0\right)$.

We can also express the required sum as $\sum_{i=1}^{2n} i^2 - \sum_{i=1}^{n} (2i)^2$

$$= \left[\frac{8n^3}{3} + \frac{4n^2}{2} + \frac{2n}{6}\right] - 4\left[\frac{n^3}{3} + \frac{n^2}{2} + \frac{n}{6}\right] = \frac{4n^3}{3} - \frac{n}{3}$$ as before.

29. Let $A = 64 - x^2$ and $B = x^2 - 36$. Then $A + B = 28$ and we seek the minimum value of $-AB$, or equivalently, the maximum value of AB. In view of formula 6.3, we take $A = B = 14$, and the minimum for $-AB$ is -196.
Alternatively, we can write $(x^2 - 64)(x^2 - 36) = x^4 - 100x^2 + 2304$ $= (x^2 - 50)^2 - 196$. Again, the minimum occurs when $x^2 = 50$.

30. Let $CX = CY = r$ (see diagram). Then $(30 - r) + (40 - r) = 50$, so $r = 10$ and $BD = 30$. Now $\cos B = 4/5$, and in triangle CBD we have $CD^2 = 40^2 + 30^2 - 2 \cdot 30 \cdot 40 \cdot (4/5) = 580$, and $CD = \sqrt{580} = 2\sqrt{145}$. See also Appendix B, (3.5).

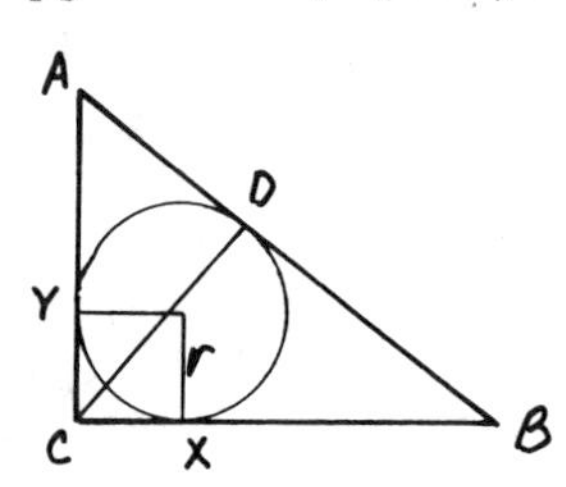

Solutions—Fall 1978

1. Let $BC = AB = x$ (see diagram). Using the Pythagorean theorem in triangle DAC shows that $DC = 18$. Then from similar triangles DAC, EBC, we have $DC:AC = EC:BC$, or $18/30 = 15/x$, so $x = 25$.

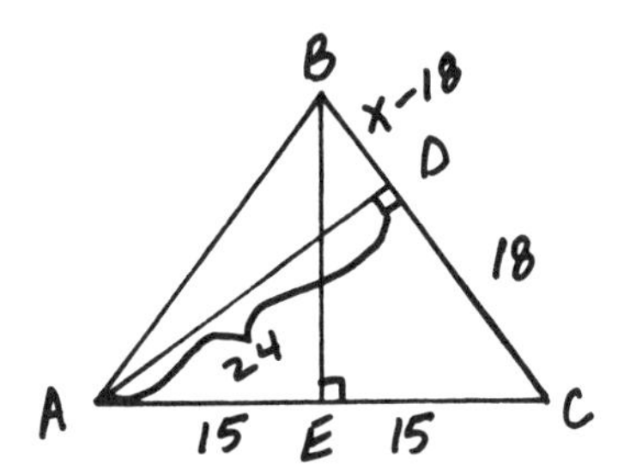

2. Since the denominators are conjugates, neither their sum nor their product contains a radical. Hence, clearing fractions gives $4x = 2x(x^2 - 1)$, which leads to $x = 0$ or $x^2 = 3$. The only integral value is 0.

3. If the number is $100h + 10t + u$, we have $h + t + u = 20$, or $\frac{1}{2}(100h + 10t + u - 16) = 100u + 10t + h$. Eliminating t, we find that $h = 2 + u + 3u/4$. Thus u must be a multiple of 4. If $u = 0$, the reversed number is not greater than 100. If $u = 8$, h is greater than 9. Hence $u = 4$, $h = 9$, and $t = 7$.

4. Let $BB_k = x$. Then in triangle ABB_k, $k = 1^2 + x^2 - 2 \cdot 1 \cdot x \cdot \cos 120°$, or $k = 1 + x + x^2$, and in triangle ABB_{3k}, we have: $3k = 1 + (x + 2)^2 - 2 \cdot 1 \cdot (x + 2) \cdot \cos 120°$ or $3k = 7 + 5x + x^2$. Eliminating k produces: $3(1 + x + x^2) = 7 + 5x + x^2$, $2x^2 - 2x - 4 = 0$, $2(x - 2)(x + 1) = 0$, $x = 2$ and $k = 1 + 2 + 4 = 7$.

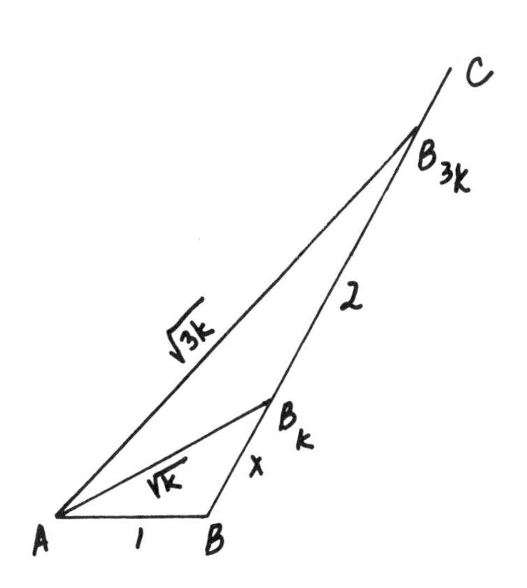

5. We have: $2^{3^5} = 2^{243}$; $3^{5^2} = 3^{25}$; $5^{3^2} = 5^9$; $2^{243} = (2^4)^{60.75}$; $3^{25} = (3^2)^{12.5}$. Clearly, $16^{60.75} > 9^{12.5} > 5^9$, as both the exponents and the bases are in decreasing order.

6. The event for which we want the probability can be represented as the union of the following disjoint events:

   P_1: $\underbrace{\text{HH...H}}_{11 \text{ times}}\text{xxxxxx}$ (the x's denote any outcome at all)

   P_2: $\text{T}\underbrace{\text{HH...H}}_{11 \text{ times}}\text{xxxxx}$

   P_3: $\text{xT}\underbrace{\text{HH...H}}_{11 \text{ times}}\text{xxxx}$

   ⋮

   The probability of the first happening is 2^{-11}, and that of each of the others is 2^{-12}. The sum of these is the desired probability, so $P = 2^{-11} + 6 \cdot 2^{-12} = 2^{-11} + 3 \cdot 2^{-11} = 2^{-11} \cdot 4 = 2^{-9} = \frac{1}{512}$.

7. The best way to attack this problem is to write out a list of the prime numbers under 100, noting the differences between successive primes on the line below:

   2 3 5 7 11 13 17 19 23 29 31 37 41 43 47 53 59 61 67 71 73 79 83 89 97
   1 2 2 4 2 4 2 4 6 2 6 4 2 4 6 6 2 6 4 2 6 4 6 8

   The situation in the problem corresponds to cases where consecutive differences are equal. This occurs only at 5 and 53. The solution in this biological context is probably unique, as the next such occurrence is at 157.

8. Calculating a few terms, it becomes clear that a_n increases exponentially with n. A good guess and (1.33) then leads to the the answer. More formally, we have $\frac{a_n}{3^n} - \frac{a_{n-1}}{3^{n-1}} = \frac{1}{3^n}$,

   so $\sum_2^k \left(\frac{a_n}{3^n} - \frac{a_{n-1}}{3^{n-1}}\right) = \sum_2^k \frac{1}{3^n}$. The sum on the left "telescopes"--consecutive terms cancel each other out--so we have:

8. (Continued)

$$\frac{a_k}{3^k} - \frac{a_1}{3^1} = \frac{\frac{1}{9}\left(1 - \frac{1}{3^{k-1}}\right)}{\frac{2}{3}} = \frac{1}{6}\left(1 - \frac{1}{3^{k-1}}\right)$$

or $\frac{a_k}{3^k} = \frac{1}{3} + \frac{1}{6}\left(1 - \frac{1}{3^{k-1}}\right)$, and $a_k = 3^{k-1} + \frac{1}{6}(3^k - 3) = \frac{3^k - 1}{2}$.

Hence $a_{1978} = \frac{1}{2}(3^{1978} - 1)$. For many more such techniques, see Spiegel (Chapter 5) or Markushevich.

9. Let $y = x + \frac{1}{x}$. Then $6y^2 - 35y + 50 = 0$, and $y = \frac{10}{3}$ or $\frac{5}{2}$.

Thus $x + \frac{1}{x} = \frac{10}{3} = 3 + \frac{1}{3}$ or $x + \frac{1}{x} = \frac{5}{2} = 2 + \frac{1}{2}$, and $x = 3, \frac{1}{3}, 2, \frac{1}{2}$.

10. We can express each product as a sum: (4.331)

$\sin 40° \sin 80° = \frac{1}{2}(\cos 40° - \cos 120°) = \frac{1}{2}(\cos 40° + \cos 60°)$

$\sin 80° \sin 160° = \frac{1}{2}(\cos 80° - \cos 240°) = \frac{1}{2}(\cos 80° + \cos 60°)$

$\sin 160° \sin 320° = \frac{1}{2}(\cos 160° - \cos 480°) = \frac{1}{2}(-\cos 20° + \cos 60°)$

Hence the required sum can be expressed as

$\frac{1}{2}(3 \cos 60° + \cos 40° + \cos 80° - \cos 20°)$; and since

$\cos 40° + \cos 80° = 2 \cos 20° \cos 60° = \cos 20°$ (see 4.323),

the required sum reduces to $\frac{1}{2}(3 \cos 60°) = \frac{3}{2} \cdot \frac{1}{2} = \frac{3}{4}$.

11. By the "chain rule" for logarithms (5.11), $(\log_7 10)(\log_{10} 3)$ $= \log_7 3 = \frac{1}{2}\log_7 9$. And since $\log_x y = \frac{1}{\log_y x}$, we can write $(1/b)(a) = \frac{1}{2}\log_7 9$ and $\log_7 9 = 2a/b$.

12. Since angle C = angle E = 90 and angle CAE = angle EAB, triangle $ACD \sim$ triangle AEB. We will use this fact to solve for BE. By the angle bisector theorem (3.31), $7:25 = CD:DB = AC:AB$. Thus we can let $AC = 7k$, $AB = 25k$, and by the Pythagorean theorem, $(7k)^2 + 32^2 = (25k)^2$. Solving shows that $k = \frac{4}{3}$ and $AB = \frac{100}{3}$, $AC = \frac{28}{3}$. Now in triangle ACD, the ratio of the legs is $AC:CD$ $= 28:21 = 4:3$, so triangle ACD has sides in the ratio 3:4:5, and triangle AEB has sides of the same ratio. Hence $BE:AB = 3:5$.

If $BE = x$, we have $\frac{x}{\frac{100}{3}} = \frac{3}{5}$, or $x = 20$.

12. (Continued)

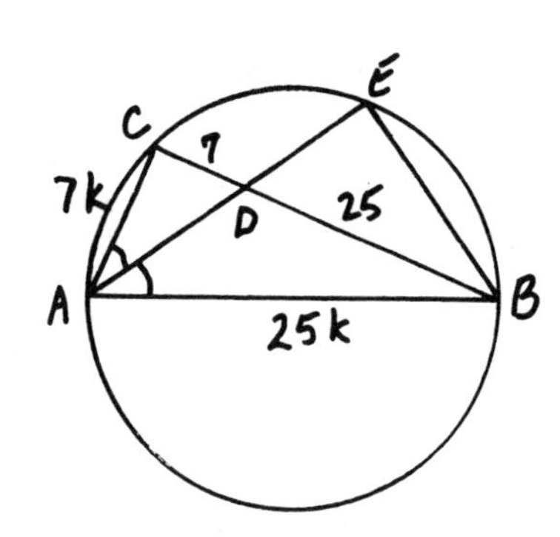

13. We need the smallest positive cube that is the sum of two squares. Listing the first few cubes, we quickly see that 125 = 100 + 25 works, and a little more experimentation shows that no smaller cube works. Thus $z^3 = 125$. For a full discussion of this Diophantine equation, see Liff.

14. The calculations for this problem are relatively simple if we use the definition of $\log_3 x$ directly:

$$x = 3^{[-2 + \log_2 100][\log_3 \sqrt{2}]} = \left[3^{\log_3 \sqrt{2}}\right]^{[2 + \log_2 100]}$$

$$= [\sqrt{2}]^{[-2 + \log_2 100]} = [2^{\frac{1}{2}}]^{[-2 + \log_2 100]} = 2^{-1 + \frac{1}{2}\log_2 100}$$

$$= 2^{-1} \cdot 2^{\log_2 10} = \tfrac{1}{2} \cdot 10 = 5.$$

15. Letting $x = y = 1, 2, 3$, etc., we have

$f(2) = f(1 + 1) = f(1) + f(1) = 2f(1)$
$f(3) = f(2 + 1) = f(2) + f(1) = 3f(1)$
$\vdots$
$f(10) = f(9 + 1) = f(9) + f(1) = 10f(1) = 30.$

In fact, we can show that for all rational numbers q, $f(q) = 3q$:

1. For natural numbers n, we can use an induction based on the above.
2. $f(0) = f(0 + 0) = f(0) + f(0) = 2f(0)$, so $f(0) = 0$.
3. $f(0) = f(1) + f(-1) = 3 + f(-1)$, so $f(-1) = -3$. Another induction gives the result for all negative integers.
4. For natural numbers n, $f\left(\frac{n}{n}\right) = 3 = f\left(\frac{1}{n} + \frac{1}{n} + \ldots + \frac{1}{n}\right) = nf\left(\frac{1}{n}\right)$, so $f\left(\frac{1}{n}\right) = 3 \cdot \frac{1}{n}$. A similar argument holds for negative integers.
5. Finally, any rational number q can be expressed as $\frac{a}{b}$, with $a \geq 0$ and $b \neq 0$. Then $f\left(\frac{a}{b}\right) = f\Big(\underbrace{\frac{1}{b} + \frac{1}{b} + \ldots + \frac{1}{b}}_{a \text{ times}}\Big) = a \cdot f\left(\frac{1}{b}\right) = 3 \cdot \frac{a}{b}$.

15. (Continued)

If we assume in addition that $f(x)$ is continuous, the result holds also for all real x.

16. The larger side lies opposite the larger angle. By the law of sines (see 3.4),

$$\frac{2}{1} = \frac{\sin(\theta + 60°)}{\sin \theta} = \frac{\sin \theta \cos 60° + \cos \theta \sin 60°}{\sin \theta} = \frac{1}{2} + \frac{\sqrt{3}}{2} \cot \theta.$$

Hence $\cot \theta = \sqrt{3}$ and $\theta = 30°$. The largest angle of the triangle is thus 90°. (In competition, the answer can easily be guessed.)

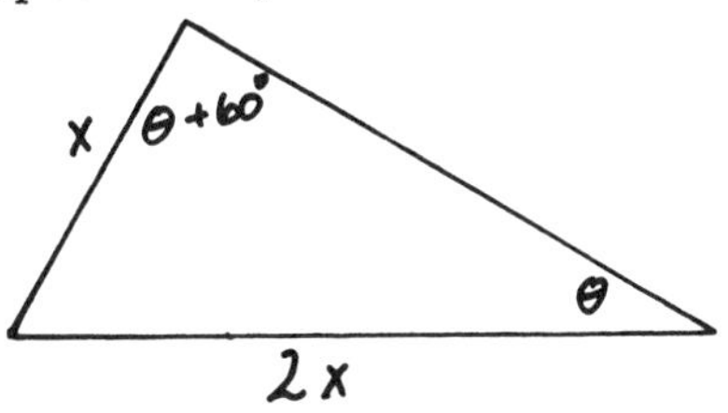

17. There can be at most one unpaired sock of each color, so if Mr. Smith packs 13 socks, at most 6 will be unpaired. Also, since the total number of socks is odd, the number of unpaired socks must also be odd and hence there cannot be more than 5 unpaired socks. Twelve socks may not be enough, since if he takes one of each color plus any 3 matching pairs he will have 12 socks, but only 3 matching pairs. The answer is 13.

18. Method 1: Labelling the points as in the diagram, we see that $AB \cdot AF = AD \cdot AH$ = area (rectangle $ABEF$) and $AC \cdot AG$ = area (square $ACEG$). Also, $AE = 2$. On the other hand, area (rectangle $ABEF$) = 4 · area(triangle ABO) = $4 \cdot \frac{1}{2} \cdot OA \cdot OB \cdot \sin$ angle BOA $= 4 \cdot 1 \cdot 1 \cdot \sin 45° = \sqrt{2}$, and area (square $ACEG$) = $\frac{1}{2} \cdot AE \cdot CG$ = 2. Hence the required product is $(\sqrt{2})^2 \cdot 2 \cdot 2 = 8$.

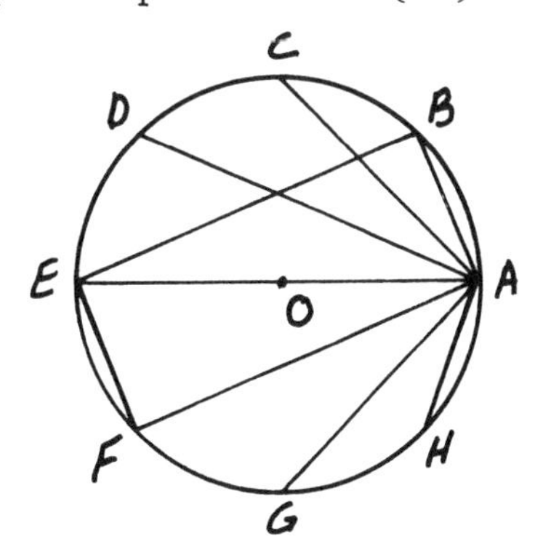

Method 2: We can solve this problem more generally using complex numbers. If n points are spaced equally about the unit circle with center at 0, and one of the points is $x_0 = 1 + 0i$, the other points are at $x_k = \text{cis}\,\frac{2k\pi}{n}$ for $k = 1$ to $n - 1$. By DeMoivre's theorem (1.6), $x^k = 1$ for each k, and all the points aside from

18. (Continued)

x_0 satisfy $\frac{x^n - 1}{x - 1} = 0$. By the factor theorem (1.311), the left side must factor into $(x - x_1)(x - x_2) \ldots (x - x_{n-1})$. On the other hand, long division shows this same polynomial to be $x^{n-1} + x^{n-2} + \ldots + x + 1$. Taking absolute values, we have (for any x): $|x - x_1||x - x_2| \ldots |x - x_{n-1}| = |x^{n-1} + x^{n-2} + \ldots + x + 1|$. Setting $x = 1$ shows $|1 - x_1||1 - x_2||1 - x_3|\ldots|1 - x_{n-1}| = |1 + 1 + 1 + \ldots + 1| = n$. The expression on the left is the required product. See also The Hungarian Problem Book I (1899 Competition), where the present situation is connected with the topic of Chebyshev Polynomials.

19. Four points form a parallelogram if and only if the sums of corresponding coordinates of opposite vertices are equal. If $A = (1,1)$, $B = (3,5)$, and $C = (-1,4)$, there are three ways we can choose a pair of vertices to be opposite:

Choosing A and B opposite puts the fourth vertex at (5,2).

Choosing A and C opposite puts the fourth vertex at (-3,0).

Choosing B and C opposite puts the fourth vertex at (1,8).

20. If x is the man's federal tax (in dollars) and y is his state tax (also in dollars), then:

$$x = .2(9800 - y)$$
$$y = .1(9800 - x).$$

These lead to the solution $(x,y) = (1800,800)$.

21. This problem can most easily be solved by trial and error. Two ways of achieving the result x^{31} in seven multiplications are given below:

$$x,\ x^2,\ x^3,\ x^5,\ x^{10},\ x^{20},\ x^{30},\ x^{31}$$

$$x,\ x^2,\ x^4,\ x^8,\ x^9,\ x^{18},\ x^{27},\ x^{31}$$

A "tree" diagram, tracing all possible paths, can be constructed which will show that seven multiplications is indeed the smallest number. For more on this rather complicated but interesting problem, see Knuth, Vol. II, p. 421.

22. Cubing both sides gives

$$(x - a) + (x - b) + 3\sqrt[3]{(x - a)(x - b)}\left[\sqrt[3]{x - a} + \sqrt[3]{x - b}\right] = x - c.$$

But the original equation says $\sqrt[3]{x - a} + \sqrt[3]{x - b} = \sqrt[3]{x - c}$, so we have have $(x - a) + (x - b) + 3\sqrt[3]{(x - a)(x - b)(x - c)} = x - c$ or $3\sqrt[3]{(x - a)(x - b)(x - c)} = -x - c + a + b$. At this point, it is not hard to see that cubing both sides will give a cubic equation in x (but not an identity!) which can have no more than 3 roots, real or complex.

To see that 3 roots are possible, take for example $a = -1$, $b = 0$, $c = 1$. We get: $3\sqrt[3]{x(x^2 - 1)} = -x$, $27x(x^2 - 1) = -x^3$, $27x^3 - 27x = -x^3$, $28x^3 - 27x = 0$ and the 3 real roots are $x = 0$, $\pm\sqrt{\frac{27}{28}}$.

23. We can generalize the problem just as easily as solving the special case given.

Let a_k be the number of lattice points on the graph of $|x| + |y| = k$. Let $x, y > 0$. For each integral value of $x < k$, $(x, k - x)$ gives a lattice point solution to $|x| + |y| = k$. There are $k - 1$ such solutions. Reflecting in the axes, and counting also the vertices of the figure, shows that $a_k = 4(k - 1) + 4 = 4k$ and $a_0 = 1$.

$$\text{Hence} \quad \sum_0^{n-1} a_k = 1 + \sum_1^{n-1} 4k = 1 + 2n(n - 1) \text{ (see 1.53)}$$
$$= 2n^2 - 2n + 1.$$

This sum gives the number of lattice points such that $|x| + |y| < n$. For $n = 5$ there are 41 solutions.

24. If we assume that x, y, and z are real, we can guess that adding the squares of the numbers has the same effect as adding the numbers themselves. This certainly works if x, y, and z satisfy the equation $t = t^2 = t^3 = t^4$. This equation is satisfied by $t = 0$ or 1, which gives 0, 1, 2, 3 as possible values of a.

More rigorously, we can take advantage of the algebraic symmetry of the situation and let x, y, and z be the roots of $t^3 - pt^2 + qt - r = 0$. Then $p = x + y + z$, $q = xy + yz + zx$, and $r = xyz$ (see 1.4). Since a is at once the sum of the roots, as well as the sum of the squares, cubes, and fourth powers of the roots of this equation, it will help to express the coefficients p, q, and r in terms of these sums. We have $p = x + y + z = a$, and $a^2 = (x + y + z)^2 = x^2 + y^2 + z^2 + 2(xy + yz + zx) = a - 2q$, so that $q = \frac{1}{2}a(1 - a)$. We next use a common trick in the theory of equations. We write:

24. (Continued)

$x^3 - px^2 + qx - r = 0$

$y^3 - py^2 + qy - r = 0$ (since x, y, z are roots of the equation)

$z^3 - pz^2 + qz - r = 0$

and add: $(x^3 + y^3 + z^3) - p(x^2 + y^2 + z^2) + q(x + y + z) - 3r = 0$,

or $a - a^2 + \frac{1}{2}a(1 - a) - 3r = 0$, which leads to $r = \frac{1}{6}(a - 1)(a - 2)$.

Last, we write $x^4 - px^3 + qx^2 - rx = 0$

$y^4 - py^3 + qy^2 - ry = 0$

$z^4 - pz^3 + qz^2 - rz = 0$

and add again. We find that $a - a^2 + \frac{1}{2}a(a^2 - a) - \frac{1}{2}a^2(a - 1)(a - 2)$ $= \frac{1}{6}a(a - 1)(a - 2)(a - 3) = 0$ so $a = 0$, 1, 2, or 3. See also problem 30, Spring 1975.

25. We have $[x^2 - x^2 + 2x - 1] = [2x - 1] = 10$, so $10 \leq 2x - 1 < 11$ and $\frac{11}{2} \leq x < 6$. Answer: $\left(\frac{11}{2}, 6\right)$ or equivalent.

26. Let the slopes of the two tangents be m and n. By symmetry, the acute angles made by either tangent and the x-axis are equal. Hence either $m = n$ or $m = -n$ and equality clearly does not hold. Since the tangents are perpendicular, we have $mn = -1$. Hence $m^2 = n^2 = 1$, and the two slopes are 1 and -1.

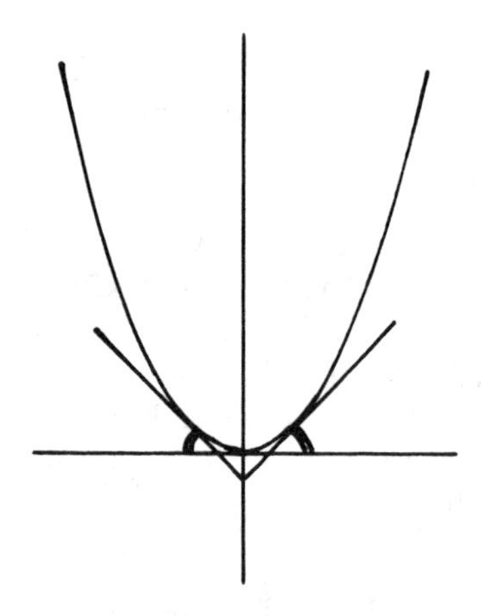

Suppose the equation of one tangent is $y = x + b$. For a line to be tangent to a parabola, the simultaneous solution of their equations must be unique. Hence we have $y = x + b$ and $y = x^2$ simultaneously, or $x^2 - x - b = 0$. For unique solution, the discriminant must vanish, or $1 + 4b = 0$, so $b = -1/4$. The equation of one tangent is thus $y = x - 1/4$, and if $x = 0$, $y = 1/4$. See also Salmon, for a discussion of geometric properties of the parabola.

27. Let $a = 1 - \sqrt{x^2 - 1} - \sqrt{x^2 + 2x + 1}$. Then $a + \frac{1}{a} = 2$, and $a^2 - 2a + 1 = (a - 1)^2 = 0$, and $a = 1$. If $1 - \sqrt{x^2 - 1} - \sqrt{(x + 1)^2} = 1$, we have $\sqrt{x^2 - 1} + \sqrt{(x + 1)^2} = 0$. This happens only when $x^2 - 1 = 0$ and $(x + 1)^2 = 0$. The only simultaneous solution is $x = -1$.

28. We will use absolute value to denote area.

Let $|ADT| = x$, $|DCT| = y$, $|CTB| = z$, $|ATB| = w$ (see diagram).

Then from the conditions in the problem, $x + y = 1$, $y + z = 6$, and $w + x = 2$. Adding, $2x + 2y + w + z = 9$. Since $2x + 2y = 2$, we have $2 + w + z = 9$, and $w + z = 7$.

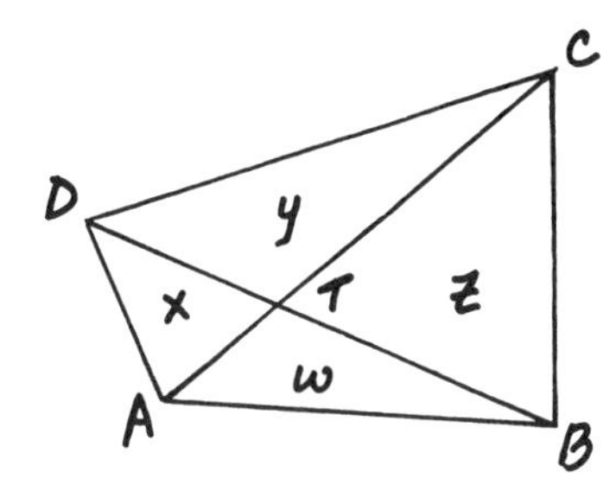

29. Let $a > b > c > 0$. Then $ab = a + b + c < 3a$, so $ab < 3a$ and $b < 3$. If $b = 1$, there is no possibility for c. Hence $b = 2$, $c = 1$, and $a = 3$.

30. We have, with a little "hindsight": $2 \cos^2 30° - 1 = \cos(2 \cdot 30°) = \cos 60°$ (see 4.21, 4.22), so $1 + \cos 60° = 2 \cos^2 30°$. Also, $\sin 60° = 2 \sin 30° \cos 30°$. Hence the given expression is equal to $2 \cos^2 30° + 2i \sin 30° \cos 30° = 2(\cos 30°)(\cos 30° + i \sin 30°) = \sqrt{3}\ \text{cis}\ 30°$. By DeMoivre's theorem (1.6), $(1 + \cos 60° + i \sin 60°)^{12} = 3^6 \text{cis} 12 \cdot 30° = 3^6 \text{cis}\ 360° = 3^6 + 0i$, and the imaginary part is 0.

A quicker solution to this problem is furnished by looking at it geometrically. On the complex plane, let A represent 1, B represent $\cos 60° + i \sin 60°$, and C represent $1 + \cos 60° + i \sin 60°$. Since complex numbers add as vectors, $\overrightarrow{OA} + \overrightarrow{OB} = \overrightarrow{OC}$, and $OBCA$ is a parallelogram. Since $|\overrightarrow{OA}| = |\overrightarrow{OB}| = 1$, $OBCA$ is a rhombus. Hence, angle $COA = 30°$, and upon raising the number C to the 12th power, the amplitude will be $12 \cdot 30° = 360°$ and the result will be real.

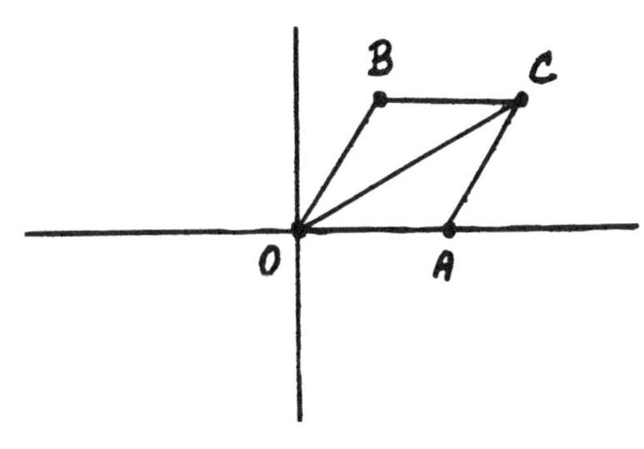

Solutions—Spring 1979

1. Let $x = \log_2 3$, $y = \log_9 4$. Then we have $2^x = 3$, or $2^2 = 3^{2/x}$. Also, $9^y = 4$, or $2^2 = 3^{2y}$. Hence $2/x = 2y$, or $xy = 1$.

2. Since $PA + PC$ is constant (see diagram 1), we need only maximize the sum $PB + PD$. A good guess in any minimum or maximum problem is an extreme point, which here would be one of the vertices of the trapezoid. It is clear that if choosing a vertex will give the maximal sum, then Y is the best possible vertex to choose. Let us place point P at Y, so that $PB = 0$. To compute PD, we extend $\overline{WZ}$ and $\overline{XY}$ to intersect at V. Then, as $WX{:}YZ = \frac{1}{2}$, we find that $VW = 2$, $VX = 3$, and we can use the law of cosines in triangle VWX to find angle WVX: $7 = 9 + 4 - 2 \cdot 2 \cdot 3 \cdot \cos\theta$, $-12\cos\theta = -6$, $\cos\theta = \frac{1}{2}$, and $\theta = 60°$. Hence $PD{:}VY = PD/6$ $= \sin 60° = \sqrt{3}/2$ and $PD = 3\sqrt{3}$. Also, if $\overline{VT}$ is the altitude to $\overline{ZY}$ in triangle ZVY, $VT = 2(PA + PC)$. Computing the area of triangle ZVY in two ways, we have (see 3.64):

$$\frac{1}{2} \cdot ZV \cdot VY \cdot \sin 60° = \frac{1}{2}\, ZY \cdot VT$$

$$\frac{1}{2} \cdot 4 \cdot 6 \cdot \frac{\sqrt{3}}{2} = \frac{1}{2} \cdot 2\sqrt{7} \cdot 2(PA + PC)$$

$$6\sqrt{3} = 2\sqrt{7}(PA + PC)$$

$$PA + PC = \frac{3\sqrt{3}}{\sqrt{7}} = \frac{3\sqrt{21}}{7}$$

and $PA + PB + PC + PD = 3\sqrt{3} + \dfrac{3\sqrt{21}}{7}$ (or equivalent).

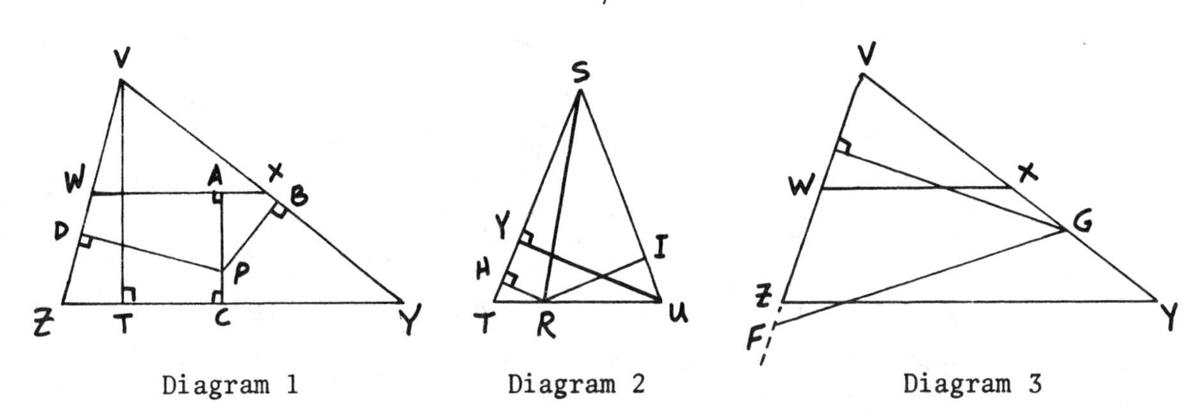

Diagram 1 Diagram 2 Diagram 3

Let us now show that placing P at point Y will indeed result in the maximal sum $PB + PD$. We first need to prove that if R is any point on the base of isosceles triangle STU (see diagram 2), then the sum of the perpendiculars from R to $\overline{ST}$ and $\overline{SU}$ is constant. This can be shown by computing the area of triangle STU in two ways. We use absolute value to denote area. On the one hand,

2. (Continued)

$|STU| = \frac{1}{2} \cdot UY \cdot ST.$

On the other hand, $|STU| = |STR| + |SRU|$

$$= \tfrac{1}{2} \cdot HR \cdot ST + \tfrac{1}{2} \cdot RI \cdot SU$$

$$= \tfrac{1}{2}(ST)(HR + RI)$$

Hence $HR + RI$ is constant and is equal to UY (the altitude to the leg of the isosceles triangle).
In the original problem, we can draw a variable line FG (see diagram 3) so that triangle VFG is isosceles. For any point P on FG, the sum of the perpendiculars $PB + PD$ is equal to the perpendicular from G to line VZ (the altitude to the leg of isosceles triangle VFG). We need to choose line FG so that this altitude is maximal. This will clearly occur when FG passes through point Y (so that P and Y must coincide).
Note that for any choice of P, we can move in a direction parallel to that of FG and keep the sum $PB + PD$ constant. The line FG is called a "level line" for the function $PB + PD$. Such level lines (or "level curves") are often useful in finding maxima or minima. In this problem, the argument given also shows that the minimum value of $PB + PD$ occurs when P is at vertex W.

3. The actual total cost of all the items was \$28.97 - \$.65 = \$28.32. The largest possible correct cost is (25)(\$1.10) = \$27.50. Hence the smallest possible error is the difference, or \$.82. Since the maximum amount of each error is .09, the number of errors is more than 82/9, or at least 10.

4. We have

$$\sin 20° + \sin 40° = 2 \sin 30° \cos 10° \text{ (see 4.321)}$$
$$= \cos 10° = \sin 80°.$$

5. $f(x,y,z) = yz(x - 3) + (2x - 5)$. Both $(x - 3)$ and $(2x - 5)$ attain maxima for $x = 2$. But then $x - 3 = -1 < 0$, so we want the factor yz to be minimized. This is achieved if $y = z = 1$, and $f(2,1,1) = 2 - 3 + 4 - 5 = -2$. That the maximum will be achieved at a corner of the cube can be predicted by noting that the expression for $f(x,y,z)$ is linear in each variable.

6. For every \$1 difference in price, Joe will go down 10¢, while Moe responds by going up 9¢. Hence the concessions are in the ratio 10:9, and the price will ultimately be reduced by $\frac{10}{19}$(\$200) to \$894.73.

7. Writing $f(x) = ax^2 + bx + c$, and substituting in turn $x = -1,1$, and 2, we find that

$$\begin{aligned} a - b + c &= -2 \\ a + b + c &= 0 \\ 4a + 2b + c &= 7. \end{aligned}$$

Solving simultaneously gives $(a,b,c) = (2,1,-3)$, and $y = f(3) = 18$.

8. We have $\dfrac{2}{\cos 2x} = \dfrac{\sin y}{\cos y} + \dfrac{\cos y}{\sin y} = \dfrac{\sin^2 y + \cos^2 y}{(\sin y)(\cos y)} = \dfrac{1}{\sin y \cos y}$,

or $\cos 2x = 2 \sin y \cos y = \sin 2y$. Thus the least positive value for $2x + 2y$ is 90°, and for $x + y$ it is 45° (see 4.11 and 4.21).

9. The first man occupies the following series of places in successive maneuvers: 1,3,5,7,9,11,12,13,14,15,16,17,18,19,20,2,4,6,8,10,1. Thus after 20 maneuvers the first man is back in place. Note that the others are also back in their original places. All the men go through the same cycle, starting at a different position.

10. For odd k, $x^k + 1$ is divisible by $x + 1$ (see 1.34).

We have $(2^{210} - 1) = (2^{105} + 1)(2^{105} - 1)$ and $2^{105} + 1 = (2^{15})^7 + 1 = (2^{21})^5 + 1 = (2^{35})^3 + 1$. The four solutions are $2^{105} + 1$, $2^{15} + 1$, $2^{21} + 1$, $2^{35} + 1$.

11. It now costs $\dfrac{60 \text{ cents/gallon}}{24 \text{ miles/gallon}} = \dfrac{60}{24}$ cents/mile $= \dfrac{5}{2}$ cents/mile to run the car. The man needs a cost of $\dfrac{70 \text{ cents/gallon}}{x \text{ miles/gallon}} = \dfrac{70}{x}$ cents/mile $= \dfrac{5}{2}$ cents/mile. Hence $140 = 5x$ and $x = 28$. He needs to get 4 more miles per gallon.

12. It can be shown that any simple planar polygon of n sides has at least $n - 3$ interior diagonals. Since the total number of diagonals is $\binom{n}{2} - n$, there can be at most $\binom{n}{2} - 2n + 3$ exterior diagonals. The first diagram shows an example of an n-gon where this maximum is actually achieved. For $n = 12$, this number is 45.

To count the minimum number of interior diagonals of a simple n-gon, let us first prove that any such figure has at least one interior diagonal. Suppose the vertices of the figure, in order, are $A_1, A_2, \ldots, A_n$ (see diagram 2). We can choose any vertex A_k and look at the angle $A_{k-1}A_kA_{k+1}$. If we find any vertex A_j of the polygon inside this angle, then there must be a closest such vertex to A_k, and then $\overline{A_kA_j}$ is an interior diagonal. If there is no such vertex A_j, then $\overline{A_{k-1}A_{k+1}}$ is an interior diagonal.

12. (Continued)

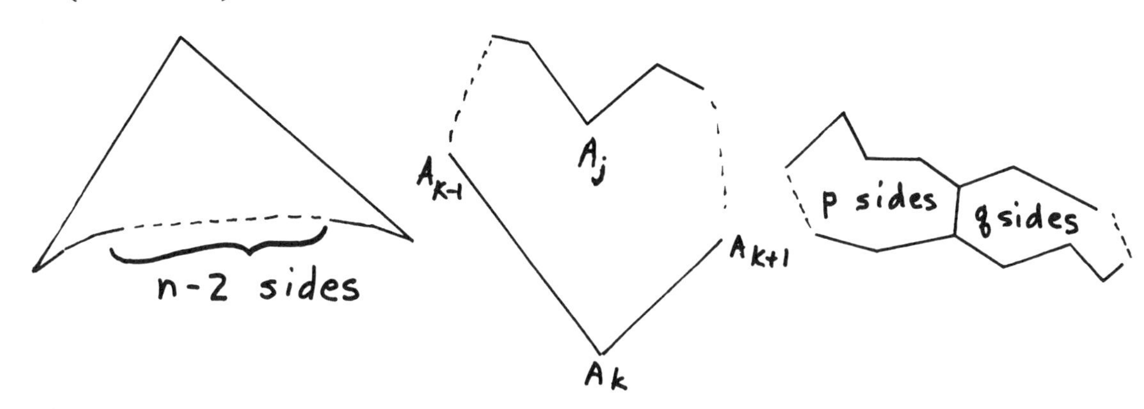

Hence any polygon must have at least one interior diagonal. We can now prove by mathematical induction that a polygon of n sides must have at least $n - 3$ interior diagonals.
For $n = 4$, the statement merely asserts the existence of one interior diagonal, which we have shown above. Suppose that for $n = 4, 5, 6, \ldots, k$ we know that an n-gon must have $n - 3$ interior diagonals. To complete the induction we must show that any polygon of $k + 1$ sides must have at least $k + 1 - 3 = k - 2$ interior diagonals. We proved above that even our $(k + 1)$-gon has at least one interior diagonal.
If we draw this in, we will separate the $(k + 1)$-gon into two polygons (diagram 3). Suppose one of these has p sides and the other has q sides. Then $p + q = k + 3$ (we count the diagonal twice as a side) and the induction hypothesis applies for these two polygons, so they must have at least $p - 3$ and $q - 3$ interior diagonals respectively. Now any interior diagonal of one of the "smaller" polygons is an interior diagonal of our $(k + 1)$-gon, so in all we must have at least $(p - 3) + (q - 3) + 1 = p + q + 1 - 6$ $= k + 3 - 5 = k - 2$ interior diagonals for the $(k + 1)$-gon. This completes the induction.
Although this reasoning is difficult to complete in the time limit given for the problem, a little experimentation can quickly lead one to the correct conjecture.

13. Dividing the first equation by the second, we find $\frac{x}{y} = 9$.
Substituting, we have $\frac{9y}{4\sqrt{y}} = \frac{9\sqrt{y}}{4} = 18$, $y = 64$, and $x = 576$.

14. One can prove that if a rectangle has four vertices on the sides of a given square and no side parallel to a side of a given square, then either the rectangle is itself a square, or each of its sides meets a side of the given square at right angles.
Since the rectangle in this problem is not a square, we have (see diagram) $AX = PX = XS$, $PQ = RS = XY$, $QY = RY = YC$, so that the perimeter of the rectangle is $PQ + QY + YR + RS + SX + XP$ $= XY + YC + YC + XY + AX + AX = 2(AC)$, or $2\sqrt{2}$. Since one of its

14. (Continued)

sides has length 1, the other side must have length $\sqrt{2} - 1$. In general, the perimeter of a rectangle inscribed in a square as shown is twice a diagonal of the square.

It is not hard to see that the largest rectangle of the type described in the problem must have one vertex on each side of the square. For if one vertex is inside the square, we can rotate the rectangle about a different vertex. Then three vertices are inside the square, and a small dilation will give us a larger rectangle than the one we started with, which still fits into the square.

A slightly longer argument will show that the sides of our rectangle must meet the sides of the square at 45° angles. For suppose $RD = a$, $QC = b$ (see diagram). Without loss of generality we may assume $a \geq \frac{1}{2}$. It is not hard to see that triangles SRD and QRC are similar, so we have:

$$\frac{a}{1 - b} = \frac{b}{1 - a}, \text{ or } a - a^2 = b - b^2 .$$

We can solve this quadratic equation for b in terms of a: $b^2 - b + a - a^2 = 0$.

$$b = \frac{1}{2}\left(1 \pm \sqrt{1 - 4(a - a^2)}\right) = \frac{1}{2} \pm \frac{1}{2}\sqrt{4a^2 - 4a + 1}$$

$$= \frac{1}{2} \pm \frac{1}{2}\left|2a - 1\right| = \frac{1}{2} \pm \left(a - \frac{1}{2}\right) \quad \left(\text{since we assumed } a \geq \frac{1}{2}\right).$$

so either $b = a$ or $b = 1 - a$. The first alternative implies that $PQ = QR$, so that the rectangle is a square. The second implies that PQ and AB meet at an angle of 45°.

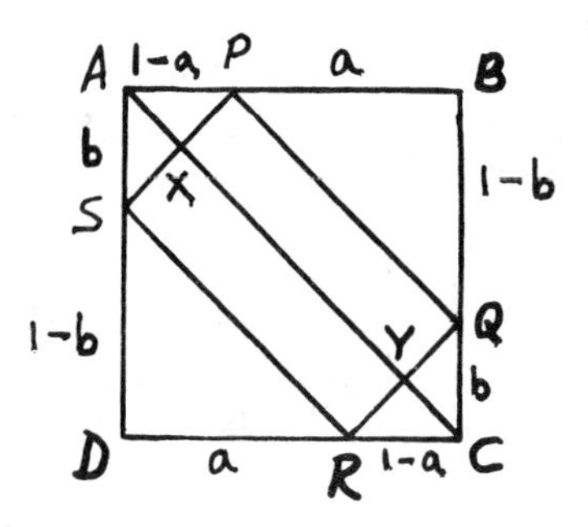

15. Using absolute value for area, we have $|ABC| = |OAC| - |OAB| - |OBC|$. Now $|OAC| = \frac{1}{2} \cdot 6 \cdot 4 \sin \text{angle } AOC = 12 \sin 120° = 6\sqrt{3}$ (see 3.64). Similarly, $|OAB| = 3\sqrt{3}$ and $|OBC| = 2\sqrt{3}$. Hence $|ABC| = 6\sqrt{3} - 5\sqrt{3} = \sqrt{3}$.

15. (Continued)

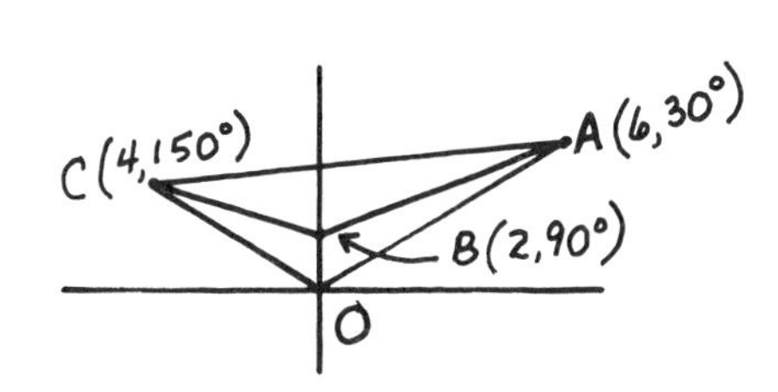

16. Let $\alpha = a + b\sqrt{2}$, where a and b are integers, and let us look at the function $f(\alpha) = f(a + b\sqrt{2}) = a - b\sqrt{2}$ (the "irrational conjugate" of intermediate algebra). Some easy, if tedious, calculations show that if α and β are numbers of the form $a + b\sqrt{2}$, then

$f(\alpha + \beta) = f(\alpha) + f(\beta)$
$f(\alpha\beta) = f(\alpha)f(\beta)$
$\alpha + f(\alpha)$ is always an integer.

Note that the second property implies that $f(\alpha^2) = [f(\alpha)]^2$, $f(\alpha^3) = [f(\alpha)]^3$, and in general that $f(\alpha^n) = [f(\alpha)]^n$ for any natural number n (a formal proof would use induction).

If $\alpha = 1 + \sqrt{2}$ (the number given in the problem), then $f(\alpha) = 1 - \sqrt{2}$ and $(1 + \sqrt{2})^{99} + (1 - \sqrt{2})^{99} = \alpha^{99} + [f(\alpha)]^{99} = \alpha^{99} + f(\alpha^{99})$ is an integer. Now $-1 < 1 - \sqrt{2} < 0$, so that $-1 < (1 - \sqrt{2})^{99} < 0$, and $(1 + \sqrt{2})^{99} + (1 - \sqrt{2})^{99} = (\sqrt{2} + 1)^{99} - (\sqrt{2} - 1)^{99}$ is the integer just below $(1 + \sqrt{2})^{99}$. But this is exactly the N referred to in the problem. Hence $N = (\sqrt{2} + 1)^{99} - (\sqrt{2} - 1)^{99}$ and $r = (\sqrt{2} - 1)^{99}$, so that $(N + r)r = [(\sqrt{2} + 1)^{99}][(\sqrt{2} - 1)^{99}] = [(\sqrt{2} + 1)(\sqrt{2} - 1)]^{99} = (2 - 1)^{99} = 1$. The function f given above is called an automorphism for the ring of numbers of the form $a + b\sqrt{2}$. For much more on this topic, see Herstein (Chapter 3, section 3).

17. If we denote the three roots by $\frac{a}{r}$, a, ar, then $\frac{a}{r} \cdot a \cdot ar = a^3 = 1$, and $a = 1$, since a, being a root, must be real (see 1.4). Also, $\frac{a}{r} + a + ar = r + 1 + \frac{1}{r} = \frac{7}{2}$, $r + \frac{1}{r} = 2 + \frac{1}{2}$, and $r = 2$ or $\frac{1}{2}$. The roots are $\frac{1}{2}$, 1, 2.

18. The central angle subtended by a side of the 60-gon is 6°, while the same angle for a 30-gon is 12°. The area of the 60-gon is 60 times that of an isosceles triangle with legs r (the common radius) and vertex angle 6°, or $60 \cdot \frac{1}{2}r^2 \sin 6°$ (see 3.64). Similarly, the area of the 30-gon is $30 \cdot \frac{1}{2}r^2 \sin 12°$. The required ratio is

$$\frac{2 \sin 6°}{\sin 12°} = \frac{2 \sin 6°}{2 \sin 6° \cos 6°} = \sec 6° \text{ (see 4.21).}$$

18. (Continued)

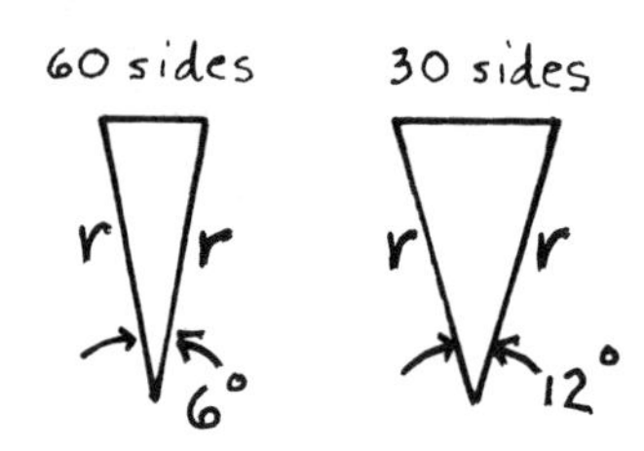

19. The expected waiting time at the stop for bus A is $\frac{1}{2} \cdot 8 = 4$ minutes, so the expected value of the time until boarding a bus is 14 minutes. Going to line B, the expected time is $5 + \frac{1}{2} \cdot 15 = 12\frac{1}{2}$ minutes, so the better choice is line B.

20. We have (1.4):

$$\begin{aligned} p + q + r &= p \\ pq + qr + pr &= q \\ pqr &= r. \end{aligned}$$

From the first equation, $q + r = 0$. From the second, $p(q + r) + qr = 0 + qr = qr = q$. Since $q \neq 0$, we must have $r = 1$. Then $q = -1$ and from the third equation, $p = -1$.

21. We have $\sin x + \cos x = \frac{1}{2}$, or $\sin^2 x + \cos^2 x + 2(\sin x)(\cos x) = 1 + 2 \sin x \cos x = \frac{1}{4}$, (see 4.11) so $\sin x \cos x = -\frac{3}{8}$. Now (1.321) $\sin^3 x + \cos^3 x = (\sin x + \cos x)(\sin^2 x - \sin x \cos x + \cos^2 x) = \frac{1}{2}(1 - \sin x \cos x) = \frac{1}{2}\left(1 + \frac{3}{8}\right) = \frac{11}{16}$.

22. We will show that angle $BFC = 90°$, so that $BF^2 + CF^2 = BC^2 = 36$. Suppose BE and CD intersect at X. The angles of quadrilateral $ADXE$ must add up to 360°. Hence, if angle $A = a°$, $a + 80 +$ angle $DXE + 100 = 360$ and angle $DXE = 180 - a$. Then angle $BXC =$ angle $DXE = 180 - a$, and in triangle BXC, angle CBX + angle $BCX + 180 - a = 180$, or $a =$ angle CBX + angle BCX.
Also, angle DCA + angle $A = 100°$ (since angle $ADC = 100°$ is an exterior angle in triangle ACD), so that angle $DCF = 50° - \frac{1}{2}a$. Similarly, angle $EBF = 40° - \frac{1}{2}a$. Now we look at triangle BCF: angle FBC + angle CBF = angle CBX + angle EBF + angle BCX + angle $DCF = a + 50 - \frac{1}{2}a + 40 - \frac{1}{2}a = 90°$. Thus angle $BFC = 90°$ and the proof is complete.

22. (Continued)

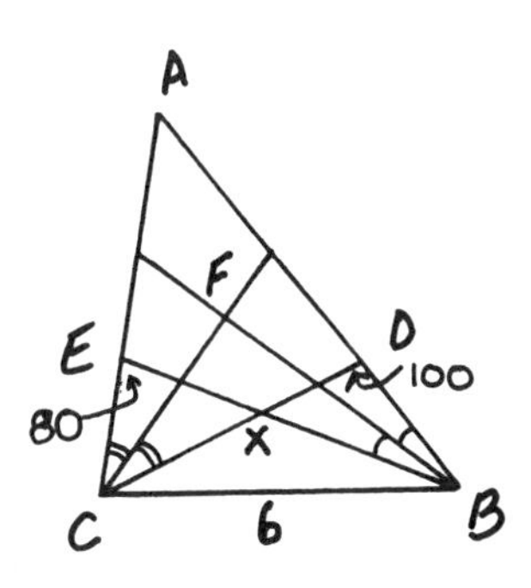

23. Let $y = x^{1/2n}$. Then $y^2 = x^{1/n}$, and the given equation can be written as $y^2 - 4y + 3 = 0$, or $(y - 1)(y - 3) = 0$. Solving for x gives $x = 1$ or $x = 3^{2n}$.

24. Let us call the smallest distance between two points in the figure a "unit." We first count triangles whose sides are parallel to the sides of the large triangle. There are nine such triangles with unit side, three such triangles with sides 2 units, and one such triangle with side 3 units. Hence there are 13 triangles with sides parallel to those of the larger triangle. But there are two more triangles whose sides are not parallel to those of the larger triangle. These are shown in the figure.

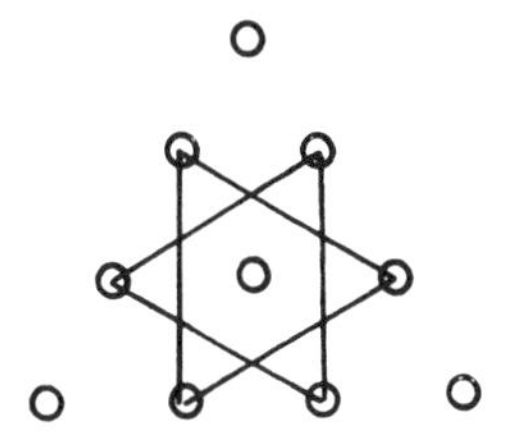

25. Let us look at the possible vowels, represented by $x + 7$. The numerical positions of A,E,I,O,U in the alphabet are, respectively, 1,5,9,15,21. Thus only $x + 7 = 9$, 15, or 21 are possible. For $x = 2$ the message is *BIHG*. For $x = 8$ it is HONM, and for $x = 14$ it is NUTS. (For a discussion of the anecdote in American history on which the problem is based, see Rawson (p. 194)).

26. In the diagram, the coordinates of P are (h,k) and those of B are (o,k). Since $PA = PB$, and P is on the given curve, we have $k^2 = 20k - 100$, which leads to $k = 10$.

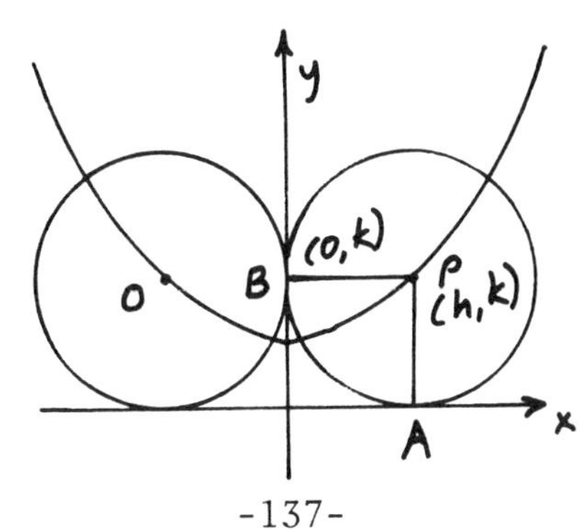

27. The given expression is equal to $x + y + z - x - x - y + x = z$ $= \min(x,y,z)$.

28. If the radius of a circle is R, then the length of a chord of a central angle of measure 2θ is given by $2R \sin \theta$ (see diagram). In the present case, we have $2R \sin \theta = 4$, $2R \sin 3\theta = 11$, where the given arcs are 2θ and 6θ. Now $\sin 3\theta = 3 \sin \theta - 4 \sin^3\theta$ (4.51), so $\frac{11}{2R} = \frac{6}{R} - \frac{32}{R^3}$, leading to $R = 8$ or -8. The radius of the circle is 8.

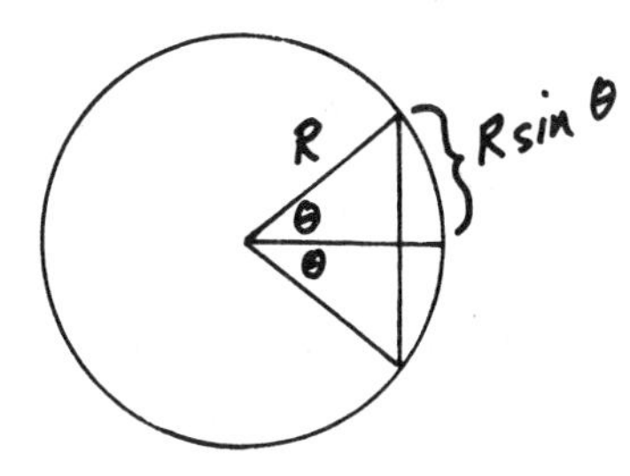

29. The numbers in question are of the form $30n + 1$. Since $3000 \leq 30n + 1 \leq 4000$, we have $99 < n \leq 133$ and there are 34 such numbers forming an arithmetic progression. The first and last are 3001 and 3991, so formula 1.512 shows their sum to be 118864.

30. The last game must be won by the Mets. Hence the Mets may win any three of the first five games and also the last game. The probability of this event is $\binom{5}{3}\left(\frac{3}{5}\right)^3\left(\frac{2}{5}\right)^2\left(\frac{3}{5}\right) = .20736$.

Solutions—Fall 1979

1. We have:

$$\frac{x + 39}{3} = \sqrt[3]{108x} + 13$$

$$x + 39 = 3\sqrt[3]{108x} + 39$$

$$x^3 = 27(108x)$$

$$x^2 = (81)(36)$$

$$x = (9)(6) = 54.$$

2. We have:

$$\frac{100A + 10B + C}{10A + C} = 9,$$ which leads to $5A + 5B - 4C = 0$.

If $5(A + B) = 4C$ then $C = 5$, so $A + B = 4$. The maximum occurs where $A = 4$ and $B = 0$, or $N = 405$.

3. We have:

$80 \cdot 45 = 3600$, so $\log 36 = \log \frac{3600}{100} = \log 3600 - \log 100$
$= a + b - 2$.

4. Using absolute value to denote area,

$|ADF| = \frac{1}{2}(AD)(AF)\sin 60° = \frac{1}{2}(1)(\sqrt{3})(\sqrt{3}/2) = 3/4$ (see 3.64),

so $|DEF| = |ABC| - 9/4$. Then $|ABC| = \frac{(1 + \sqrt{3})^2\sqrt{3}}{4} = \frac{6 + 4\sqrt{3}}{4}$,

so $|DEF| = \frac{-3 + 4\sqrt{3}}{4}$, and $(a,b) = \left(-\frac{3}{4}, 1\right)$. See problem 34, Spring 1976.

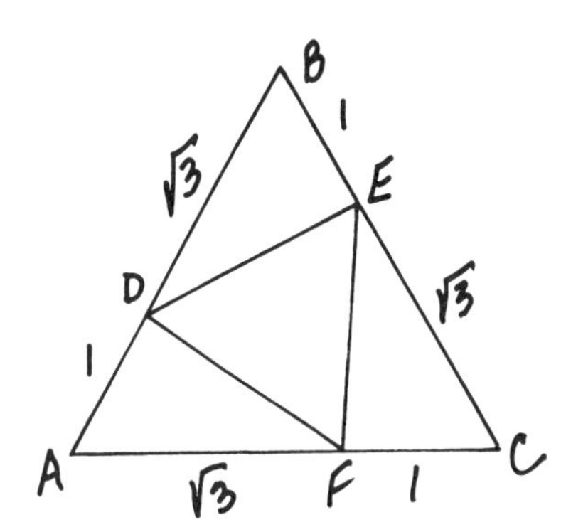

5. $\cos A = -\sqrt{1 - \sin^2 A}$

$= -\sqrt{(1 + \sin A)(1 - \sin A)}$

$= -\sqrt{1 + 24/25)(1 - 24/25)}$

$= -\frac{7}{25}$.

$\sin \frac{A}{2} = \sqrt{\frac{1 - \cos A}{2}} = \sqrt{\frac{1 + 7/25}{2}}$

$= \sqrt{\frac{16}{25}} = \frac{4}{5}$. (See 4.11, 4.26.)

6. If we let $a = \sqrt{3x + 1} + \sqrt{2 - x}$, we are looking for solutions to the equation $a^3 + a^2 + a = 14$. One real solution that we can get by inspection is $a = 2$. From this we find:

$$\sqrt{3x + 1} + \sqrt{2 - x} = 2, \text{ or } \sqrt{3x + 1} = 2 - \sqrt{2 - x}.$$

Squaring and simplifying gives us $4x - 5 = -4\sqrt{2 - x}$. Squaring again yields $16x^2 - 24x - 7 = (4x - 7)(4x + 1) = 0$, and $x = \frac{7}{4}, -\frac{1}{4}$. A check shows the value $x = \frac{7}{4}$ to be extraneous.

6. (Continued)

The equation $a^3 + a^2 + a - 14$ also has two complex roots:

$a^3 + a^2 + a - 14 = (a - 2)(a^2 + 3a + 7) = 0$ when $a = 2$ or $a = -\frac{3}{2} \pm i\sqrt{19}/2$. We can show that these complex roots do not lead to real values for x. For suppose $\sqrt{3x + 1} + \sqrt{2 - x} = p + qi$, where $q \neq 0$. Then, as the square root of a real number is either positive, zero, or "pure" imaginary, either $p = 0$, or $p = \sqrt{3x + 1}$ and $qi = \sqrt{2 - x}$, or $p = \sqrt{2 - x}$ and $qi = \sqrt{3x + 1}$. In each case, $p \geq 0$, which is not true of the complex roots we have found.

7. Let x = the sixth term. Then $1 + 2 + 4 + 8 + 16 + 32 = 63$, $63 = \frac{6}{2}(1 + x)$ (see 1.512), and $x = 20$.

8. There are $\binom{5}{2} = 10$ ways in which to choose the two numbers that are to be in their correct places. The other three numbers must be in incorrect places. The first of the three empty places can be filled in two ways and the remaining two in only one way. Since the total number of ways to fill the five spaces is $5!$, the answer is $\frac{10 \cdot 2}{5!} = \frac{1}{6}$.

9. $S = \log (1/2)(2/3)(3/4)\ldots(99/100)$

$= \log 1/100$

$= -2.$

10. Without loss of generality, we may assume that both lines pass through the origin, that $OP = 7$, and that $AP = 1$ (see diagram). Then the Pythagorean theorem shows that $OB = 7\sqrt{2}$ and $AO = 5\sqrt{2}$. Using the law of cosines in triangle OAB, we find $36 = 50 + 98 - 140 \cos 2\theta$, so $\cos 2\theta = 112/140 = 4/5$. Using 4.24, $\sin \theta = \sqrt{\frac{1 - \cos 2\theta}{2}} = \sqrt{\frac{1}{10}} = \frac{\sqrt{10}}{10}$. Then $\cos \theta = \frac{3\sqrt{10}}{10}$ and $\tan \theta = \frac{1}{3}$.

The slope of line L is $\tan(\alpha + \theta) = \frac{\tan \alpha + \tan \theta}{1 - \tan \alpha \tan \theta}$

$= \frac{\frac{1}{7} + \frac{1}{3}}{1 - \frac{1}{7} \cdot \frac{1}{3}} = \frac{10}{20} = \frac{1}{2}$. (4.341) A similar argument will show that if $0 \leq m \leq n$, then the tangent of the acute angle x formed by two lines with slopes m and n is given by $\tan x = \frac{n - m}{1 + mn}$.

10. (Continued)

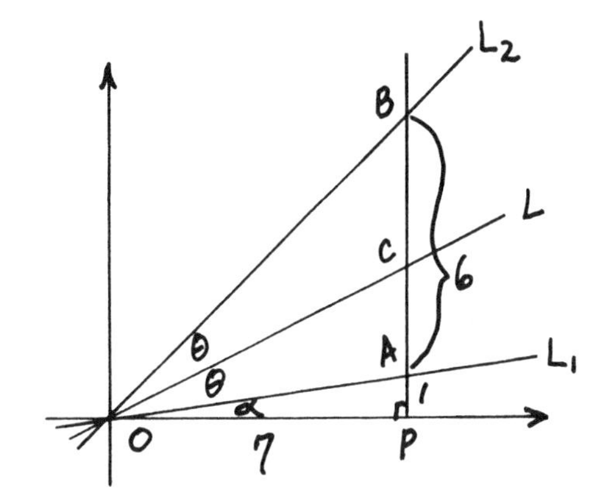

11. $\dfrac{1 + \cot A}{\cot A} = \dfrac{1}{\cot A} + 1 = \tan A + 1.$

$\dfrac{1 + \cot B}{\cot B} = \tan B + 1 = \tan\,(225° - A) + 1 \quad (4.342)$

$= \dfrac{1 - \tan A}{1 + \tan A} + 1$, so the product is

$(\tan A + 1)\left(\dfrac{1 - \tan A}{1 + \tan A} + 1\right) = 2.$

12. $x^3 + y^3 = (x + y)^3 - 3xy(x + y)$. Also, $2xy = (x + y)^2 - (x^2 + y^2)$.
Then $c = a^3 - 3a \cdot \dfrac{a^2 - b}{2} = \dfrac{3ab - a^3}{2}$ or equivalent equation.

13. Let $[x]$ denote the greatest integer not exceeding x. There are $[343/7] = 49$ factors of 7 in 343!. There are $[343/7^2] + [343/7^3] = 7 + 1$ additional factors of 7, since each multiple of 7^2 or 7^3 contributes more 7's. The sum is $49 + 7 + 1 = 57$.

14. Using formula 4.341, we get $\dfrac{x + x^2}{1 - x^3} = x$, or $x^4 + x^2 = 0$, so

$x^2(x^2 + 1) = 0$, so the single real solution is $x = 0$.

15. $A = \dfrac{1}{2}y^2 \sin\theta$ and $\sin\theta$ is maximum when $\theta = 90°$. This gives

$A = \dfrac{y^2}{2}$. (see 3.64)

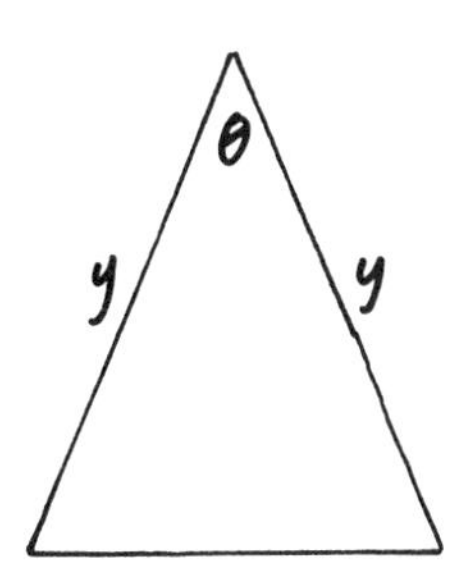

16. $\frac{(x - 3y)(x + y)}{x - 2y + 1} = 0$ and $\frac{(x - 3y)(x - y)}{2x - 3y - 2} = 0$, so $(x - 3y)(x + y)$ $= 0$ and $(x - 3y)(x - y) = 0$, and $x = 3y$ satisfies both equations. Setting $x - y$ and $x + y$ equal to zero leads only to $x = y = 0$, which is included in the above. Thus all simultaneous solutions are given by $x = 3y$, PROVIDED:

$$\begin{aligned} x - 2y + 1 &\neq 0 \\ 3y - 2y + 1 &\neq 0 \\ y &\neq -1 \\ x &\neq -3 \end{aligned} \qquad \begin{aligned} 2x - 3y - 2 &\neq 0 \\ 6y - 3y - 2 &\neq 0 \\ y &\neq 2/3 \\ x &\neq 2 \end{aligned}$$

17. The number of unit squares touching the sides is $2x + 2y - 4$. This provides for half the area of the rectangular region. Thus $\frac{1}{2}xy = 2x + 2y - 4$, or $xy - 4x - 4y + 16 = (x - 4)(y - 4) = 8$. The only factors of 8 are 1, 2, 4, and 8. Therefore, $x - 4 = 1, 2, 4,$ or 8, while $y - 4$ is equal to 8, 4, 2, or 1, respectively. Thus $(x,y) = (5,12), (6,8), (8,6)$ or $(12,5)$, and $xy = 48$ or 60.

18. We have
$$P(x) = Q_1(x)(x - 2) + 2 \qquad \text{(i)}$$
$$P(x) = Q_2(x)(x + 2) - 2 \qquad \text{(ii)}$$

We need to find a and b where

$$P(x) = Q_3(x)(x^2 - 4) + ax + b \qquad \text{(iii)}$$

From (i) and (iii), $P(2) = 2 = 2a + b$.
From (ii) and (iii), $P(-2) = -2 = -2a + b$.
Solving simultaneously, we get $a = 1$, $b = 0$, so the remainder is x.

19. The greatest power of 5 to divide 127! is $\left[\frac{127}{5}\right] + \left[\frac{127}{5^2}\right] + \left[\frac{127}{5^3}\right]$ $= 25 + 5 + 1 = 31$. A much higher power of 2 divides 127!. Therefore, 10^{31} divides 127! to make it end in 31 zeroes.

20. Let $y = \sqrt{x} + \sqrt{\frac{1}{x}}$. Then $\frac{y - \sqrt{2}}{y} = \frac{y}{y + 3\sqrt{2}}$

$$y^2 = y^2 + 2y\sqrt{2} - 6$$

$$y = \frac{3}{\sqrt{2}}, \text{ so } \sqrt{x} + \sqrt{\frac{1}{x}} = \frac{3}{\sqrt{2}}.$$

Squaring leads to $2x^2 - 5x + 2 = 0$, which gives $x = 2, \frac{1}{2}$.

21. We have:

$$6\left(\frac{n!}{(n-2)!}\right)^2 = 5\,\frac{(n^2)!}{(n^2-2)!}$$

$$6n^2(n-1)^2 = 5n^2(n^2-1).$$

Since $n \neq 0$, $n \neq 1$, $6(n-1) = 5(n+1)$, and $n = 11$.

22. The given equation is

$$\frac{\sin^2 x}{\cos^2 x} - \frac{4\sin x}{\cos x} + 1 = 0.$$

Multiplying by $\cos^2 x$, we get:

$\sin^2 x - 4(\sin x)(\cos x) + \cos^2 x = 0$, or $1 - 4(\sin x)(\cos x) = 0$, or $\sin 2x = \frac{1}{2}$. (See 4.21.)

23. We have:

$$\left.\begin{aligned} xyz^3 &= 2^3 \cdot 3 \\ xy^3z &= 3^3 \cdot 2 \\ x^3yz &= 2 \cdot 3 \end{aligned}\right\}$$

Multiplying the above together gives:

$x^5y^5z^5 = 3^5 \cdot 2^5$, or $xyz = 6$.

Dividing each equation by $xyz = 6$, we have:

$z^2 = 4$, $y^2 = 9$, $x^2 = 1$, and all three can be positive or two can be negative, giving $(1,3,2)$, $(-1,3,-2)$, $(-1,-3,2)$, and $(1,-3,-2)$.

24. Since $FG = GH$ and m angle $GFH = 60$, triangle FGH is equilateral. With x and s as in the diagram, in triangle AGF, $\frac{x}{\sin 120°} = \frac{s}{\sin 45°}$ implies $x^2 = \frac{3}{2}s^2$. (See 3.4.) The required ratio is

$$\frac{(x^2/4)\sqrt{3}}{6(s^2/4)\sqrt{3}} = \frac{x^2}{6s^2} = \frac{(3/2)s^2}{6s^2} = \frac{1}{4}.$$

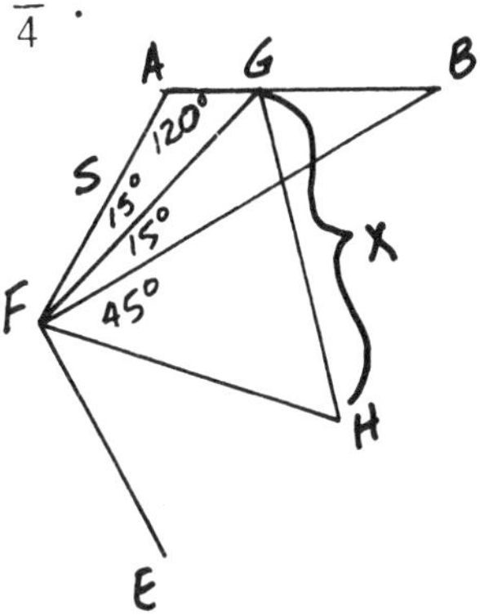

25. Let N be the base. Since $\frac{1}{5} = .333\ldots$, multiplying both sides by N produces $\frac{N}{5} = 3.333\ldots$. Then $\frac{N-1}{5} = 3$, so $N = 16$. Alternatively, we can use 1.523.

26. Let $a = \sqrt{x+y}$ and $b = \sqrt{x-y}$ (note that $a > 0$ and $b > 0$). Then $a^2 + a = 12$ and $b^2 + b = 6$, or $(a-3)(a+4) = 0$

$$(b-2)(b+3) = 0, \quad \text{so}$$

$a = 3$ and $b = 2$. Therefore, $x + y = 9$

$$x - y = 4, \text{ so } x = 13/2,$$

$$y = 5/2.$$

27. The probabilities of success on the first question for the four students are $\frac{3}{5}$, $\frac{1}{2}$, $\frac{2}{5}$, and $\frac{1}{4}$. The probability that they all fail to answer the first question correctly is $\frac{2}{5} \cdot \frac{1}{2} \cdot \frac{3}{5} \cdot \frac{3}{4} = \frac{9}{100}$. Thus the probability of at least one success is $1 - \frac{9}{100} = \frac{91}{100}$.

28. The given equation is equivalent to

$$\log_N x = \log_N 2 + \log_N 4 + \log_N 6 + \log_N 8 + \log_N 10$$

$$= \log_N 2 \cdot 4 \cdot 6 \cdot 8 \cdot 10, \text{ so } x = 3840. \quad \text{(See 5.12.)}$$

29. If $x = 1$, then $y^2 + 6y + 6 = 0$ and $y = -3 \pm \sqrt{3}$.
If $x = -1$, then $y^2 + 2y = 0$ and $y = 0$ or -2.
The points which are furthest apart are $(-1,0)$ and $(1,-3-\sqrt{3})$.
The distance between them is $\sqrt{16 + 6\sqrt{3}}$, so $p + q = 22$.

30. By the "extended law of sines" (see 3.4),

$$\frac{BC}{\sin A} = 2R = \frac{1}{BH} = \frac{1}{\sqrt{1 - \frac{x^2}{4}}}$$

(since $BH = \sin A$). This gives

$$R = \frac{1}{\sqrt{4 - x^2}}, \text{ so that}$$

$$\pi R^2 = \frac{\pi}{4 - x^2} \text{ and } (a,b) = (4,-1).$$

30. (Continued)

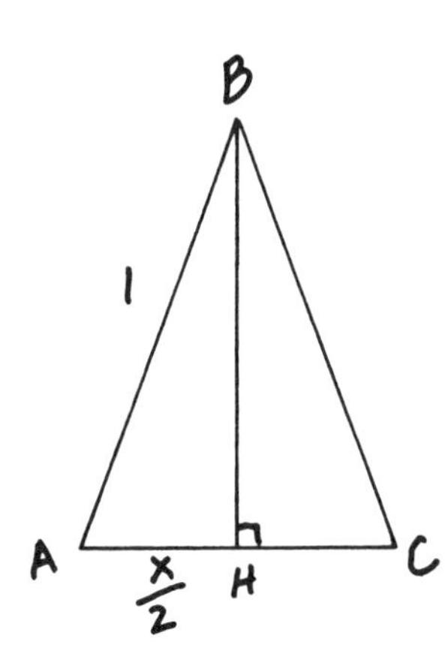

Solutions—Spring 1980

1. There are $5! = 120$ seating arrangements altogether. There are $(4 \cdot 2) \cdot 3 \cdot 2 \cdot 1 = 48$ arrangements in which the Bronx and Canton coaches sit together. Therefore, there are $120 - 48 = 72$ arrangements as required in the problem.

2. $x^3 + x + 2x^4 + 4x^2 + 2 = x(x^2 + 1) + 2(x^2 + 1)^2$

   $= (x^2 + 1)(x + 2(x^2 + 1)) = (x^2 + 1)(2x^2 + x + 2)$.

3. If $(a - 1)a(a + 1) = 6$, then $a^3 - a - 6 = 0$, so $(a - 2)(a^2 + 2a + 3) = 0$, which has one real and two complex solutions. Letting $a = \sqrt{x}$, the complex solutions for a do not lead to real solutions for x. Therefore, we need $1 \cdot 2 \cdot 3 = 6$, and $a = 2$, so $x = 4$.

4. Rationalizing the denominators, we have

   $(\sqrt{2} - \sqrt{1}) + (\sqrt{3} - \sqrt{2}) + (\sqrt{4} - \sqrt{3}) + \ldots + (\sqrt{25} - \sqrt{24}) = \sqrt{25} - \sqrt{1} = 4$.

   This is called a "telescoping sum."

5. Squaring: $\sin^2 40° + \cos^2 40° + 2(\sin 40°)(\cos 40°) = 1 + \sin x$. (See 4.21.)

   $$1 + \sin 80° = 1 + \sin x$$

   $$x = 80°$$

6. Since two tangents drawn to a circle from the same external point are equal in length, $BT = BC = 4$ so $AT = 1$. In right triangle ATO, $1^2 + r^2 = (3 - r)^2$ so $r = \frac{4}{3}$.

6. (Continued)

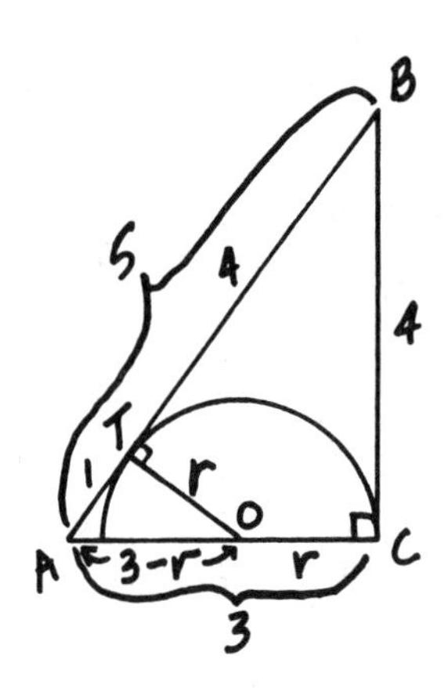

7. If x represents the number of students, $x(x - 1) = 870 = 30 \cdot 29$, so $x = 30$.

8. $x^8 - 2x^4 - 120 = (x^4 - 12)(x^4 + 10) = 0$. Solving for x^2, we find that $\sqrt{12}$ is the only positive real value. Thus,

$$\text{Area} = \frac{\sqrt{12} \cdot \sqrt{3}}{4} = \frac{3}{2}.$$

9. A number that is divisible by 22 is divisible by both 11 and 2. Hence, if the number is $100h + 10t + u$, u must be even. For $u = 2, 4, 6, 8$, t must be 9, 7, 5, 3 respectively. Using criteria for division by 11, we find 792, 374, 638 as the only possible numbers.

10. By 1.81 and 1.82, the roots of the polynomial are $\pm\sqrt{-7}$ and $\pm\sqrt{3}$. Hence, by 1.311, the polynomial must be divisible by $x^2 + 7$ and $x^2 - 3$, or by $x^4 + 4x^2 - 21$. Since its coefficients have no common factor greater than 1, and the polynomial is of degree 4, it can only be $x^4 + 4x^2 - 21$.

11. Adding: $\dfrac{x + y}{x^2 + y^2} = 3$. Dividing, $\dfrac{x}{y} = \dfrac{1}{2}$, or $y = 2x$. Substitution gives $\dfrac{3x}{5x^2} = 3$, so $x = \dfrac{1}{5}$, $y = \dfrac{2}{5}$ ($x = 0$ yields no solution).

12. Triangle $O_1O_2O_3$ is a 3-4-5 right triangle. Using absolute value for area, $|O_1O_2O_3| = 6$, $|AO_3C| = \frac{1}{2} \cdot 3^2 \sin\alpha = (9/2)(3/5) = 27/10$. $|BO_2C| = \frac{1}{2} \cdot 2^2 \sin\beta = (4/2)(4/5) = 16/10$. (See 3.64.) $|AO_1B| = \frac{1}{2} = 5/10$. $|ABC| = |O_1O_2O_3| - \Big[|AO_3C| + |BO_2C| + |AO_1B|\Big]$ $= 6 - (27 + 16 + 5)/10 = 12/10 = 6/5$. See problem 34, Spring 1976.

12. (Continued)

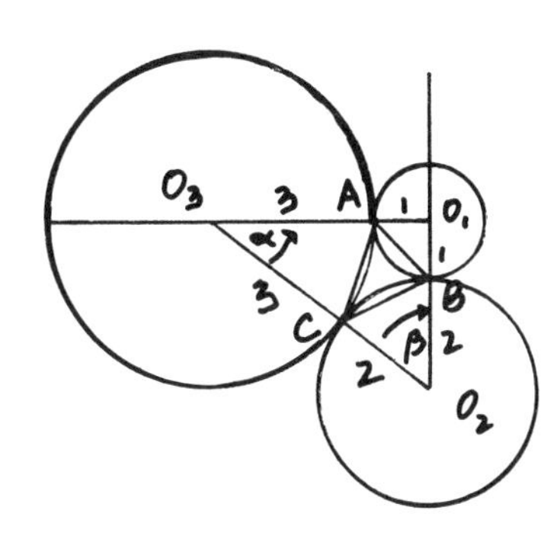

13. Since $\log_a b = \dfrac{1}{\log_b a}$, $\log_2 x + \log_2 x = 4$, so $\log_2 x = 2$ and $x = 2^2 = 4$. (See 5.12.)

14. We could use De Moivre's theorem (1.6) and convert back to rectangular form, or:

$4x^4 + 1 = 4x^4 + 4x^2 + 1 - 4x^2 = 0$

$(2x^2 + 1)^2 - (2x)^2 = 0$

$(2x^2 + 2x + 1)(2x^2 - 2x + 1) = 0$

Setting each factor equal to zero and using the quadratic formula gives $x = \pm\frac{1}{2} \pm \frac{1}{2}i$.

15. Let r = the radius. Then $\overline{OP} = r + 5$, $\overline{PQ} = 6$, $\overline{OQ} = 7 - r$, so $(7 - r)^2 + 36 = (r + 5)^2$, or $r = 5/2$.

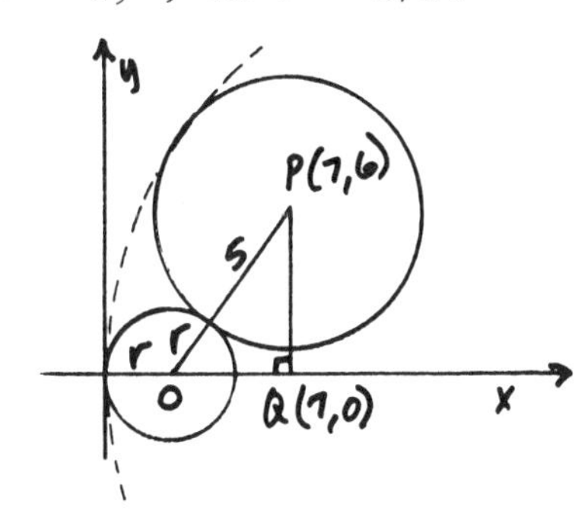

16. The outcome of choosing from urn B is either two balls of different colors (with probability 2/3) or two black balls (with probability 1/3). In the first instance, the probability of drawing a white ball from urn A becomes 3/4, and in the second instance it becomes 1/2. Thus the probability of drawing a white ball from urn A is (2/3)(3/4) + (1/3)(1/2) = (1/2) + (1/6) = 2/3. A tree diagram makes this solution quite clear:

16. (Continued)

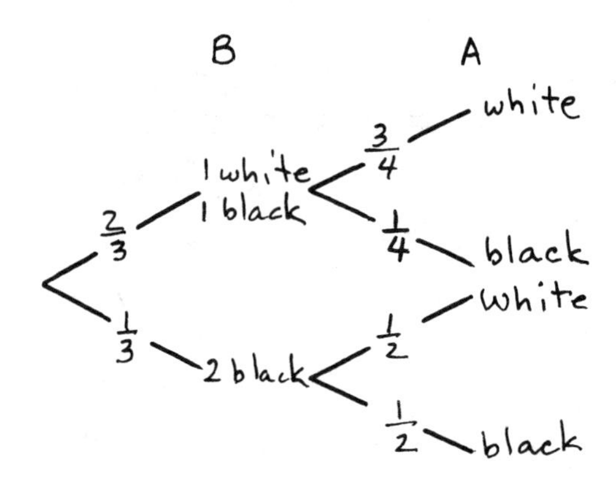

17. $\cos^2 A + 2 \sin A = 1 - \sin^2 A + 2 \sin A = 1$ (see 4.11), so $\sin A(2 - \sin A) = 0$, and $A = 0°, 180°$.

18. If the sides are $2x - y$, $2x$, and $2x + y$, the semiperimeter is $3x$. Using 3.61 and 3.62, $(3x)(3) = 126$, so $x = 14$. From Hero's formula $42 \cdot 14 \cdot (14 + y)(14 - y) = 126^2$, or $196 - y^2 = 27$, or $y = 13$. Thus $2x - y = 15$.

19. For any value of n, $n^5 - 1 = (n - 1)(n^4 + n^3 + n^2 + n + 1)$, so that $n^5 - 1$ is not prime unless one of these factors equals 1. If n is a natural number, this happens only when $n = 2$. (See 1.33.)

20. The graph of $y = f(x)$ must cross the x-axis once between -4 and 0 and again between 0 and 4. By 1.81 the third root must also be real. See the "Intermediate Value Theorem" in any calculus book.

21. If the numbers are x and y, we have: $\frac{x + y}{2} - \sqrt{xy} = 50$, so $x + y - 2\sqrt{xy} = 100$, or $(\sqrt{x} - \sqrt{y})^2 = 100$, and $\sqrt{x} - \sqrt{y} = 10$.

22. We have: $[Q(x)]^2 - [P(x)]^2 = 2x^2 + 1$, or $[Q(x) + P(x)] \cdot [Q(x) - P(x)] = (2x^2 + 1) \cdot (1)$. So either:

(i)		(ii)
$Q(x) + P(x) = 2x^2 + 1$	or	$Q(x) + P(x) = 1$
$Q(x) - P(x) = 1$		$Q(x) - P(x) = 2x^2 + 1$
and		and
$Q(x) = x^2 + 1$, $P(x) = x^2$.		$Q(x) = x^2 + 1$, $P(x) = -x^2$.

The solution obtained from (ii), however, does not fit the conditions of the problem. Notice how closely the arithmetic of polynomials resembles the arithmetic of integers.

23. If the roots of $4x^2 + ax + b = 0$ are r and s, then $r = -\frac{1}{2}$ and $s = \frac{3}{2}$, so the equation is $\left(x + \frac{1}{2}\right)\left(x - \frac{3}{2}\right) = x^2 - x - \frac{3}{4} = 0$, or $4x^2 - 4x - 3 = 0$ (see 1.4).

24. We have: $(a_1 + a_2 + \ldots + a_{n-1}) + a_n = n^2 a_n$

$$(n - 1)^2 a_{n-1} = (n^2 - 1)a_n$$

so that $a_n = \frac{n - 1}{n + 1} a_{n-1}$ for $n > 1$.

Thus $a_2 = (1/3)(1/2) = 1/(3 \cdot 2)$, $a_3 = (2/4)\,[1/(3 \cdot 2)] = 1/(4 \cdot 3)$, $a_4 = (3/5)\,[1/(3 \cdot 4)] = 1/(4 \cdot 5)$, and $a_{50} = 1/(50 \cdot 51)$, so that:

$$a_1 + a_2 + \ldots + a_{50} = 50^2 \cdot \frac{1}{50 \cdot 51} = \frac{50}{51}.$$

25. There are three ways to color card 1. Then there are two possibilities for coloring card two. This leaves only one color for card three, and the coloring of the first three cards (or even the first two) then determines the color of all the others. Thus there are $3 \cdot 2 \cdot 1 = 6$ colorings altogether.

26. $721 = 1 \cdot 721 = 7 \cdot 103$, so $x^3 - y^3 = (x - y)(x^2 + xy + y^2) = 1 \cdot 721 = 7 \cdot 103$ (see 1.322). By inspection, if $x, y > 0$, the second factor on the left is the larger, so either:

(i)	or	(ii)
$x - y = 1$		$x - y = 7$
$x^2 + xy + y^2 = 721$		$x^2 + xy + y^2 = 103.$

Substitution leads to:

$y^2 + y - 240 = 0$	$y^2 + 7y - 18 = 0$
$y = 15, -16$	$y = 2, -9.$

The ordered pairs of <u>positive</u> integers are (16, 15) and (9, 2).

27. $\log_4(\log_3(\log_2 x)) = 0$ when $\log_3(\log_2 x) = 1$, or $\log_2 x = 3$, or $x = 8$.

28. Let S = area of triangle ABC, $x = AB = DC$, and $y = BD$. Then $y(x + y) = x^2$, or $x^2 - xy - y^2 = 0$, so $\left(\frac{x}{y}\right)^2 - \left(\frac{x}{y}\right) - 1 = 0$.

Solving for $\frac{x}{y}$ and rejecting the negative root gives $\frac{x}{y} = \frac{1 + \sqrt{5}}{2}$.

28. (Continued)

Since the triangles have equal altitudes to line AD,

$$\frac{\text{Area of triangle } ABC}{\text{Area of triangle } CBD} = \frac{S}{2\sqrt{5}} = \frac{x}{y} = \frac{1 + \sqrt{5}}{2}.$$ This leads to

$S = 5 + \sqrt{5}$, and $(a,b) = (5,1)$.

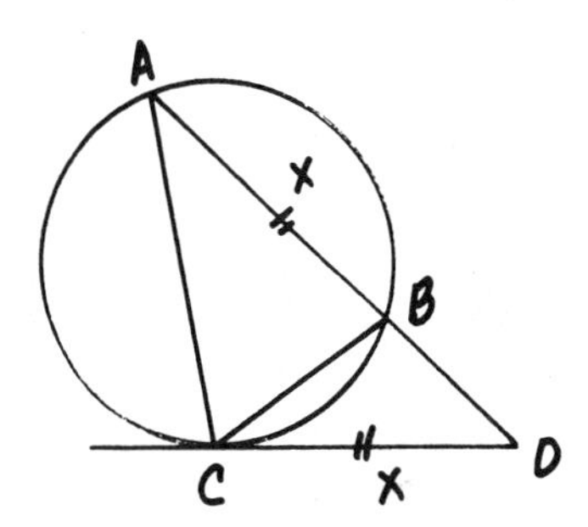

29. Let the original triangle be ABC and let $y = k$ intersect AB, AC, and the y-axis at D, E, and P respectively. Then triangle ADE ~ triangle ABC, and since their areas are in the ratio 1:2, corresponding altitudes are in the ratio $1:\sqrt{2}$. Hence, if O is the origin, $AP:AO = 1:\sqrt{2}$, or $(6 - k)/6 = 1/\sqrt{2}$, so that $k = 6 - 3\sqrt{2}$.

Note: In the limit, as $\overline{BC}$ approaches 0 in length, the line bisecting the area of the triangle does not approach the perpendicular bisector of $\overline{AO}$(!).

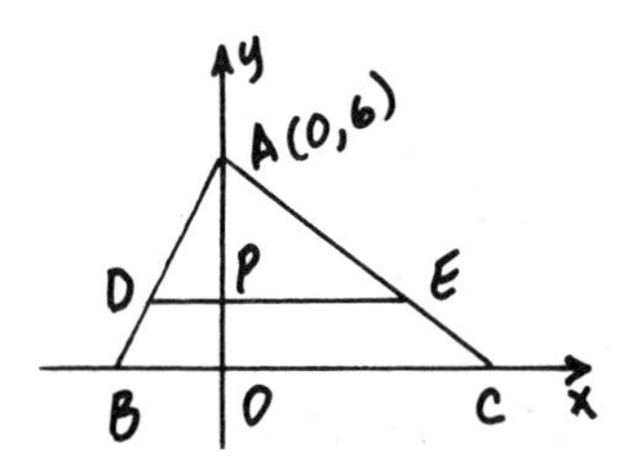

30. If the man has n children, then $(n + n^2)^2 - n^4 = 12(n + n^2 - 3)$, or $2n^3 - 11n^2 - 12n + 36 = 0$. Using the rational root theorem (1.7), we find $(n - 6)(2n^2 + n - 6) = (n - 6)(2n - 3)(n + 2) = 0$. Only $n = 6$ satisfies the conditions of the problem.

We can avoid the use of (1.7) by writing the original equation in the form $(n + n^2)^2 - 12(n + n^2) + 36 = (n^2 + n - 6)^2 = n^4$, so that $n^2 + n - 6 = \pm n^2$.

Solutions—Fall 1980

1. Cost = $m + (k - 12)s = m + ks - 12s$.

2. If we let $y = x + 3$, then $x = y - 3$, and $P(y) = (y - 3)^2 + 7(y - 3) + 4 = y^2 + y - 8$, so that $P(x) = x^2 + x - 8$, and $(a,b,c) = (1,1,-8)$.

3. Altogether, $6 \cdot 5/2 = 15$ games were played. Clearly, the winning team could not have won more than 5 games. If this team had won four games (or fewer), then two teams would have to win the same number of games. Hence, the winning team must have won five games.

4. By the law of cosines, $AC^2 = 7^2 + 8^2 - 2 \cdot 7 \cdot 8 \cdot \cos 120°$. $BD^2 = 7^2 + 8^2 - 2 \cdot 7 \cdot 8 \cdot \cos 60°$. The difference is $4 \cdot 7 \cdot 8 \cdot \cos 60° = 112$. Compare 3.84.

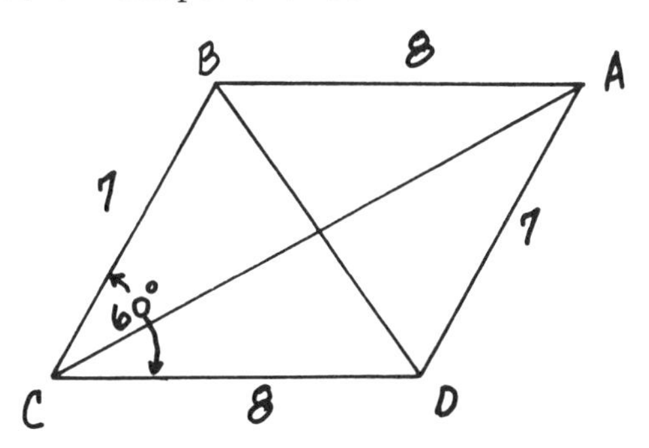

5. By trial and error, $17 = 11 + 3 + 3 = 5 + 5 + 7 = 3 + 7 + 7$. For an interesting discussion of the sense in which the statement about odd integers is "known," see, for example, Sierpinski.

6. Let $x = \log_{1/125} 25 \sqrt[3]{25}$. Then $\left(\frac{1}{125}\right)^x = 25 \cdot 25^{1/3} = 25^{4/3}$,

$5^{-3x} = 5^{8/3}$, so $-3x = \frac{8}{3}$, and $x = -\frac{8}{9}$.

7. $f(x) = Q(x) \cdot [x - a] + R = \frac{1}{3} Q(x) \cdot 3[x - a] + R$. The remainder is not affected (!). Therefore, $(g,h) = \left(\frac{1}{3}, 1\right)$.

8. Let $\overline{AE}$ intersect $\overline{FG}$ at Y, and let $FG = s$, so $EY = s\sqrt{3}/2$. Clearly $AE = 4$. From similar triangles AYG, AEC, we have:

$$\frac{4}{3} = \frac{4 - s\sqrt{3}/2}{s/2} = \frac{8 - s\sqrt{3}}{s}$$

$4s = 24 - 3s\sqrt{3}$, $s = 24/(4 + 3\sqrt{3}) = \frac{-24}{11}(4 - 3\sqrt{3})$, and $(a,b) = (4,-3)$.

8. (Continued)

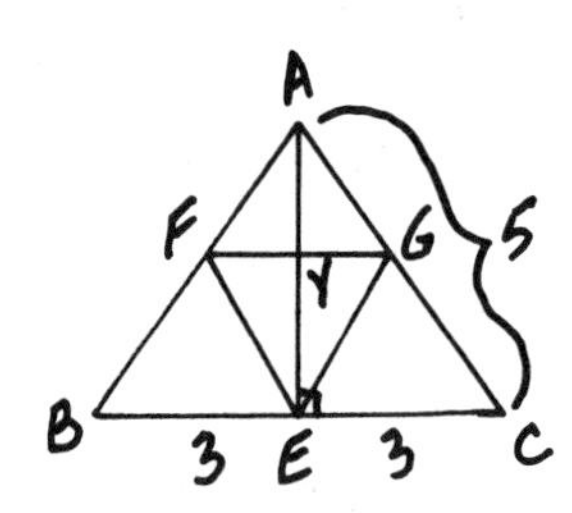

9. $n = (a + 2/b)(b + 2/a) = ab + 4 + 4/ab$. But $ab = q$, so $n = q + 4 + 4/q$, or equivalent.

10. We have $a + (a + 2) + (a + 4) + \ldots + (a + 2n - 2) = 105$, or, summing the arithmetic progression, $(n/2)(2a + 2n - 2) = n(a + n - 1) = 105 = 3 \cdot 5 \cdot 7$. Since both n and $a + n - 1$ are integers, we can have $n = 1, 3, 5, 7, 15, 21, 35, 105$ only. The last four lead to negative values for a. The other values of n lead to $a = 105, 33, 17, 9$, respectively. Hence the smallest possible value for a is 9. (See 1.512.)

11. Let $\log x = A$, $\log y = B$, so $\log xy = A + B$. Then $\log (A + B) = \log A + \log B = \log AB$, or $A + B = AB$, and $B = A/(A - 1)$. For integer solutions, we need $A - 1$ to divide A, so $A = B = 2$, and $(x,y) = (100,100)$.

12. Extend $\overline{AM}$ its own length to D. $AM = \sqrt{70}$, so $AD = 2\sqrt{70}$. Then $ACDB$ is a parallelogram, so (see problem 4, Fall 1980) $AD^2 + BC^2 = 2AB^2 + 2AC^2$. (See 3.84.) If $BC = x$, then $x^2 + 280 = 296$, so $BC = \sqrt{18} = 3\sqrt{2}$.

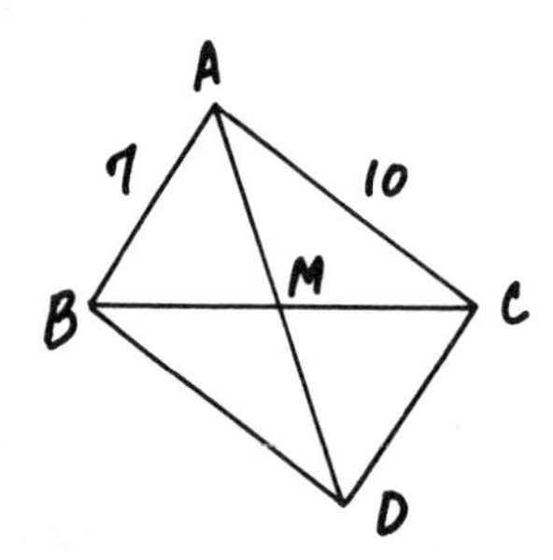

13. Using mass points, we can assign weights of 2, 1, 1 to A, B, C, respectively. Then F has a weight of 3, D of 2, and E of 4. We find $FE/EB = 1/3$ and $AF/FC = 1/2$. The required sum is then 5/6.

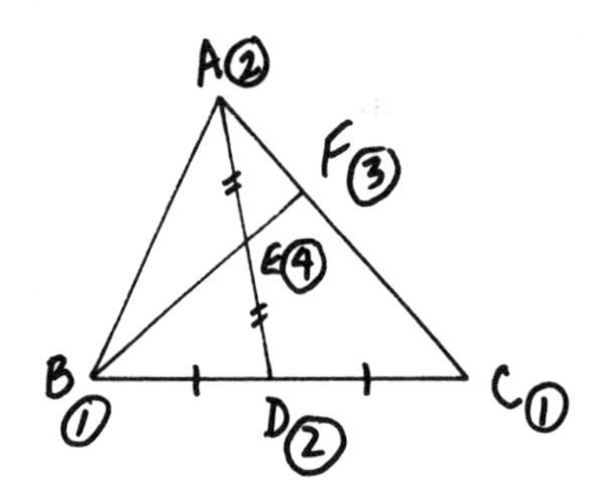

14. The form of the numbers is $2^n \cdot 9^n - 1 = 18^n - 1$. Since the polynomial $x - 1$ always divides the polynomial $x^n - 1$ without remainder, $17 = 18 - 1$ must divide $18^n - 1$ for all n. To see that 17 is the greatest common divisor, take $n = 2$ and $n = 3$. The two numbers $323 = 17 \cdot 19$ and $5831 = 17 \cdot 343$ have no greater common divisor than 17, so 17 must be the greatest common divisor for the entire set (see 1.33).

15. The least common multiple of 3, 4, 5, 6, and 7 is 420. Adding each of these numbers to the common multiple will produce multiples of the number added, so $m = 423$. A smaller value for m would make $m - 3$ a smaller common multiple for the five numbers.

16. Let D be the foot of the altitude from B on AC. Then $AD = c \cos A$ and $DC = a \cos C$. Hence $\dfrac{\cos A}{a} + \dfrac{\cos C}{c} = \dfrac{c \cos A + a \cos C}{ac} = \dfrac{b}{2b} = \dfrac{1}{2}$.

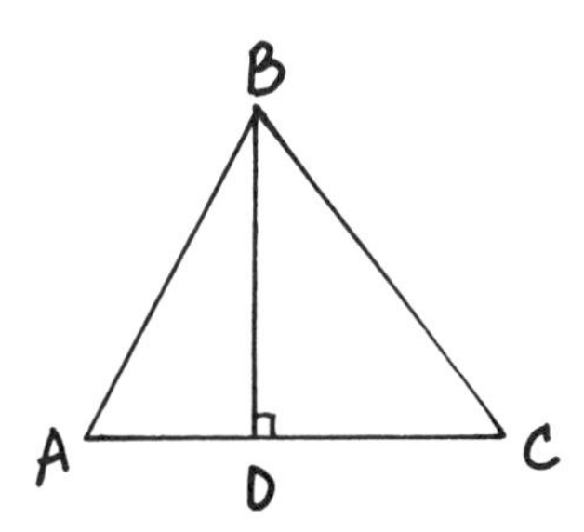

17. If the three angles are $3x$, $4x$, and $5x$, then $3x + 4x + 5x = 180$, and $x = 15$. The required ratio is $\dfrac{\sin 45°}{\sin 60°} = \dfrac{\sqrt{2}}{\sqrt{3}} = \dfrac{\sqrt{6}}{3}$.

18. Let $OB = \frac{1}{2} AB = r$, $CM = \frac{1}{2} CD = s$, and $OM = CT = h$. Then from the given areas, $rh = 150$, $sh = 120$, so $r:s = 5:4$, and triangle OMC is a 3-4-5 right triangle. Let $r = 5x$, $s = 4x$, and $h = 3x$. Then $TB = r - s = x$, and $CB = \sqrt{CT^2 + TB^2} = \sqrt{10x^2} = x\sqrt{10}$. To find x, note that $rh = 15x^2 = 150$, so $x^2 = 10$ and $CB = 10$.

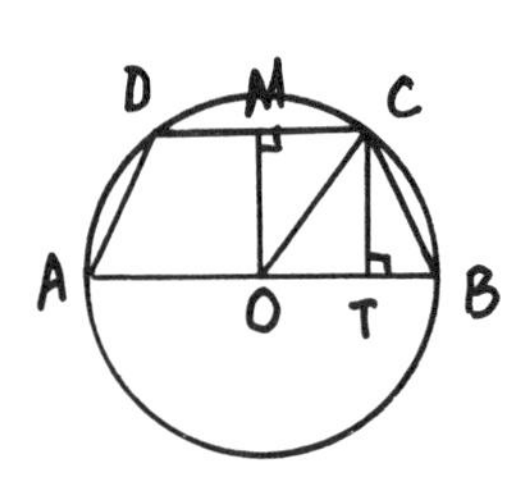

19. We have: $\sin x = \frac{\cos x}{\sin x}$, so $\sin^2 x = 1 - \cos^2 x = \cos x$, or $\cos^2 x + \cos x - 1 = 0$, and $\cos x = \frac{-1 \pm \sqrt{5}}{2}$. But $\frac{-1 - \sqrt{5}}{2} < -1$, so the only possible value for $\cos x$ is $\frac{-1 + \sqrt{5}}{2}$.

20. Suppose the runners meet at point R. Let $PQ = d$ and $QR = a$. Then, if B increases his speed by x MPH, we have the time equations: $\frac{d - a}{9} = \frac{d + a}{12}$ and $\frac{d + a}{9 + x} = \frac{d - a}{12}$. From the first equation, we find that $\frac{d + a}{d - a} = \frac{4}{3}$, and from the second we find that $\frac{d + a}{d - a} = \frac{9 + x}{12}$, so $\frac{9 + x}{12} = \frac{4}{3}$, and $x = 7$.

21. We want $x + y + 2\sqrt{xy} = 10 + \sqrt{q}$, so $x + y = 10$ and $xy = \frac{q}{4}$. Thus we are maximizing the product of two numbers, given that their sum is constant (see 6.3). We would want to make the two numbers equal, but this is contrary to the conditions of the problem. Thus the best we can do is to let $x = 6$, $y = 4$, giving $q = 96$ (letting $x = 4$ and $y = 6$ gives the same answer).

22. Draw $\overline{EX} || \overline{AG}$. Then if $AD = 2a$, $EX = \frac{AD + BC}{2} = \frac{3a}{2}$, and since triangle $EXF \sim$ triangle GDF, $XF:FD = EX:DG = 3:4$, so $CF:FD = 10:4 = 5:2$.

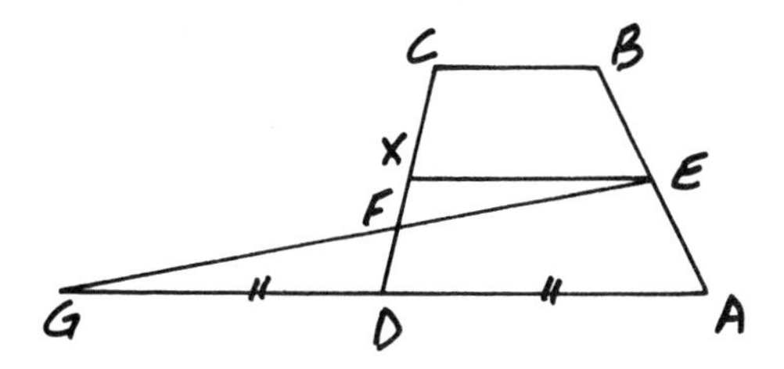

23. $-1 < \log x < 1$ for $.1 < x < 10$, so possible values for x lie within these bounds. A sketch of the graphs of the two functions shows three points of intersection.

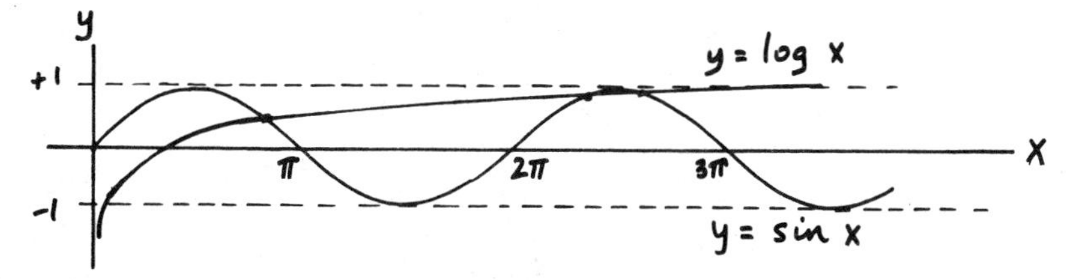

24. The subject is right about the jack 1/3 of the time, about the queen 4/11 of the time, about the king 4/10 of the time, and about all three (1/3)(4/11)(4/10) = 8/165 of the time.

25. Since D is the midpoint of base AB of the isosceles triangle, we have $BD = AD = 13/2$. And since E lies on a circle with diameter $\overline{AB}$, $DE = DB = 13/2$. Segments AC and DE are sometimes said to be antiparallel (see, for instance, Kay).

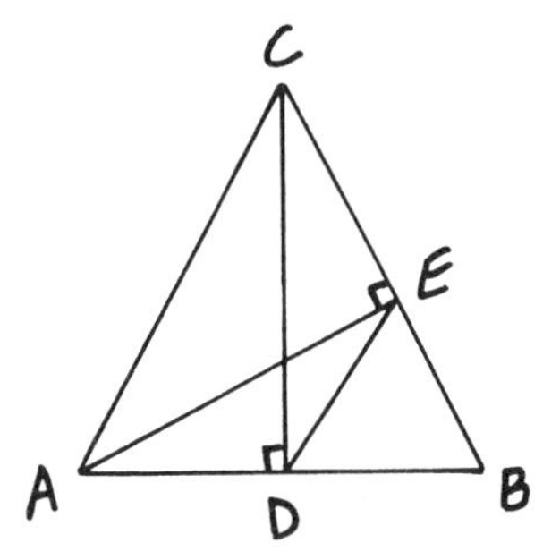

26. Let $x = \sqrt[3]{\sqrt{28} + 6} - \sqrt[3]{\sqrt{28} - 6}$.

Then $x^3 = (\sqrt{28} + 6) - (\sqrt{28} - 6) - 3\sqrt[3]{28 - 36}\left(\sqrt[3]{\sqrt{28} + 6} - \sqrt[3]{\sqrt{28} - 6}\right)$

so $x^3 = 12 - 3\sqrt[3]{-8} \cdot x$. Thus $x^3 = 12 + 6x$, or $x^3 - 6x - 12 = 0$,

so $(a,b,c) = (0,-6,-12)$.

27. $(\cos 3x)(\cos x) + (\sin 3x)(\sin x) = \cos(3x - x) = \cos 2x$, so $\cos 2x = \frac{3\pi}{5}$, and $0 \leq 2x \leq \pi$, so $2x = 3\pi/5$, $x = 3\pi/10$. (See 4.314.)

28. Suppose there are a white balls and b black balls in urn A. Then the probability of picking a white ball from urn A and then picking a white ball from urn B is $(a/8)(1/2) = a/16$. Similarly, the probability of picking a black ball from urn A and then a white ball from urn B is $b/24$. Thus we have $(a/16) + (b/24) = 3/8$, and $a + b = 8$. These two simultaneous equations have solution $(2,6)$, so $a = 2$.

29. $\frac{1}{x + y} = \frac{1}{x} + \frac{1}{y}$, so $(x + y)^2 = xy$ and $x^2 + xy + y^2 = 0$. Thus $\left(\frac{x}{y}\right)^2 + \frac{x}{y} + 1 = 0$, and $\frac{x}{y} = \frac{-1 \pm \sqrt{1 - 4}}{2} = \frac{-1 \pm i\sqrt{3}}{2}$. Note that

$\frac{x}{y} = \omega$ or ω^2, the complex cube roots of 1.

30. Let $S = \sin 10° \sin 50° \sin 70° = \cos 80° \cos 40° \cos 20°$. Then

$$\begin{aligned} 8S(\sin 20°) &= 8 \sin 20° \cos 20° \cos 40° \cos 80° \\ &= 4 \sin 40° \cos 40° \cos 80° \\ &= 2 \sin 80° \cos 80° \qquad \text{(using 4.21)} \\ &= \sin 160° = \sin 20°, \text{ so } S = 1/8. \end{aligned}$$

Solutions—Spring 1981

1. $a = rb$, so $\dfrac{a+b}{a-b} = \dfrac{rb+b}{rb-b} = \dfrac{r+1}{r-1}$.

2. If the area of triangle ABC is 1 and the area of triangle CDE is A, we have:

$$\frac{A}{1-A} = \frac{1-A}{1}$$

which leads to $A^2 - 3A + 1 = 0$, whence $A = \dfrac{3 \pm \sqrt{5}}{2}$. Since the ratio is clearly less than 1, only the negative sign is valid.

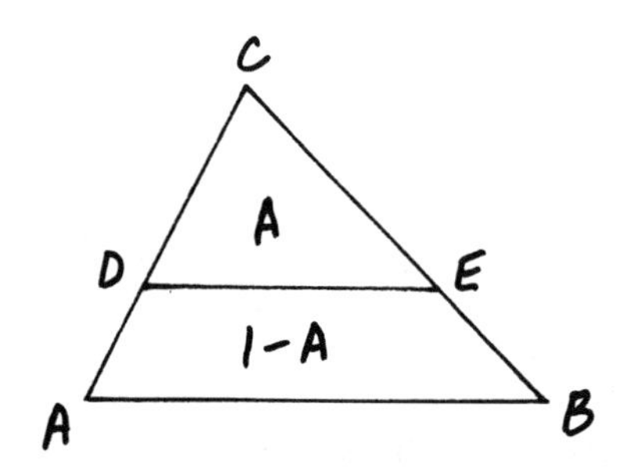

3. The discriminant of the equation is $p^2 - 4q$, and this must be positive or 0. Thus $p^2 - 4q = 1 - 4q \geq 0$, or $q \leq \frac{1}{4}$, so q must -1, and p can be either +1 or -1.

4. The diagram shows the large circle and one smaller circle. In the right triangle shown, $\dfrac{r}{r+1} = \sin\dfrac{\pi}{7} = a$, so that $r = \dfrac{a}{1-a}$.

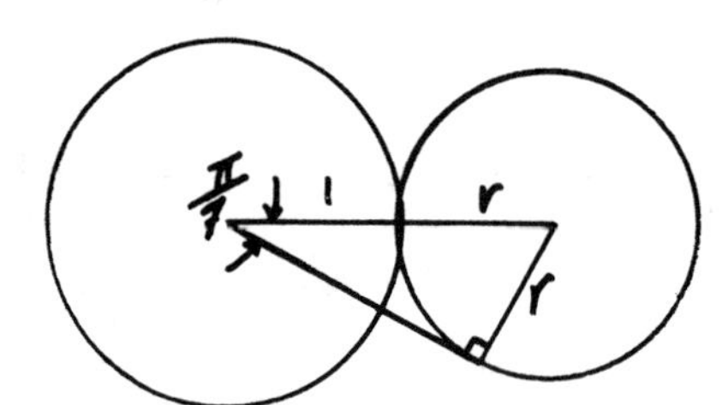

5. We have: $\left(a + \dfrac{1}{a}\right)^2 = 3^2$, so $a^2 + \dfrac{1}{a^2} + 2 = 9$, and $a^2 + \dfrac{1}{a^2} = 7$.

Hence $\left(a - \dfrac{1}{a}\right)^2 = a^2 + \dfrac{1}{a^2} - 2 = 7 - 2 = 5$, and $\left|a - \dfrac{1}{a}\right| = \sqrt{5}$.

6. Note that if the number of heads is equal to the number of tails, the man will have returned to his starting point. And if the number of heads does not equal the number of tails, their difference must be even (since the numbers add up to 6), and thus must be at least 2. Hence we need only calculate the probability of <u>not</u> getting 3 heads and 3 tails. This is

$$1 - P(3H,3T) = 1 - \binom{6}{3}(1/64) = 44/64 = 11/16.$$

7. If each student who passed saw only ten out of eleven failures, and each student who failed saw every failing grade but his own, there could have been exactly eleven failures. There could not be fewer, since then a failing student could not see more than nine failing grades.

8. The diagram shows the circumscribed circle. The inscribed circle has radius $OM = 1$. It is easy to see that angle $MOA = \pi/7$. Also, since O, the center of the circle, is on the bisector of angle RAQ (by symmetry), we see that angle $OAP = (1/2)$ angle RAQ $= \pi/14$. Hence: $x = OA \sin \pi/14$

$$= \frac{1}{\cos \pi/7} \cdot \sin \pi/14$$

$$= \frac{\sin \pi/14}{\cos 2\pi/14} = \frac{b}{1 - 2b^2} \quad \text{(using 4.22).}$$

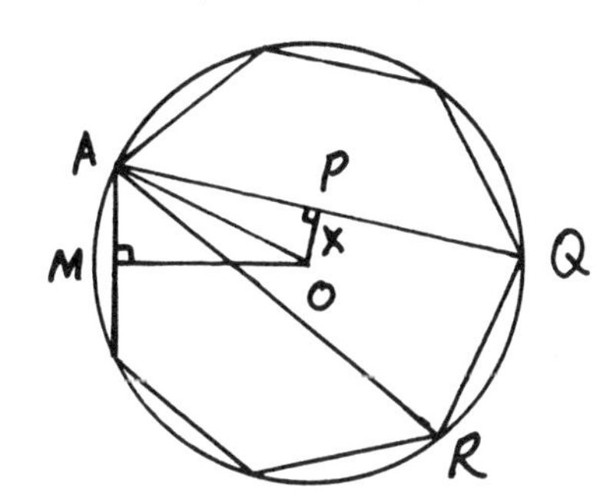

9. We have $x^r = \dfrac{A}{1 + A}$, $x^s = \dfrac{B}{1 + B}$ and $x^{r+s} = x^r \cdot x^s$, so

$$C = \frac{\dfrac{A}{(1 + A)} \cdot \dfrac{B}{(1 + B)}}{1 - \dfrac{AB}{(1 + A)(1 + B)}} = \frac{AB}{1 + A + B}.$$

10. If I is Irwin's age and G is Gilbert's, we have $I - G = 8$ and $I + 1 = nG$. Hence $nG - 1 - G = (n - 1)G - 1 = 8$, or $9 = (n - 1)G$. This means that G is a factor of 9 and can only equal 1, 3, or 9.

11. $(a + bi)^2 = b + ai$, so $a^2 - b^2 + 2abi = b + ai$, $a^2 - b^2 = b$ and $2ab = a$. From the second equation, $b = 1/2$, and from the first, $a = \sqrt{3}/2$. Thinking of the complex numbers in polar form helps to motivate this solution.

12. The four points determine in all six line segments. If two of their midpoints coincide, the endpoints of their segments form a parallelogram, and none of the other midpoints can coincide (some of them lie on parallel lines, while others are midpoints of segments with a common endpoint; if these last coincided, the other endpoints of the segments would also coincide). The minimum number of midpoints is thus five.

13. $c^2 = x^2 + 2xy + y^2 = (x^2 - 2xy + y^2) + 4xy = a^2 + 4b$, and $c > 0$, so $c = \sqrt{a^2 + 4b}$.

14. In m minutes, the minute hand will move $6m°$ around the face of the clock, while the hour hand will move $(1/2)m°$. Thus, at 12:10, the hands make an angle of $60° - 5° = 55°$. They will make this angle again shortly before 1:00. If this time is p minutes after 12:10, we have:

$$5 + \frac{1}{2}p + (300 - 6p) = 55$$ (since this time the hour hand is clockwise ahead of the minute hand).

This leads to $p = 500/11 = 45\frac{5}{11}$.

15. Compare Problem 12, above. The four points determine six line segments. For any selection of three or more of these segments, at least two segments will have a common endpoint. Hence no three (or more) midpoints can coincide. Let us next see how many pairs of midpoints can coincide.
Call the points, from left to right, A_1, A_2, A_3, and A_4. The only possible pairs of segments without common endpoints are A_1A_2 and A_3A_4, A_1A_3 and A_2A_4, and A_1A_4 and A_2A_3. The first pair is disjoint, so they cannot share a midpoint. Suppose M were the common midpoint of A_1A_3 and A_2A_4. Then M must be between A_2 and A_3, on the overlap of the two segments. If $A_2M < MA_3$, then $MA_4 = A_2M < MA_3$, and A_4 is to the left of A_3, contrary to assumption. If $A_2M > MA_3$, then $A_1M = MA_3 < A_2M$, and A_1 is to the right of A_2, contrary to assumption. If $A_2M = MA_3$, then $MA_4 = MA_3 = A_2M$, and two of the three points A_2, A_3, and A_4 must coincide.
Hence only A_1A_4 and A_2A_3 can share a midpoint. This will happen, for example, if the four points are equally spaced, and five is the minimal number of midpoints. How can this be generalized?

16. The existence of a block of length 2 implies the existence of a 1 before or after the block (or in both places). If the block occurs first, we have 001xx, which can happen four ways. If it appears in the next two places, we have 1001x, or two possibilities. If it appears in the third and fourth place, we have x1001, for two more possibilities. If it occurs last, we have xx100, for four more possibilities. But we have counted twice any arrangement which has two blocks of 2: the only one in this case is 00100. Thus there are 11 possibilities altogether. A general solution to this complicated sort of problem is given in Feller's Probability.

17. $\frac{5}{4+5} + \frac{2}{5+2} + \frac{1}{8+1} + \frac{1}{x+1} = 1$, so $x = 20$.

18. Let the vertices of the triangle be at (0,0), (4,0), and (0,3). If (x,y) is the point where the minimum occurs, we must have a minimal value for $x^2 + y^2 + (x - 4)^2 + y^2 + x^2 + (y - 3)^2 = 3x^2 - 8x + 16 + 3y^2 - 6y + 9$. We can minimize each quadratic separately, hence their sum, by taking $x = 4/3$, $y = 1$ (think of the formula for the turning point of a parabola). This gives 50/3 for the minimum value.
Note: It can be shown, by generalizing the above, that the minimum value always occurs at the centroid of the triangle. It is interesting to note (and challenging to prove) that if X is any point in the plane of triangle ABC and G is the centroid of triangle ABC, then $XA^2 + XB^2 + XC^2 - 3XG^2 = \frac{a^2 + b^2 + c^2}{3}$.

19. Let $y = \sqrt[4]{x}$, so that $y^2 = \sqrt{x}$. Then $y^2 + y = 12$, so $y = 3, -4$. The latter is extraneous, so $\sqrt[4]{x} = 3$ and $x = 81$.

20. Using Ptolemy's theorem, we have (3.91) $7x + 7y = 7 \cdot 8$, so $x + y = 8$. Also, angle $APC = 120°$, so in triangle APC, the Law of Cosines gives $x^2 + y^2 + xy = 49$. These two equations lead to $xy = 15$, so $x = 3$ and $y = 5$.

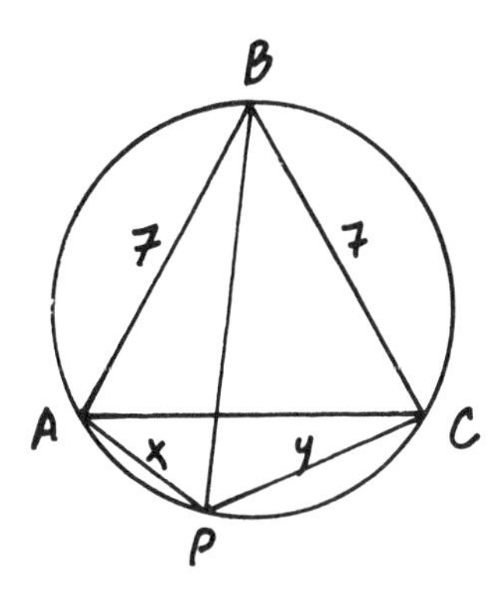

21. One could use Hero's formula (3.61), but a glance at the diagram will show why this particular triangle was used. The shortest altitude is drawn to the longest side (can you prove this?), so its length is 15.

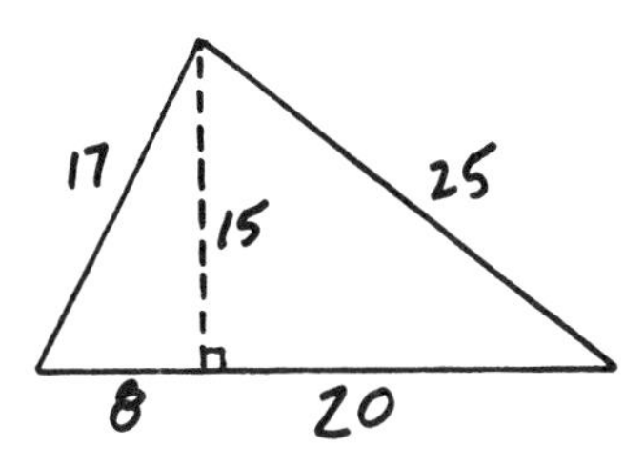

22. The probability that the cables don't all break in one hour is $1 - (2/3)^m$, where m is the number of cables. So $1 - (2/3)^m \geq 4/5$, or $(2/3)^m \leq 1/5$, and $m \geq 4$. At least 4 cables are needed.

23. There are many proofs of this result. A very simple one uses mass points, assigning weights of 1 to each vertex of the tetrahedron (see Appendix B). A more straightforward proof follows.

    Let P be the midpoint of edge $\overline{CD}$ of regular tetrahedron $ABCD$.

    Then $\overline{AP}$, $\overline{BP}$ are medians in triangles ADC, BCD respectively. Let X and Y be the respective centroids of these two triangles. Then $AX:XP = BY:YP = 2:1$, and A, B, X, Y, and P are all coplanar. Hence $AB||XY$. If AY and BX intersect at M, then M is on the plane through A, B, and P, triangle $ABM \sim$ triangle YXM, and $AM:MY = AB:XY = BM:MX = 3:1$. Also, triangle $APB \sim$ triangle XPY, so $AB:XY = AP:XP = 3:1$.
    Note that this proof, as well as the mass points proof, holds for any tetrahedron, and not just a regular one. In a regular tetrahedron, it shows that the altitudes as well as the medians are coincident (since the medians are perpendicular to the sides to which they are drawn). For more tetrahedron problems, see Aref and Wernick (Chapter 8).

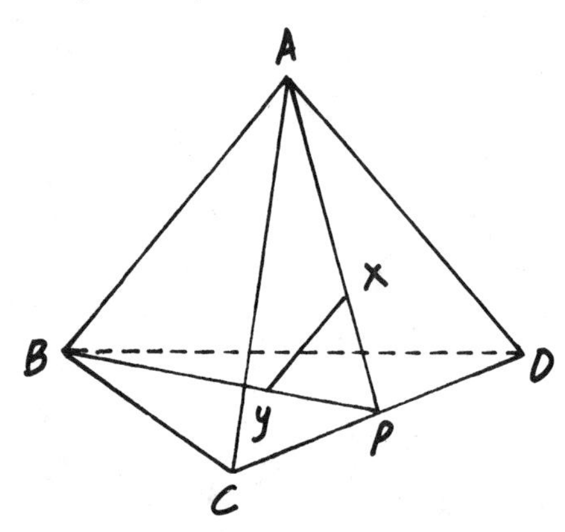

24. The equation is $3MN = 4(M - 2)(N - 2)$, since the unpainted region is a block whose dimensions are 4, $(M - 2)$, and $(N - 2)$. Rearranging the equation, we have $3MN = 4MN - 8(M + N) + 16$, or $(M - 8)(N - 8) = 48 = 1 \cdot 48 = 2 \cdot 24 = 4 \cdot 12 = 6 \cdot 8$. The possible values for (M,N) are (56,9), (32,10), (24,11), (20,12), and (16,14). The smallest N is 9.

25. If John's age is J and Mary's age is M, then $J = rM$. We want the ratio $M:[M - (J - M)] = \frac{M}{2M - J} = \frac{M}{2M - rM} = \frac{1}{2 - r}$. Note that if $r < 1$ or $r > 2$ there would be no time such as that referred to by the problem.

26. AC is a diameter of the circle, so $AC = 5\sqrt{2}$. By the Pythagorean theorem, $PC^2 = AC^2 - AP^2 = 34$. Now we can use (3.91) Ptolemy's theorem in quadrilateral $PADC$. $4 \cdot 5 + 5\sqrt{34} = (PD)5\sqrt{2}$, so $PD = \dfrac{20 + 5\sqrt{34}}{5\sqrt{2}} = \sqrt{17} + 2\sqrt{2}$.

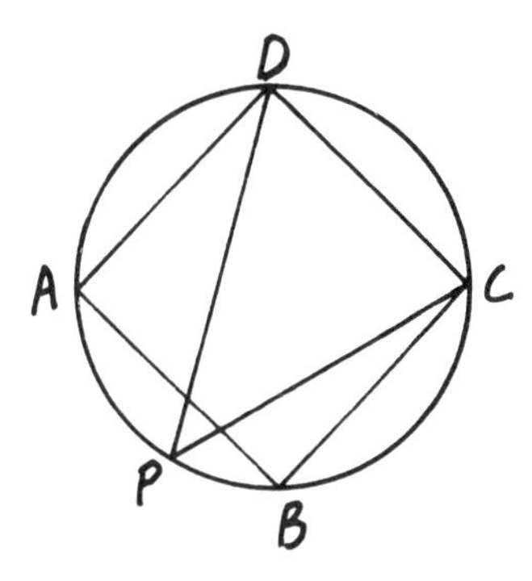

27. It is not hard to see that all such polynomials are given by $f(x) = x^k$ for k an integer greater than 3. To prove this, note first that the lead coefficient of $f(x)$ must be 1, and suppose that $f(x)$ included other terms as well. Then we can write $f(x) = x^n + q(x)$, where $q(x) \neq 0$, and if the degree of q is m and that of f is n, we have $m < n$.
Hence $x^{2n} + q(x^2) = x^{2n} + 2x^n q(x) + [q(x)]^2$, or $q(x^2) = q(x)(2x^n + q(x))$. Comparing the degrees of both sides, we find $2m = m + n$, or $m = n$, which is impossible.
Note: If we drop the condition that degree $(f) > 3$, the only new solutions would be $f(x) = 0, 1, x, x^2, x^3$.

28. The centers of the four spheres form a regular tetrahedron of edge 2. The circumcenter of this tetrahedron divides each median, which is also an altitude, in the ratio 1:3 (see problem 23, Spring 1981). If x is a length of an altitude,

$$x^2 + \frac{1}{3} = 3, \text{ so } x = \sqrt{8/3}.$$

The radius of the large sphere is equal to the radius of the sphere circumscribed about the tetrahedron plus the radius of one smaller sphere, or $1 + \frac{3}{4}\sqrt{8/3} = \frac{2 + \sqrt{6}}{2}$. See also problem 10, Fall 1976, or Coxeter.

28. (Continued)

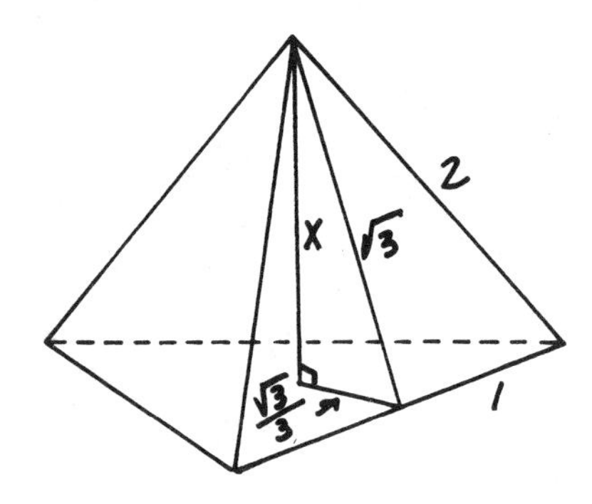

29. To get to (3,1), the particle must move three times along the x-direction and once in the y-direction. The probability of this event is $4p^3(1 - p)$. Similarly, the probability of passing through (1,3) is $4p(1 - p)^3$. Thus $16 \cdot 4p(1 - p)^3 = 4p^3(1 - p)$, or $16(1 - p)^2 = p^2$. This leads to $4 - 4p = p$ and $p = 4/5$.

30. By dropping altitudes from C and D and using the 30-60-90 triangles and the rectangle that result, it follows that $AB = 8$. From similar triangles ECD, EAB, $DE:EB = CE:EA = 1:4$. Thus $BG:EB = AF:EA = 1:4$, so $EG:EB = EF:EA = 3:4$, and the triangles EFG, EAB are similar. Finally, $FG:AB = x:8 = EG:EB = 3:4$, so $x = 6$.

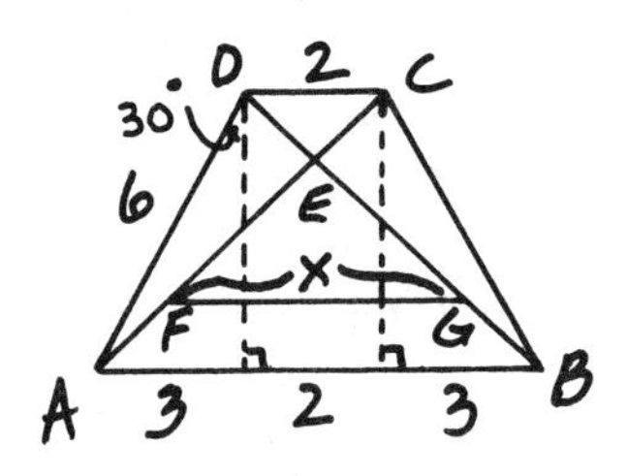

Solutions—Fall 1981

1. If one Munchie bar costs M¢, the students had $M - 7$ and $M - 2$ cents. Together they had $2M - 9$ cents, and $2M - 9 < M$, so $M < 9$. If $M < 8$, the first student could not have needed all 7¢, so $M = 8$.

2. $$y = 2^x + \frac{1}{2^x} = \left(\sqrt{2^x} - \frac{1}{\sqrt{2^x}}\right)^2 + 2 \text{ and since } \left(\sqrt{2^x} - \frac{1}{\sqrt{2^x}}\right)^2 \geq 0$$

then $y \geq 2$. Since $y = 2$ when $x = 0$, we have that the smallest value of y is 2 (or use 6.1 and note that $2^0 = 1$).

3. We can inscribe the regular pentagon in a circle. Then arc $\overset{\frown}{BA} = \overset{\frown}{DE}$ $= \frac{360°}{5} = 72°$. Angle BXA is the average of these two arcs, which is also 72°.

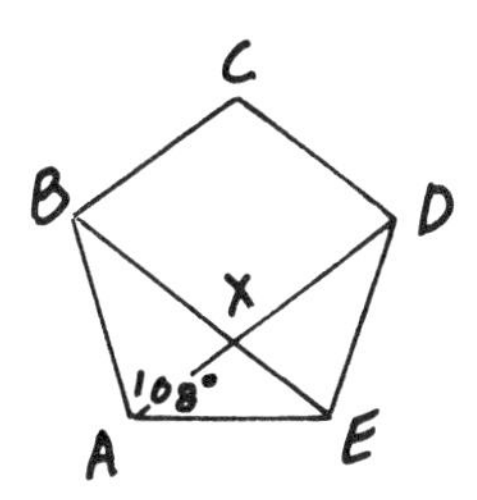

4. $x^6 - 64 = (x^3)^2 - 8^2 = (x^3 + 8)(x^3 - 8)$

$= (x - 2)(x^2 + 2x + 4)(x + 2)(x^2 - 2x + 4)$

(See 1.321, 1.322.)

5. Method 1: The two "pieces" of the original parallelogram are two trapezoids with equal altitudes. Hence their areas will be equal if the sums of their bases are equal. Since $AB = CD$ (see figure), we have simultaneously: $x + y = q + w$

$x + q = y + w$

Subtracting, we find $y - q = q - y$, so $y = q$.
Adding, we find $2x + y + q = 2w + y + q$, so $x = w$.
Since $AE = FC = 3$, $DF = 5$.

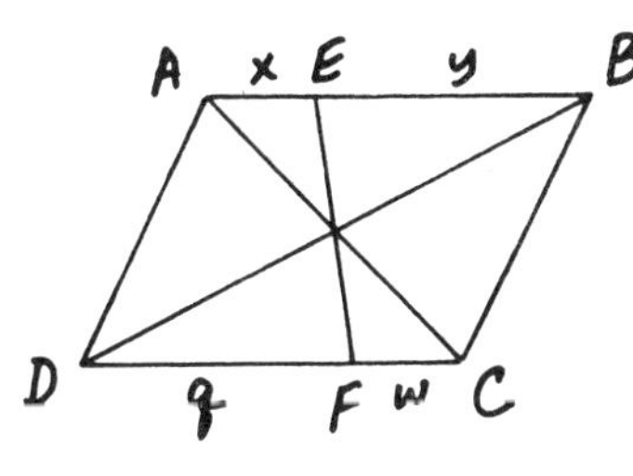

Method 1

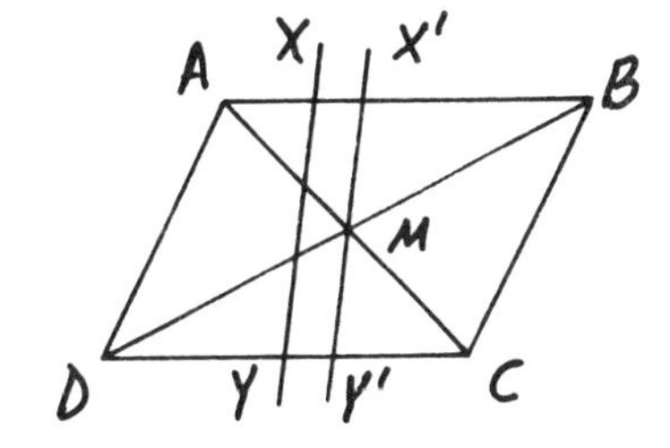

Method 2

Method 2: Line EF must pass through the center of the parallelogram, so $DF = EB = 5$.
If it's not clear to you right away that EF must contain the center symmetry of the parallelogram and you think that line XY, for instance, might do the job, draw $X'Y' || XY$ through the center M. It's easy to prove that area($AX'Y'D$) = area($X'BCY'$) = $\frac{1}{2}$ area ($ABCD$). But area($AXYD$) is clearly less than area($AX'Y'D$).

6. $xf(x) + f(1 - x) = x^2 + 2$. Replacing x by $1 - x$ we get $(1 - x)f(1 - x) + f(x) = x^2 - 2x + 3$. We solve these two simultaneous equations for the unknown $f(x)$:

$$x(1 - x)f(x) + (1 - x)f(1 - x) = (x^2 + 2)(1 - x)$$
$$f(x) + (1 - x)f(1 - x) = x^2 - 2x + 3.$$

6. (Continued)

Subtracting:

$$[-x^2 + x - 1]f(x) = -x^3 - 1 \text{ and}$$

$$f(x) = \frac{x^3 + 1}{x^2 - x + 1} = x + 1. \quad \text{(See 1.321.)}$$

7. In any right triangle ABC, if M is the midpoint of hypotenuse AB, then $CM = AM = BM$. Therefore, angle MCB = angle ABC = 30° and triangle BXC is a 30-60-90 triangle, so $BX = \frac{1}{2} BC = \frac{1}{2} \cdot \frac{\sqrt{3}}{2} \cdot AB$ $= 3\sqrt{3}$.

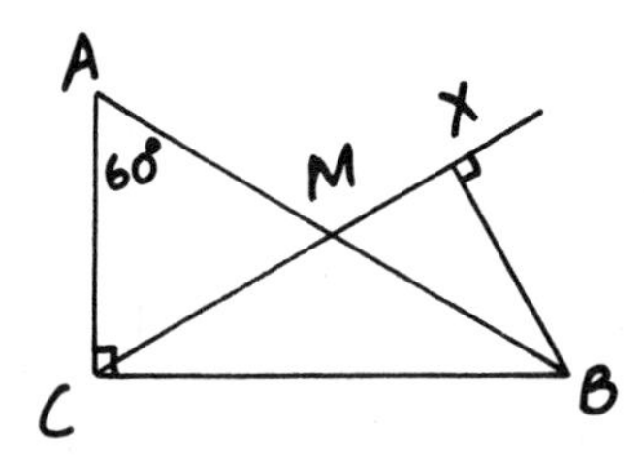

8. $2 \cdot 4^x + 6^x = 9^x$

$2^{2x+1} + 2^x \cdot 3^x - 3^{2x} = 0$

$(2^{x+1} - 3^x)(2^x + 3^x) = 0$

$2^{x+1} = 3^x$ $\quad$ ($2^x = -3^x$ is not possible)

$\left(\frac{2}{3}\right)^x = \frac{1}{2}$ $\quad$ so $x = \log_{\frac{2}{3}} \frac{1}{2}$ and $a = \frac{1}{2}$

9. $1000027 = (100^3 + 3^3) = (100 + 3)(10000 - 300 + 9) = (103)(9709)$(see 1.321)

$= 103 \cdot 7 \cdot 1387$ (and 1387 has no prime factor between 100 and 1000)

10. The sum is an infinite geometric progression whose limit is:

$$\frac{a}{1 - r} = \frac{1/10}{1 - 1/10} = \frac{1/10}{9/10} = \frac{1}{9} \text{ gram.} \quad \text{(See 1.523.)}$$

Alternatively, the weight of the sack can be expressed in decimal notation as $.1111\ldots = \frac{1}{9}$ gram.

11. If we represent the numbers as $(n - 5)$, $(n - 4), \ldots, (n - 1)$, n, $(n + 1)$, $(n + 2), \ldots, (n + 5)$, $(n + 6)$, then their sum is $12n + 6$ $= 4(3n + 1) + 2$, so the remainder is always 2.

12. Any line bisecting the common area must pass through the midpoint of the centerline, which is (2,4). Since (2,4) satisfies $14x + 3y = k$, we have $14 \cdot 2 + 3 \cdot 4 = k$, and $k = 40$.

13. $$AB^2 = \frac{1}{\cos^2 15} + \frac{1}{\sin^2 15} = \frac{\sin^2 15 + \cos^2 15}{\cos^2 15 \sin^2 15} = \frac{1}{\cos^2 15 \sin^2 15}$$

$$= \frac{4}{4(\sin^2 15)(\cos^2 15)} = \frac{4}{\sin^2 30} = \frac{4}{\frac{1}{4}} = 16,$$ so $AB = 4$. (See 4.11, 4.21.)

14. Since only one number is chosen, the required probability is the sum of the probabilities of picking 3,6,9,... .

This is simply $\sum_{k=1}^{\infty} \frac{1}{2^{3k}}$, a geometric progression whose sum is

$$\frac{\frac{1}{8}}{1 - \frac{1}{8}} = \frac{1}{7}.$$ (See 1.523.)

15. Set $(a + bi)^2 = a^2 - b^2 + 2abi = -3 + 4i$. Then $a^2 - b^2 = -3$ and $ab = 2$. A good guess gives the answer, or substitution shows that $a^4 + 3a^2 - 2 = (a^2 + 4)(a^2 - 1) = 0$. Since a is real, $a = \pm 1$. Taking $a = +1$ leaves $b = 2$.

16. By the law of cosines, $x^2 = 75 + 4 - 4 \cdot 5\sqrt{3} \cdot \frac{\sqrt{3}}{2} = 49$ so $BC = 7$.

Then by the extended Law of Sines (3.4), $2R = \frac{BC}{\sin A} = \frac{7}{\sin 30°}$,

so $R = 7$. Without using the extended law of sines directly, we can draw diameter COA'. Then angle A' = angle $A = 30°$ and angle $A'BC = 90°$, so $A'C = 2R = 2BC$ and again $R = 7$.

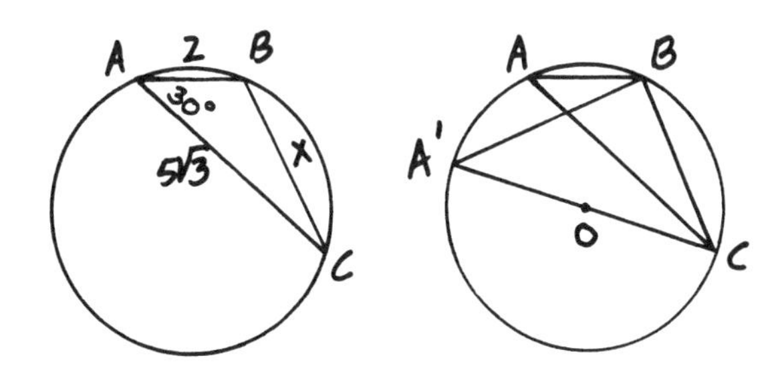

17. Use 4.4, or:

If $a = \text{Arctan}\,\frac{3}{5}$, $b = \text{Arctan}\,\frac{3}{4}$

then $x = \tan(a + b) = \dfrac{\tan a + \tan b}{1 - (\tan a)(\tan b)} = \dfrac{\frac{3}{5} + \frac{3}{4}}{1 - \frac{9}{20}} = \dfrac{27}{11}$.

(See 4.341.)

18. The given number is slightly less than 10^{12} and is in fact $10^{12} - 4096 = 10^{12} - 2^{12}$. This factors as $2^{12} \cdot (5^{12} - 1)$ and

$$5^{12} - 1 = (5^3 - 1)(5^3 + 1)(5^2 + 1)(5^4 - 5^2 + 1)$$
$$= (5 - 1)(5^2 + 5 + 1)(5 + 1)(5^2 - 5 + 1)(5^2 + 1)(5^4 - 5^2 + 1)$$
$$= 2^2 \cdot 31 \cdot 2 \cdot 3 \cdot 3 \cdot 7 \cdot 2 \cdot 13 \cdot 601,$$

thus the given number $= 2^{16} \cdot 3^2 \cdot 7 \cdot 13 \cdot 31 \cdot 601$. (See 1.321 and 1.322.)

19. Since $AC \| WZ$, we know that triangle $DPZ \sim$ triangle DMC and both are isosceles. Hence $DP = PZ$. In the same way, we find $BQ = QY$. Since $ZY = PQ$, $PZ + ZY + YQ = DP + PQ + QB = DB$. Thus the perimeter of the parallelogram is equal to the sum of the diagonals of the rectangle, or 20.

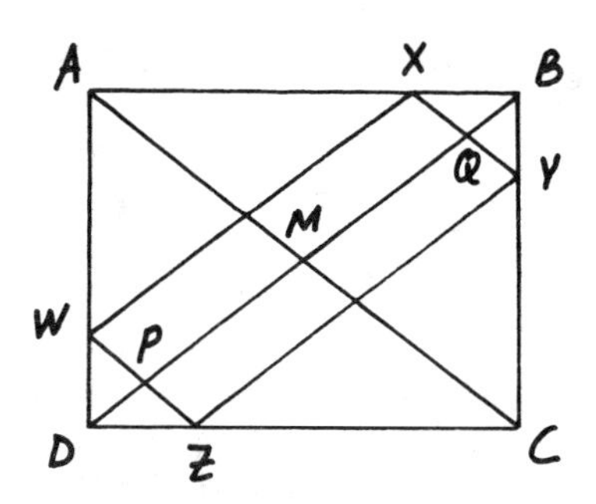

20. There are 6^5 possible outcomes for this experiment. There are 6 favorable outcomes: 12345, 12346, 12356, 12456, 13456, 23456. Hence the required probability is $\frac{6}{6^5} = \frac{1}{6^4} = \frac{1}{1296}$.

21. The number must be self-descriptive, in the sense that it counts the number of letters in its own name. For larger numbers, the English names are too short. Hence, the number must be small. By trial and error, it is not hard to see that only "FOUR" fits its own description. For another example of self-reference, read this sentence.

22. $\sin A = \frac{4}{5}$, so $\tan A = \frac{4}{3}$; $\cos B = \frac{5}{13}$, so $\tan B = \frac{12}{5}$.

Now $\tan C = \tan[180 - (A + B)] = -\tan(A + B) = -\frac{\tan A + \tan B}{1 - \tan A \tan B}$

$$= -\frac{\frac{4}{3} + \frac{12}{5}}{1 - \frac{48}{15}}$$

$$= \frac{56}{33} \qquad \text{(see 4.341)}$$

23. It is not hard to see that neither $\sqrt{x + 1}$ nor $\sqrt{x + 34}$ can be irrational, and in fact both must be integers.
Set $x + 1 = m^2$
$x + 34 = n^2$.
Then $n^2 - m^2 = 33$ or $(n + m)(n - m) = 33 = 1 \cdot 33 = 3 \cdot 11$.
Equating factors gives $(m,n) = (16,17)$ or $(4,7)$, leading to $x = 15$ or 255.

24. Let $N = 20 + t$. Then $M = (20 + t)^2 + 20 + t - 16$
$= t^2 + 41t + 404$ is a multiple of 101
Therefore $t(t + 41)$ must be a multiple of 101.
Since 101 is prime, either $t = 101$ or $t + 41 = 101$ and $t = 60$.
Hence $N = 80$.

25. If x grams of gold were cut off, then $\frac{6 - x}{6} = \frac{x}{12}$, $6x = 72 - 12x$, so $x = 4$.

26. If the triangle is ABC, with angle $C = 90°$, then by 3.31, $AC:CB = 2:5$. If the altitude is CD, we need the ratio $AD:BD$. From similar triangles, $\frac{AD}{CD} = \frac{AC}{CB} = \frac{2}{5}$ and $\frac{CD}{BD} = \frac{AC}{BC} = \frac{2}{5}$. Multiplying these two equations, we find $\frac{AD}{CD} \cdot \frac{CD}{BD} = \frac{AD}{BD} = \frac{2}{5} \cdot \frac{2}{5} = \frac{4}{25}$.

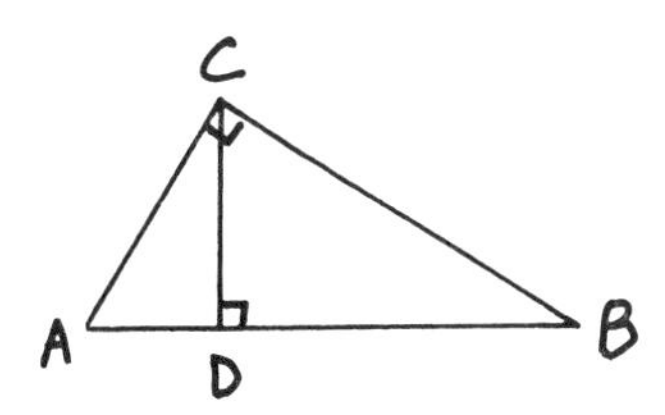

27. The first few triangular numbers are 1,3,6,10,15,21,28,36,45,... It is not hard to guess (and to prove by induction) that two out of every three triangular numbers are multiples of 3. These are the elements of the sequence with subscripts of the form $3k - 1$ or $3k$. Since $n = 100$ is not of this form, there are $\frac{2}{3} \cdot 99 = 66$ multiples of 3 in the set described.

28. We have $x = \frac{n}{n - 10}$, so $n - 10$ must divide n. If $n - 10$ divides n, it also divides $n - (n - 10) = 10$. Hence $n - 10 = \pm 1, \pm 2, \pm 5, \pm 10$, and $n = 11, 12, 15, 20, 9, 8, 5, 0$. Now if $n < 10$, then $n - 10 < 0$ and $x < 0$. Hence possible values of x are 11, 6, 3, 2. The largest is 11.

29. We have: $\sin\theta + 2\sin\theta\cos\theta = \cos\theta + 1 - 2\sin^2\theta$

$$2\sin^2\theta + \sin\theta - 1 = \cos\theta - 2\sin\theta\cos\theta$$

$$(2\sin\theta - 1)(\sin\theta + 1) = \cos\theta(1 - 2\sin\theta)$$

If $2\sin\theta - 1 \neq 0$, this means that $\sin\theta + \cos\theta = -1$.

Squaring, $\sin^2\theta + \cos^2\theta + 2\sin\theta\cos\theta = 1 + \sin 2\theta = 1$ and

$\sin 2\theta = 0$. No acute angles satisfy this. (See 4.21 and 4.22.)

If $2\sin\theta - 1 = 0$, $\sin\theta = \frac{1}{2}$ and $\theta = 30°$.

30. First note that $BC^2 = AB^2 + AC^2 - 2 \cdot AB \cdot AC \cdot \cos A = 25 + 49 - 35 = 39$. Also, since angle ANX = angle $AKX = 90°$, quadrilateral $ANXK$ is cyclic: that is, X is on the circle circumscribing triangle ANK. If the radius of this circle is t, then by the extended Law of Sines, $\frac{NK}{\sin A} = 2t$, or $NK = t\sqrt{3}$. Hence NK is maximal when t is maximal. But, since $\overline{AX}$ is a diameter of the circle circumscribing triangle ANK and is also a chord of the circle circumscribing triangle ABC, its largest value occurs when $\overline{AX}$ is a diameter of the latter circle. In this case, $NK = BC = \sqrt{39}$. (See 3.92, 3.4.)

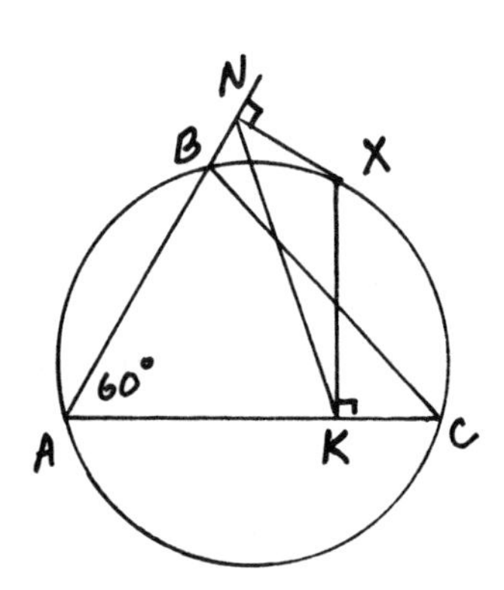

Solutions—Spring 1982

1. In base ten notation, squares of integers can end only in 0, 1, 4, 5, 6, or 9. If the difference of two squares ends in the digit 2, the two units digits of the squares themselves must be 6 and 4 or 1 and 9 (with a "carry" necessary to subtract). In either case, the units digit of the sum of the squares is 0.

2. $3087 = 3^2 \cdot 7^3$, so we need the sum $1 \cdot 3^0 + 1 \cdot 3^1 + 1 \cdot 3^2 + 7 \cdot 3^0 + 7 \cdot 3^1 + 7 \cdot 3^2 + 7^2 \cdot 3^0 + 7^2 \cdot 3^1 + 7^2 \cdot 3^2 + 7^3 \cdot 3^0 + 7^3 \cdot 3^1 + 7^3 \cdot 3^2$. Factoring, we find this is $(1 + 3 + 3^2)(1 + 7 + 7^2 + 7^3) = 13 \cdot 400 = 5200$. For a full discussion, see Dudley (or any book on number theory).

3. The sum of a set of integers is equal to their average times the number of integers. If M is the average of John's previous tests and N is the number of previous tests, we have
$$90(N + 1) = MN + 97$$
$$87(N + 1) = MN + 73.$$
Subtracting, $3(N + 1) = 3N + 3 = 24$ and $N = 7$.

4. Using absolute value to denote area, we have:

$$\frac{|\text{triangle } XBY|}{|\text{triangle } ABC|} = \frac{\frac{1}{2}XB \cdot BY \sin B}{\frac{1}{2}AB \cdot BC \sin B} = \frac{4}{5} \cdot \frac{1}{5} = \frac{4}{25} = \frac{|ZYC|}{|ABC|} = \frac{|AXZ|}{|ABC|}.$$

(See 3.64.) Hence $|\text{triangle } AXZ| + |\text{triangle } BXY| + |\text{triangle } CYZ| = \frac{12}{25}|\text{triangle } ABC|$, so $|\text{triangle } XYZ| = \frac{13}{25}|\text{triangle } ABC|$.

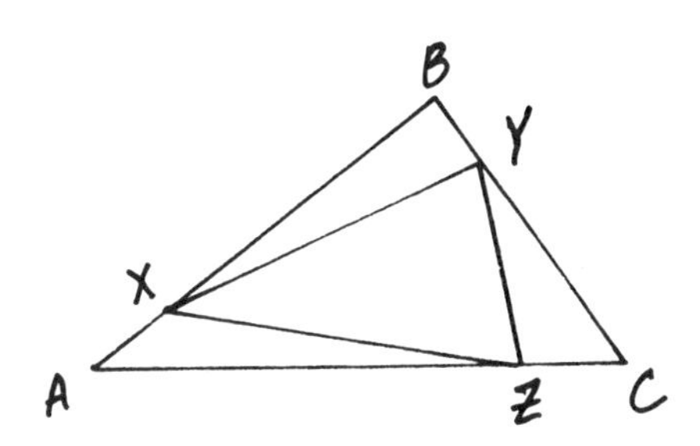

5. If a, b, c are the numbers of 1¢, 2¢, and 5¢ stamps purchased, respectively, then $a + 2b + 5c = 100$ and $a = 10b$. Thus $12b + 5c = 100$, or $c = 20 - \frac{12b}{5}$, and b must be a multiple of 5. If $b \geq 10$, $\frac{12b}{5} \geq 24$, and $c < 0$. Hence $b = 5$, $a = 50$, and $c = 8$.

6. We have $(x + y)^2 - 2(x + y) = (x + y - 2)(x + y) = 0$ (i), and
$(x + y)(x - y) - 2(x - y) = (x - y)(x + y - 2) = 0$ (ii)
From (i), either $x + y = 2$ or $x = -y$. If $x + y = 2$, (x,y) satisfies (ii) as well. If $x = -y$, we find $x = y = 0$ from (ii), which doesn't satisfy the conditions of the problem. A similar analysis starting with (ii) shows that $x + y = 2$.

7. The answer (6) is easily read from a good diagram. For a discussion of the general case, see Polya: Mathematics and Plausible Reasoning, Vol. I. There it is shown that n lies create $\frac{n(n - 1)}{2} + n + 1$ regions. It is not hard to see that $2n$ of these are unbounded.

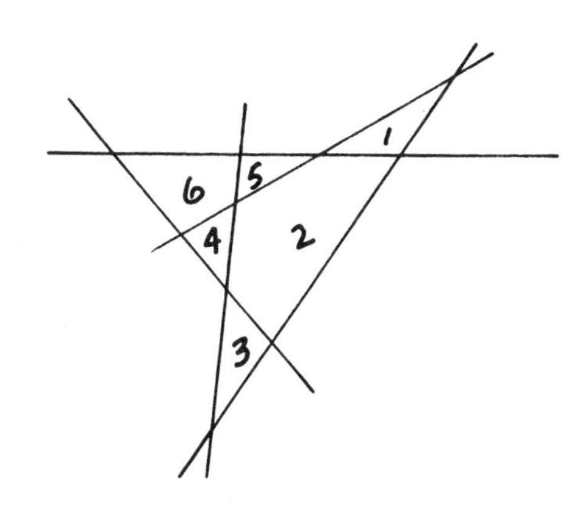

8. $\frac{500}{125} = 4$, so we can multiply 5^3 by any power of 2 up to 2^2 and get one of the required numbers. This sum will be $5^3(1 + 2 + 2^2)$ $= 5^3(2^3 - 1) = 125 \cdot 7 = 875$. Continuing, $\frac{500}{25} = 20 > 2^4$, so more such numbers are (see 1.522) $5^2(1 + 2 + 2^2 + 2^3 + 2^4) = 5^2(2^5 - 1)$ $= 25 \cdot 31 = 775$. $\frac{500}{5} = 100 > 2^6$, so we must add in $5(1 + 2 + \ldots + 2^6) = 5 \cdot 127 = 635$ and of course $500 > 2^8$ and $1 + 2 + 2^2 + \ldots + 2^7 + 2^8 = 2^9 - 1 = 511$. The sum is $875 + 775 + 635 + 511 = 2796$.

9. Let angle ADX = angle $AXD = a$, angle BAX = angle $BXA = b$, angle CBX = angle $CXB = c$. Since angle D and angle C are supplementary, $a + (180 - 2c) = 180$, so $a = 2c$. Since angle D and angle DAB are supplementary, $a + (180 - 2a + b) = 180$, so $b = a$. Summing the angles about point X, we have $a + b + c = a + a + (a/2) = 180$, so $a = 72°$.

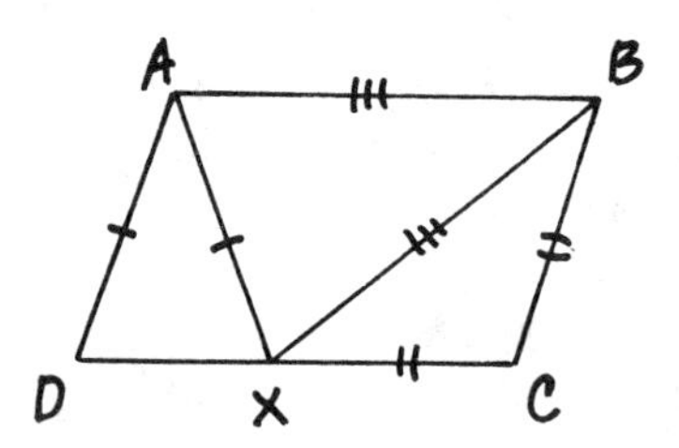

10. $x^4 + 64 = x^4 + 16x^2 + 64 - 16x^2 = (x^2 + 8)^2 - (4x)^2$.
$P(x) = x^2 + 8$.
The trick of adding and subtracting the same quantity is difficult to see, until one notices how similar it is to the process of completing the square.

11. We have $\dfrac{\cos 2b}{\sin^2 2b(\cot^2 b - \tan^2 b)} = \dfrac{\cos 2b}{\sin^2 2b} \cdot \dfrac{\sin^2 b \cos^2 b}{\cos^2 b - \sin^2 b}$

$= \dfrac{\sin^2 b \cos^2 b}{4\sin^2 b \cos^2 b} = \dfrac{1}{4}$. (See 4.21 and 4.22.)

12. Let $AX{:}XB = m/n$. Then we can write $AX = mx$, $XB = nx$, $AB = (m + n)x$, $CZ = my$, $ZA = ny$, $CA = (m + n)y$.
Then $\dfrac{|\text{triangle } AXZ|}{|\text{triangle } ABC|} = \dfrac{\frac{1}{2}AX \cdot AZ \sin A}{\frac{1}{2}AB \cdot AC \sin A} = \dfrac{mn}{(m + n)^2}$. Similarly,

$\dfrac{|\text{triangle } XBY|}{|\text{triangle } ABC|} = \dfrac{|\text{triangle } YZC|}{|\text{triangle } ABC|} = \dfrac{mn}{(m + n)^2}$. Thus $\dfrac{7}{25}$

$= \dfrac{|\text{triangle } XYZ|}{|\text{triangle } ABC|}$

$= \dfrac{|\text{triangle } ABC| - |\text{triangle } AXZ| - |\text{triangle } XBY| - |\text{triangle } YCZ|}{|\text{triangle } ABC|}$

$= 1 - \dfrac{3mn}{(m + n)^2} = 1 - \dfrac{3\frac{m}{n}}{\left(1 + \frac{m}{n}\right)^2}$.

Writing $m/n = k$, we have $\dfrac{18}{25} = \dfrac{3k}{(1 + k)^2}$ or $6k^2 - 13k + 6 = 0$
and $k = \dfrac{3}{2}$ or $\dfrac{2}{3}$. Since the problem requires $k < 1$, $k = \dfrac{2}{3}$.

13. $(\sqrt{a} + \sqrt{b})^2 = 20 + \sqrt{384}$, so $a + b = 20$
$2\sqrt{ab} = \sqrt{384} = 2\sqrt{96}$
$ab = 96$ and $(a,b) = (8,12)$.

14. Let BX bisect angle ABC. Then angle $XBC = 36° =$ angle ABX, and angle $BXC = 72° =$ angle BCX. Hence triangle BXC is isosceles, and it is similar to triangle ABC. Let $AB = AC = a$, $BC = BX = XA = b$.
Then $\dfrac{XC}{BC} = \dfrac{BC}{AB}$ or $\dfrac{a - b}{b} = \dfrac{b}{a}$. If $\dfrac{a}{b} = r$, $r - 1 = \dfrac{1}{r}$: $r^2 - r - 1 = 0$,

$r = \dfrac{1 \pm \sqrt{5}}{2}$. But $r > 0$, so $r = \dfrac{1 + \sqrt{5}}{2}$. For much more on this "golden triangle," see Huntley or Coxeter.

14. (Continued)

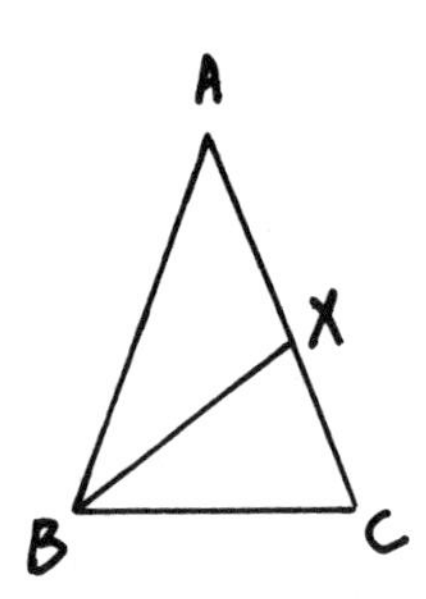

15. Color the first row red, white, blue. There are 2 possible choices for square 4 (white or blue). After this, all colors are determined. There are 6 permutations of the first row, hence $6 \times 2 = 12$ colorings of the square.

1 R	2 W	3 B
4	5	6
7	8	9

16. Method 1: Use undetermined coefficients (see Hall and Knight, Higher Algebra).

Method 2: We have, in general:

$$\begin{aligned}(a^2 + b^2)^2 &= a^4 + 2a^2b^2 + b^4 \\ &= a^4 - 2a^2b^2 + b^4 + 4a^2b^2 \\ &= (a^2 - b^2)^2 + (2ab)^2.\end{aligned}$$

In the present case, we can let $a = x + 1$, $b = x$, so $(a^2 - b^2) = R(x) = 2x + 1$, and $2ab = Q(x) = 2x(x + 1) = 2x^2 + 2x$.

17. We have $x^2 + y^2 = 1$ and $x^2 + (y + 1)^2 = a^2$

$$\begin{aligned}x^2 + y^2 + 2y + 1 &= a^2 \\ 2y + 2 &= a^2 \\ y &= \frac{a^2 - 2}{2}\end{aligned}$$ or equivalent.

17. (Continued)

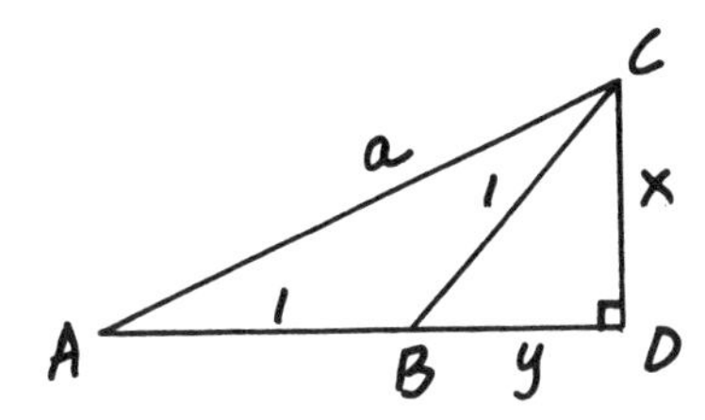

18. By the distributive law, the required sum is:

$$\left(1 + \frac{1}{2} + \frac{1}{4} + \frac{1}{8} + \dots\right)\left(1 + \frac{1}{3} + \frac{1}{9} + \frac{1}{27} + \dots\right)\left(1 + \frac{1}{5} + \frac{1}{5^2} + \dots\right).$$

Since all series are geometric (see 1.522),

$$\left(\frac{1}{1 - \frac{1}{2}}\right)\left(\frac{1}{1 - \frac{1}{3}}\right)\left(\frac{1}{1 - \frac{1}{5}}\right) = \frac{2}{1} \cdot \frac{3}{2} \cdot \frac{5}{4} = \frac{15}{4}.$$

19. We have $\frac{n!}{(n - r)!} = 6 \cdot \frac{n!}{(n - r)!r!}$. Therefore, $r! = 6$ and $r = 3$.

20. $2\sqrt{5} \approx 4.48$, $\sqrt{21} \approx 4.6$ so the answer is at least 8. But is it as much as 9? We have $(2\sqrt{5} + \sqrt{21})^2 = 4 \cdot 5 + 21 + 4\sqrt{105} = 41 + 4\sqrt{105}$ and $(4\sqrt{105})^2 = 16 \cdot 105$. $4^2 \cdot 10^2 < 16 \cdot 105 < 4^2 \cdot 11^2$ so $40 < 4\sqrt{105} < 44$ and $81 < 41 + 4\sqrt{105} < 85$ so $9 < 2\sqrt{5} + \sqrt{21} < \sqrt{85} < 10$ and the answer is 9.

21. If his salary was S before the first year, it was $\left(1 + \frac{p}{100}\right)S$ after the increase. After the decrease, it was $\left(1 + \frac{p}{100}\right)S\left(1 - \frac{p}{100}\right)$ $= \left(1 - \frac{p^2}{100^2}\right)S$. The percent of decrease is

$$\frac{S - S\left(1 - \frac{p^2}{100^2}\right)}{S} \times 100 = \left(1 - 1 + \frac{p^2}{100^2}\right)100 = \frac{p^2}{100}.$$

22. $2^{22} + 1 = (2^{11} + 1)^2 - 2^{12} = (2^{11} + 1 + 2^6)(2^{11} + 1 - 2^6)$
$= 2113 \cdot 1985 = 2113 \cdot 397 \cdot 5$

23. We want to reproduce the situation of problem 14, Spring 1982. Since $108 = \frac{3}{2} \cdot 72$, we can do this by drawing the trisectors of angle C. If these are CX, CY (see diagram), then angle XCB = angle $CXB = 72°$, so we can let $CB = XB = a$. If $AB = b$, then $AX = AB - XB = b - a$. Also, angle ACX = angle $CAX = 36°$, so triangle $ACX \sim$ triangle ABC and $\frac{AC}{AX} = \frac{AB}{AC}$, or $\frac{a}{b - a} = \frac{b}{a}$. Hence if $r = \frac{b}{a}$, we find $\frac{b - a}{a} = \frac{b}{a} - 1 = \frac{a}{b}$, or $r - 1 = \frac{1}{r}$. Thus $r^2 - r - 1 = 0$ and $r = \frac{1 \pm \sqrt{5}}{2}$. Since $b > a$, $r > 1$ and only the "+" sign applies.

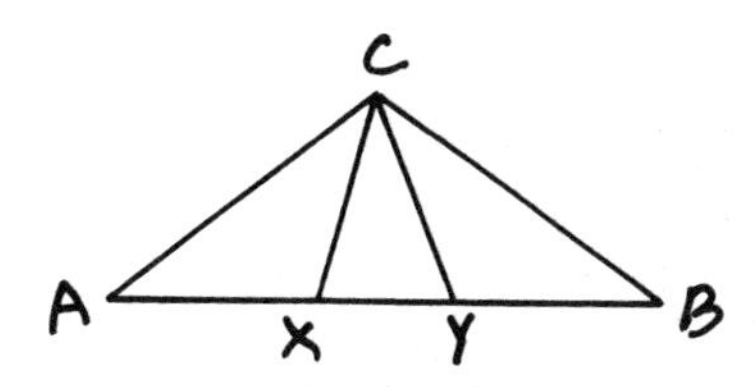

24. First note that $\tan B = \frac{1}{3}$ (see diagram). Hence $\tan 2B = \frac{2 \tan B}{1 - \tan^2 B}$ $= \frac{2}{3} \div \frac{8}{9} = \frac{3}{4}$. Now $\tan(A + 2B) = \frac{\tan A + \tan 2B}{1 - \tan A \tan 2B}$

$$= \frac{\frac{1}{7} + \frac{3}{4}}{1 - \frac{3}{28}} = \frac{\frac{4}{28} + \frac{21}{28}}{\frac{25}{28}} = 1$$ (see 4.341) and $A + 2B = 45°$.

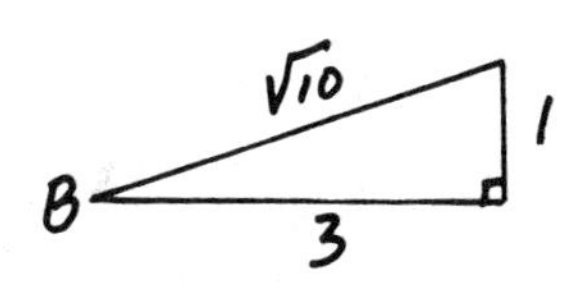

25. There are 100 integers in the given set. Let us find how many are multiples of squares. If $d(n)$ is the number of integers divisible by n, we have $d(4) = 25$; $d(9) = 11$; $d(25) = 4$; $d(36) = 2$; $d(49) = 2$. Now multiples of 16 are counted among those of 4, and multiples of 49 cannot be multiples of any other square. But $d(4 \text{ or } 9) = d(4) + d(9) - d(4 \text{ and } 9) = d(4) + d(9) - d(36)$. $d(4 \text{ or } 25) = d(4) + d(25) - d(4 \text{ and } 25) = d(4) + d(25) - d(100)$.
Hence we must subtract two multiples of both 4 and 9 and one multiple of both 4 and 25 (the other squares are too large to necessitate this)(see also 7.1), and $d(\text{some square}) = 22 + 11 + 4 + 2 = 39$, and there are 61 square-free numbers. Note that we have included "1" in our definition of square-free. For more on this topic, see Greger.

26. If $m = \sqrt[3]{20 + 14\sqrt{2}}$, $n = \sqrt[3]{20 - 14\sqrt{2}}$, then $x = m + n$.
Hence $x^3 = m^3 + n^3 + 3mn(m + n) = 40 + 3mnx =$
$40 + 3x(\sqrt[3]{400 - 2 \cdot 196}) = 40 + 3x\sqrt[3]{8} = 40 + 6x$, so $x^3 - 6x = 40$.

27. The equation can be solved by writing: $(\sin x + \sin 3x) + \sin 2x$
$= 2 \sin 2x \cos x + \sin 2x$
$= 2 \sin x \cos x \,(2 \cos x + 1) = 0$,
so $\sin x = 0$ and $x = 0, \pi$,
or $\cos x = 0$ and $x = \frac{\pi}{2}, \frac{3\pi}{2}$, (see 4.321).
or $\cos x = -\frac{1}{2}$ and $x = \frac{2\pi}{3}, \frac{4\pi}{3}$ and there are 6 solutions.

28. Let B be the center of a circle of radius 5. Then, looking at segments of intersecting chords, $(4 + 5)(1) = 4x$ so $x = \frac{9}{4}$.

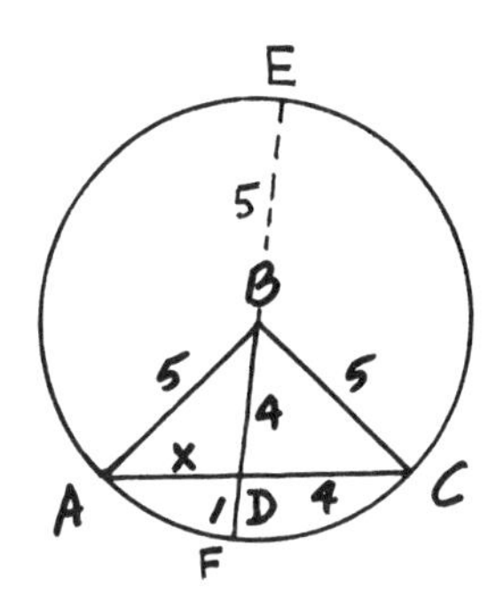

29. The probability of getting exactly a heads in b tosses of a fair coin is given by $\binom{b}{a} 2^{-b}$. Here we have $\binom{35}{k} 2^{-35} = \binom{36}{k+1} 2^{-36}$

or $\frac{35!}{k!(35 - k)!2^{35}} = \frac{36!}{(k + 1)!(36 - k - 1)!2^{36}}$ or $1 = \frac{36}{2(k + 1)}$

or $k = 17$.

30. Draw $DF \| AE$. Then since D is the midpoint of AB, F is the midpoint of BE, and also $DF = \frac{1}{2}AE = CD$. Letting angle FDB = angle EAB = angle $EAC = x$, we have angle $CDF = 90 - x$, angle CBD = angle CAD $= 2x$, angle CFD = angle $DCF = 3x$ (it's an exterior angle in triangle FDB). In triangle CDF, $90 - x + 3x + 3x = 180$. Therefore $x = 18$ and angle $ACB = 6x = 108°$.

30. (Continued)

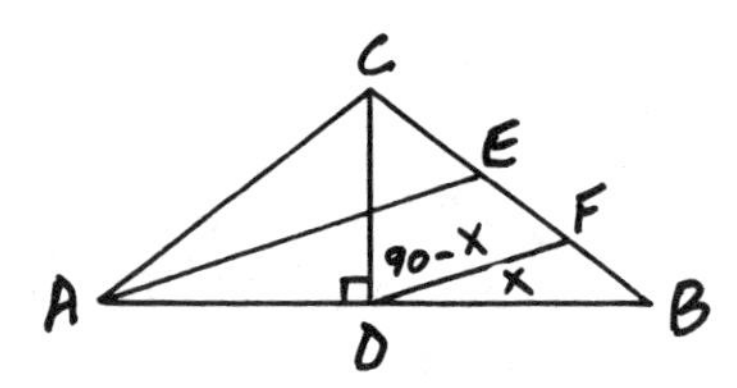

Solutions—Fall 1982

1. One could solve the quadratic resulting from multiplying both sides by x, or merely note that $x + \frac{1}{x} = 3 + \frac{1}{3}$ so $x = 3, \frac{1}{3}$. There cannot be more roots as the equation will be quadratic.

2. The following argument shows that there is only one such ordered triple. The second condition given can be written as $(a + b)(a - b) = c$. This implies $a + b = c$ and $a - b = 1$, since c is prime. Then $a = b + 1$, so $a = 3$ and $b = 2$ (since 2 and 3 are the only consecutive integers that are both prime). Finally, $c = a + b = 5$, and the unique ordered triple is (3,2,5).

3. If the original number is k, we have added $100n - 10n - n = 89n$ to the number by multiplying by n. That is, $nk = k + 89n$, or $(n - 1)k = 89n$. But 89 is prime and $n - 1$ is relatively prime to n, so $n - 1$ must be 1, $n = 2$, and $k = 178$.

4. $$\left(\sqrt[5]{\sqrt{18} + \sqrt{2}}\right)^2 = \sqrt[5]{(\sqrt{18} + \sqrt{2})^2} = \sqrt[5]{20 + 2\sqrt{36}} = \sqrt[5]{32} = 2.$$

5. If $\overline{TU}$ is the common tangent, draw $\overline{O_2X}$ parallel to $\overline{TU}$. Then (see diagram) $t^2 = 144 - 16 = 128$, so $TU = \sqrt{128} = 8\sqrt{2}$.

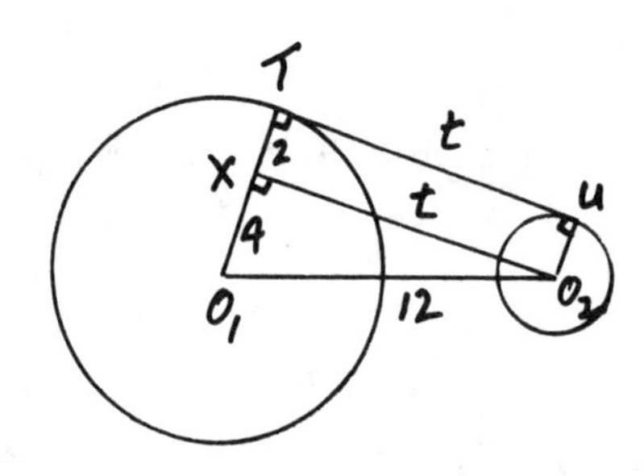

6. For each marble, there are 3 choices for an urn to place it in. This gives 3^{10} possible arrangements--including some where an urn will end up empty. If we choose any one urn to stand empty, there are 2^{10} arrangements of marbles in the other two urns. Hence

6. (Continued)

$3 \cdot 2^{10}$ arrangements leave an empty urn, and this number must be subtracted. However, we have now subtracted off twice the 3 arrangements in which two urns are empty, so we must add back 3 (see 7.3). This gives $3^{10} - 3 \cdot 2^{10} + 3$ arrangements.

7. If the numbers are A and B, we have $AB = 588000 = 2^5 \cdot 3 \cdot 5^3 \cdot 7^2$. We need to "distribute" these factors over the numbers A and B as evenly as possible. Doing this, we find that $GCD(A,B) = 2^2 \cdot 5 \cdot 7 = 140$.

8. Compare with problem 1, Fall 1982. Let $y = \sqrt{\dfrac{x+4}{x-1}}$. Then

$y + \dfrac{1}{y} = \dfrac{5}{2}$, so $y = 2$ or $y = \dfrac{1}{2}$. If $y = 2$, $\dfrac{x+4}{x-1} = 4$ and $x = \dfrac{8}{3}$.

If $y = \dfrac{1}{2}$, $\dfrac{x-1}{x+4} = 4$ and $x = -\dfrac{17}{3}$.

9. Since $x \cdot \dfrac{5}{x} = 5$ (a constant), the minimal value is achieved when

$x = \dfrac{5}{x}$, or $x = \sqrt{5}$, and $x + \dfrac{5}{x} = 2\sqrt{5}$ (see 6.4). For more such

problems, see Beckenbach and Bellman, or Korovkin.

10. The two figures have a common apothem. Since the area of any regular polygon is $\frac{1}{2}ap$, the ratio of the areas is the ratio of perimeters of the two figures in question. If the side of the octagon is s, then the side of the square is $s(1 + \sqrt{2})$ (see diagram). Hence the required ratio is $\dfrac{8s}{4s(1+\sqrt{2})} = \dfrac{2}{1+\sqrt{2}} = 2\sqrt{2} - 2$.

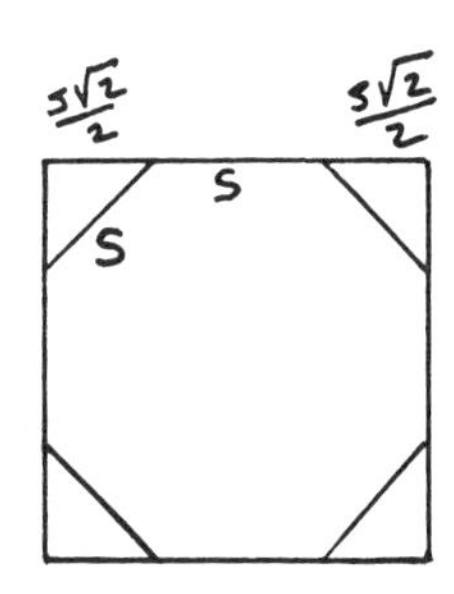

11. If T is the required point of intersection, let $O_1T = x$, $O_2T = y$. Then $x + y = 12$, and from similar right triangles O_1PT, O_2QT,

$\dfrac{x}{6} = \dfrac{y}{2}$, so $x = 3y$. Hence $x = 9$ and $y = 3$.

The distance from circle $O_2 = 3 - 2 = 1$.

11. (Continued)

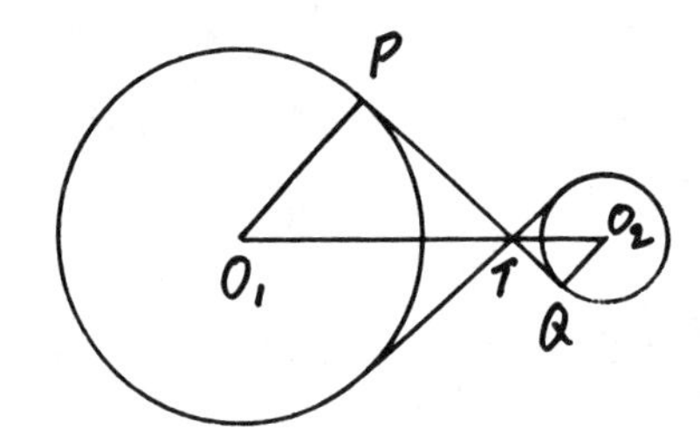

12. We have $0 = (a + b + c)^2 = a^2 + b^2 + c^2 + 2(ab + bc + ca)$, so

$$\begin{aligned}(a^2 + b^2 + c^2)^2 &= 4(ab + bc + ca)^2 \\ &= 4(a^2b^2 + b^2c^2 + c^2a^2 + 2a^2bc + 2ab^2c + 2abc^2) \\ &= 4(a^2b^2 + b^2c^2 + c^2a^2 + 2abc(a + b + c)) \\ (*) \qquad &= 4(a^2b^2 + b^2c^2 + c^2a^2).\end{aligned}$$

But also, $(a^2 + b^2 + c^2)^2 = a^4 + b^4 + c^4 + 2(a^2b^2 + b^2c^2 + c^2a^2)$.

Hence (**) $a^4 + b^4 + c^4 = (a^2 + b^2 + c^2)^2 - 2(a^2b^2 + b^2c^2 + c^2a^2)$

$$= 2(a^2b^2 + b^2c^2 + c^2a^2).$$

Comparing (*) with (**), we find that $\dfrac{(a^2 + b^2 + c^2)^2}{a^4 + b^4 + c^4} = 2$, a constant. See Hall and Knight, Higher Algebra (Chapter 34) on symmetric functions.

13. Note that $2^1 = 2$, $2^2 = 4$, $2^3 = 8$, $2^4 = 16$, and thereafter the last last digits repeat in a cycle or 4. Hence 25 of the numbers given end in the digit 6.

14. Let $y = \log_4 x$. Then $\dfrac{1}{y} = \log_x 4$ (see 5.12), and $y + \dfrac{1}{y} = \dfrac{17}{4} = 4 + \dfrac{1}{4}$.

Hence $\log_4 x = 4$ or $\log_4 x = \dfrac{1}{4}$ and $x = 256$ or $\sqrt[4]{4} = \sqrt{2}$.

15. $s = \sqrt{\dfrac{(4 + x)(1 + x)}{x}}$ is minimal when s^2 is minimal.

$s^2 = \dfrac{4}{x} + x + 5$ is minimal when $s^2 - 5 = \dfrac{4}{x} + x$ is minimal.

$\dfrac{4}{x} \cdot x = 4$, a constant, so the minimal value of their sum occurs when $\dfrac{4}{x} = x$, or $x = 2$, and $s = \sqrt{\dfrac{18}{2}} = 3$ (see 6.4). Also see references given in problem 9, Fall 1982.

16. Let the radius of the third circle be x. From similar triangles OPA, PQB, we have $\frac{OP}{PA} = \frac{PQ}{QB}$, or $\frac{x+9}{x-9} = \frac{16+x}{16-x}$. Cross multiplying gives $2x^2 = 288$, so $x = 12$. Or, using 1.1, we have: $\frac{2x}{x-9} = \frac{32}{16-x}$ or $\frac{x}{x-9} = \frac{16}{16-x}$. Then $\frac{x}{9} = \frac{16}{x}$, and $x = 12$. For useful exercises on ratio and proportion, see Hall and Knight, Higher Algebra (Chapters 1 and 2).

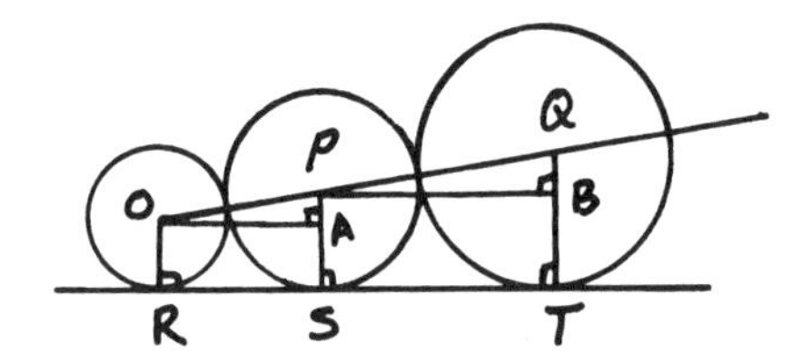

17. $2a + 3b = 31$. Since $a - b$ is a multiple of 4, a and b are either both odd or both even: in fact, it is easy to see that they must both be odd. Looking at the equation modulo 3 we have $2a \equiv 1$, so $a \equiv 2 \bmod 3$. The first few possibilities are $a = 5,11,17$. However, $2 \cdot 17 = 34 > 31$, and $a = 5$ gives $b = 7$ and $a - b = -2$, which is not a multiple of 4. Hence $(a,b) = (11,3)$.

18. Let the sides of the triangle be $a = b - d$, b, $c = b + d$, with $b = 2$, and let the inradius be r. The semiperimeter $s = (a + b + c)/2 = 3b/2 = 3$. Also, if the points of contact of the incircle with the sides are X, Y, and Z (see diagram), we have $BX = BZ = s - b = 1$. Now OBX is a 30-60-90 triangle, so $OX = \sqrt{3}/3 = r$, and the area is $rs = \sqrt{3}/3 \cdot 3 = \sqrt{3}$ (see 3.5 and 3.62).

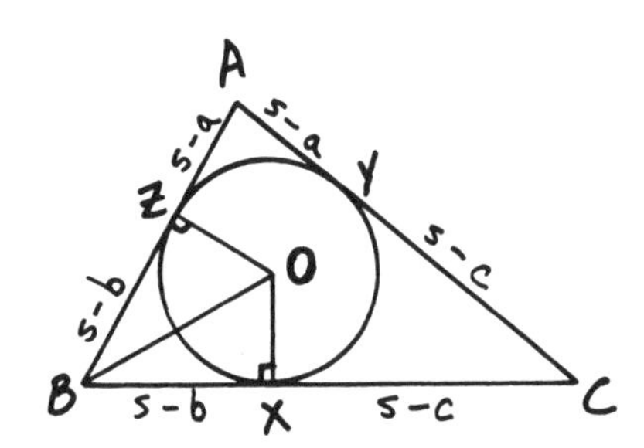

19. Let $\tan\theta = x$. Then $x + \frac{1}{x} = 3 + \frac{1}{3}$, so $\tan\theta = 3, \frac{1}{3}$ and (see 4.12) $\sec\theta = \pm\sqrt{\tan^2\theta + 1} = \pm\sqrt{10}$ or $\frac{\pm\sqrt{10}}{3}$. Since $0 < \theta < \frac{\pi}{2}$, $\sec\theta$ is positive.

20. Substitution gives $2a - 3b = 8$. Hence $(a,b) = \left(\frac{56}{11}, \frac{8}{11}\right)$,
$a + 4b = 8$.
and another substitution gives $k = -66$.

21. $\frac{(4 + x)(1 + x)}{x} = \frac{4 + 5x + x^2}{x} = x + \frac{4}{x} + 5.$

This is minimal when $x + \frac{4}{x}$ is minimal. Now $x \cdot \frac{4}{x} = 4$, a constant, so $x + \frac{4}{x}$ is smallest when $x = \frac{4}{x}$ or $x = 2$, and $\frac{(4 + x)(1 + x)}{x} = 9$ (see 6.4 and problem 9, Fall 1982).

22. By Fermat's "little" theorem (see Dudley), $p \nmid 10 \to 10^{p-1} \equiv 1 \bmod p$, so $p|10$. The largest such prime is 5.

23. Let the radius $= x$. Then (see diagram) $OX + PY = RT = 12$. From right triangles OPX and PQY,

$\sqrt{(x + 1)^2 - (x - 1)^2} + \sqrt{(9 + x)^2 - (9 - x)^2} = 12$

$\sqrt{4x} + \sqrt{36x} = 12$

$8\sqrt{x} = 12$ and $x = \frac{9}{4}$.

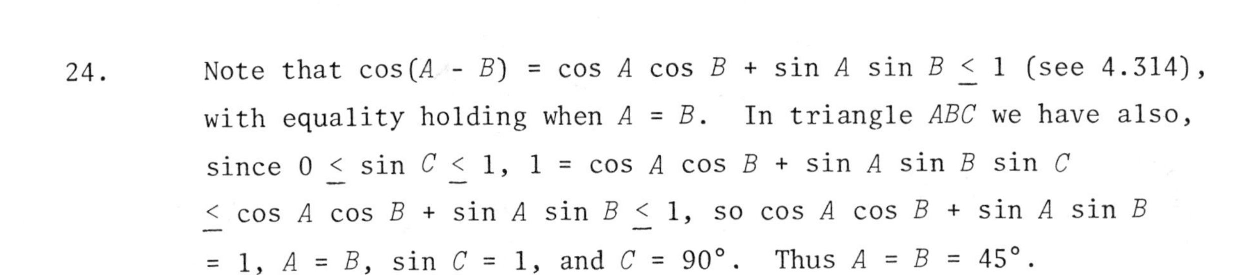

24. Note that $\cos(A - B) = \cos A \cos B + \sin A \sin B \leq 1$ (see 4.314), with equality holding when $A = B$. In triangle ABC we have also, since $0 \leq \sin C \leq 1$, $1 = \cos A \cos B + \sin A \sin B \sin C$ $\leq \cos A \cos B + \sin A \sin B \leq 1$, so $\cos A \cos B + \sin A \sin B$ $= 1$, $A = B$, $\sin C = 1$, and $C = 90°$. Thus $A = B = 45°$.

25. $$\sqrt{\frac{a}{b}\sqrt{\frac{b}{a}\sqrt{\frac{a}{b}}}} = \sqrt{\frac{a}{b}\sqrt{\sqrt{\frac{b^2}{a^2}\cdot\frac{a}{b}}}} = \sqrt{\frac{a}{b}\sqrt[4]{\frac{b}{a}}} = \sqrt{\sqrt[4]{\frac{a^4}{b^4}\cdot\frac{b}{a}}}$$
$$= \sqrt[8]{\frac{a^3}{b^3}} = \left(\frac{a}{b}\right)^{3/8}.$$

26. We can write $1 + 3x + 5x^2 + 7x^3 + \ldots$ as follows:

$$\begin{aligned} & 1 + x + x^2 + x^3 + \ldots = \frac{1}{1-x} \\ + \quad & 2(x + x^2 + x^3 + \ldots) = \frac{2x}{1-x} \\ + \quad & 2(x^2 + x^3 + \ldots) = \frac{2x^2}{1-x} \\ + \quad & \vdots \end{aligned}$$

$$1 + 3x + 5x^2 + \ldots = \frac{1 + 2x + 2x^2 + \ldots}{1-x} = \frac{-1 + 2(1 + x + x^2 + \ldots)}{1-x}$$

$$= \frac{-1 + 2\left(\frac{1}{1-x}\right)}{1-x} = \frac{1+x}{(1-x)^2}.$$ (See 1.523.) Letting $x = \frac{1}{3}$, we obtain the value 3.

27. Let $(4 + \sqrt{15})^x = u$. Then $(4 + \sqrt{15})^x + (4 - \sqrt{15})^x = u + \frac{1}{u}$. As in problems 9, 15, and 21, Fall 1982, $u + \frac{1}{u}$ is least when $u = \frac{1}{u}$, so $u = 1$ $(x = 0)$ and its value then is 2. Note $u + \frac{1}{u}$ $= 2 + \frac{(u-1)^2}{u} \geq 2$ (see 6.1 and problem 9, Fall 1982).

28. The sum of the digits must be 9. The largest sum possible with no digits 3 is 8, and the smallest with three digits 3 is 10. Hence there must be at least one 3 and at most two. If there is only one 3, the other digits must be 2's: there are 4 such numbers. If there are two digits 3, the other digits must be 1 and 2. There are 12 of these, so altogether there are 16 such numbers.

29. N is of the form $10^x - 1$. By Fermat's "little" theorem (see Dudley), $10^{16} - 1$ is a multiple of 17. We must show that no smaller exponent will work. Let us first check divisors of 16: $10^2 \equiv -2$, $10^4 \equiv 4$, and $10^8 \equiv -1$. Now if $10^y \equiv 1 \bmod 17$ and $y < 16$, let the G.C.D. of y and 16 be d. Then we can find m and n such that $ym + 16n = d$, or $10^{ym} \cdot 10^{16n} = 10^d$, or $1 \equiv 10^d$. This is impossible, as d is a divisor of 16.

30. Multiplying out, we have: $\sin^2 3\theta - \sin 3\theta \cos \theta = \sin^2\theta - \sin \theta \cos 3\theta$ or: $\sin^2 3\theta - \sin^2\theta = \sin 3\theta \cos \theta - \sin \theta \cos 3\theta$ or: $(\sin 3\theta + \sin \theta)(\sin 3\theta - \sin \theta) = \sin 3\theta \cos \theta - \sin \theta \cos 3\theta$.

Using the formula for the sums and differences of sines (on the left) and for the sine of a difference (on the right), we have (see 4.321, 4.322, 4.312): $(2 \sin 2\theta \cos \theta)(2 \cos 2\theta \sin \theta)$
$= \sin(3\theta - \theta)$
$4 \sin \theta \cos 2\theta \sin 2\theta \cos \theta = \sin 2\theta$
$\sin 2\theta(2 \sin 2\theta \cos 2\theta - 1) = 0$, so either (i) $\sin 2\theta = 0$, or (see 4.21) (ii) $2 \sin 2\theta \cos 2\theta - 1 = \sin 4\theta - 1 = 0$. From (i) we find $2\theta = 0, \pi$ and $\theta = 0, \pi/2$.
From (ii) we find $4\theta = \frac{\pi}{2}, \frac{5\pi}{2}$ and $\theta = \frac{\pi}{8}, \frac{5\pi}{8}$. For a deeper discussion, see Honsberger, Ingenuity in Mathematics (essay 16).

Solutions—Spring 1983

1. $1234321 = 11 \cdot 112211$. It is not hard to factor this last number. In fact, $112211 = 11 \cdot 10^4 + 22 \cdot 10^2 + 11 = 11(10^4 + 2 \cdot 10^2 + 1)$
$= 11(10^2 + 1)^2$
$= 11 \cdot 101^2$.
The square root, then, is $11 \cdot 101 = 1111$.

2. Take $AB = 4$, so that $AP = 3$ and $PB = 1$. Using absolute value for area, we have $|APS| = |PBQ| = |QCR| = |RDS| = \frac{3}{2}$, and $|ABCD| = 16$, so that $|PQRS| = 16 - 4|APS| = 16 - 6 = 10$, and $|PQRS|:|ABCD| = 5:8$. Or: use the Pythagorean Theorem in triangle APS.

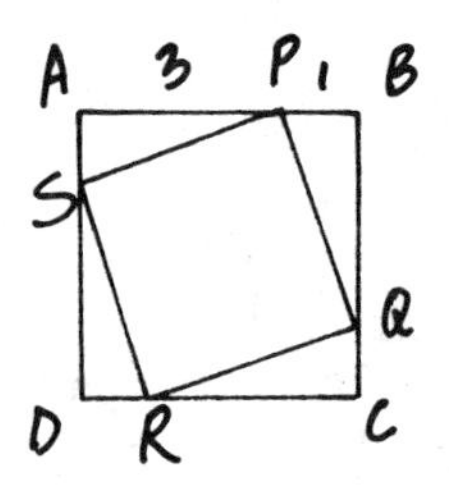

3. Between 1 P.M. and 2 P.M. the hand travelled 2π radians. Between 2 P.M. and 2:30 it travelled π radians, and between 2:30 and 2:35 it travelled $\frac{\pi}{6}$ radians. Altogether, it travelled $\frac{19\pi}{6}$ radians. Since the hand travelled clockwise (by definition), we can write the answer by the usual convention as $-\frac{19\pi}{6}$.

4. There are $\binom{6}{2} = 15$ ways to form a first subset. From the remaining four elements, there are $\binom{4}{2} = 6$ ways to form a second subset. The last subset is whatever is left. Hence there are $15 \cdot 6 \cdot 1 = 90$ ways of choosing a first, second, and third subset. Since we have counted each combination of three subsets $3! = 6$ times, the number of divisions is $\frac{90}{6} = 15$.

5. If the roots of $x^2 - px + q = 0$ are r and s, then $p = r + s$ and $q = rs$. Hence $p = r^2s^2 = q^2$ and $q = r^2 + s^2 = (r + s)^2 - 2rs = p^2 - 2q$. Substituting, we find $q = q^4 - 2q$, or $q^3 = 3$, so $q = \sqrt[3]{3}$ and $p = \sqrt[3]{9}$.

6. We can write the given sum, modulo 11, as: $4^N + 5^N + (-5)^N + (-4)^N$. For odd N this is clearly 0. For even N, the sum is congruent to $2 \cdot 4^N + 2 \cdot 5^N \equiv 2(4^N + 5^N) \equiv 2(4^N + 16^N) \equiv 2(2^{2N} + 2^{4N}) \equiv 2^{2N+1}(1 + 2^{2N})$. For this to be a multiple of 11, $2^{2N} + 1$ must be a multiple of 11. But then $4^N \equiv -1 \pmod{11}$. A quick check shows that this never happens, so the answer is 6.

7. $$\left(\sqrt[3]{\sqrt{75} - \sqrt{12}}\right)^{-2} = \left(\sqrt[3]{5\sqrt{3} - 2\sqrt{3}}\right)^{-2} = \frac{1}{\left(\sqrt[3]{3\sqrt{3}}\right)^2} = \frac{1}{\left(\sqrt[3]{\sqrt{3}^{\,3}}\right)^2}$$
$$= \frac{1}{(\sqrt{3})^2} = \frac{1}{3}$$

8. In the diagram, G is the centroid and $\overline{AA'}$ is a median. Then $AG = 2GA'$, and from similar triangles AGX, $A'GY$, $AX = 2A'Y$, so $A'Y = 5$. Then, in trapezoid $BCWZ$, $\overline{A'Y}$ is a median, so if $CW = x$, $2 \cdot 5 = x + 6$, and $x = 4$. Or: set up a system of coordinates in which one of the axes coincides with ℓ, and use the theorem that the coordinates of the centroid are the averages of the coordinates of the vertices.

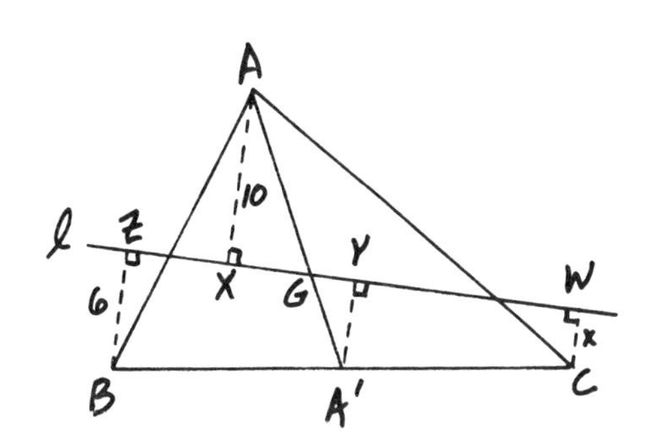

9. $A = 1 + \frac{1}{A}$, so $A - \frac{1}{A} = 1$. Also, $B = 2 + \frac{1}{B}$, so $B - \frac{1}{B} = 2$.

Thus $A + B - \left(\frac{1}{A} + \frac{1}{B}\right) = 1 + 2 = 3$. See also Olds.

10. $(\sin A + \cos A)^2 = 1 + 2 \sin A \cos A = 1 + \frac{120}{169} = \frac{289}{169}$.

Therefore, $\sin A + \cos A = \frac{17}{13}$ (see 4.11).

11. Suppose $\frac{AOB}{AB} = k$. Then $100A + B = 10Ak + Bk$, or $100A - 10Ak$ $= Bk - B = B(k - 1)$. Since the left side is a multiple of 5, the right side must also be. And since the right side is positive, so is the left, and $1 < k < 10$. First suppose $5|k - 1$. Then $k = 6$ and $40A = 5B$, or $B = 8A$. Hence $A = 1$, $B = 8$ and $AB = 18$. Now suppose $5|B$. Then $B = 5$, and we have $10A(10 - k) = 5(k - 1)$, or $2A(10 - k) = k - 1$, and $A = \frac{k - 1}{2(10 - k)}$. Since the denominator is even, $k - 1$ must be even and k odd. Trying $k = 3,5,7,9$ ($k = 1$ is impossible), we find only $AB = 15,45$ corresponding to $k = 7,9$, respectively.

12. It is clear that triangle $ABC \sim$ triangle DEC and that $\overline{EK}$ bisects angle DEC. By the angle bisector theorem (see 3.31), then $AD:DC$ $= AB:BC = 2:3 = DK:KC$. If $KC = 3x$, then $DK = 2x$ and $AD = 3x + 1$. Hence $\frac{3x + 1}{5x} = \frac{2}{3}$, and $x = 3$. Hence $AC = 8x + 1 = 25$.

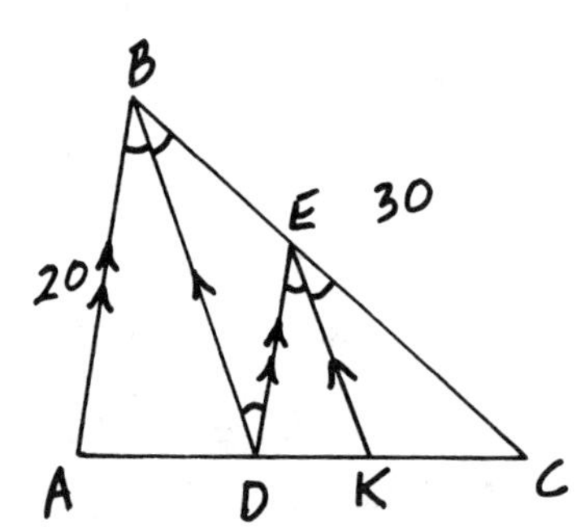

13. If he paid C dollars for the goods, he sold them for $1.2C$ instead of $1.3C$, or for 12/13 the price he should have. Hence he must have thought that each "yard" was 12/13 as long as it actually was. Thus if the "yardstick" was x units long, $12x/13 = 36$, and $x = 39$.

14. Let $AW = x$, $WB = y$. Using absolute value for area, $|ABCD|$ $= (x + y)^2$, while $|AWZ| = |BXW| = |XYC| = |YZD| = \frac{1}{2}xy$. Hence

$|WXYZ| = (x + y)^2 - 2xy = x^2 + y^2$. We know that $\frac{x^2 + y^2}{(x + y)^2} = \frac{5}{8}$.

Letting $\frac{x}{y} = r$, $\frac{r^2 + 1}{(r + 1)^2} = \frac{5}{8}$, leading to $3r^2 - 10r + 3$

14. (Continued)

$= (3r - 1)(r - 3) = 0$, and $r = 3, \frac{1}{3}$. Since $AW < WB$, $r = \frac{1}{3}$.

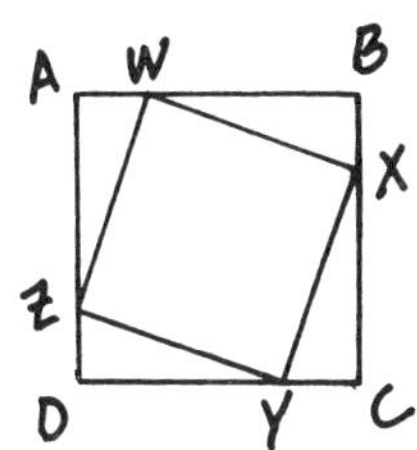

15. The fractions described are all contained in the sequence 1/24, 2/24, 3/24, . . ., 23/24. Hence there are 23 of them.

16. Since the function to be minimized involves addition and the condition given involves multiplication, we will try using the arithmetic-geometric mean inequality (see 6.2). Here, we have $\frac{20x^2 + 5y^2}{2} \geq \sqrt{20x^2 \cdot 5y^2} = \sqrt{100x^2y^2} = 10xy = 70$, so the minimum for $20x^2 + 5y^2$ is 140, occurring when $20x^2 = 5y^2$, or $y = 2x$. See problem 9, Fall 1982.

17. Since $10^4 < 61224 < 10^5$, we have $10^8 < 61224^2 < 10^{10}$, and 61224^2 has 9 or 10 digits. Now $\sqrt{10^9} = 10^4\sqrt{10} < 10^4 \cdot 4$, while $61224 > 6 \cdot 10^4 > 4 \cdot 10^4 > \sqrt{10^9}$. Hence $61224^2 > 10^9$ and has 10 digits.

18. Let us first assume that each face of a polyhedron with the maximal number of interior diagonals is a triangle. If such a polyhedron has F faces, E edges, and V vertices, then it has $\binom{V}{2} - E$ interior diagonals. Hence we must minimize E. In such a polyhedron, $3F$ counts each edge twice, or $3F = 2E$. Using Euler's formula $(V + F = E + 2)$, we find $\frac{2E}{3} - E + V = 2$ or $E = 3V - 6$. Hence the maximal number of interior diagonals is $\binom{V}{2} - 3(V - 2)$ $= \frac{(V - 3)(V - 4)}{2}$. This can be achieved, for example, by two convex pyramids sharing a common base that is a $(V - 2)$-gon. For $V = 10$, the answer is 21. The reason we choose polyhedra with triangular faces is that if a face has more than three vertices, some of the diagonals will fall on the faces, and so will not be interior diagonals. See Coxeter (Section 10.3).

19. If the factory actually produced N fleebles per day, it produced $10N$ fleebles altogether. It was supposed to produce $\frac{4N}{5}$ fleebles per day, for 12 days, or $\frac{48N}{5}$ fleebles. Hence $\frac{48N}{5} + 42 = 10N = \frac{50N}{5}$, or $42 = \frac{2N}{5}$ and $N = 105$.

20. If Arcsin $\frac{5}{13} = \theta$, then $\sin\theta = \frac{5}{13}$ and $\cos(90° - \theta) = \frac{5}{13}$, and Arccos $\frac{5}{13} = 90° - \theta$. Hence Arcsin $\frac{5}{13}$ + Arccos $\frac{5}{13}$ $= \theta + 90° - \theta = 90°$.

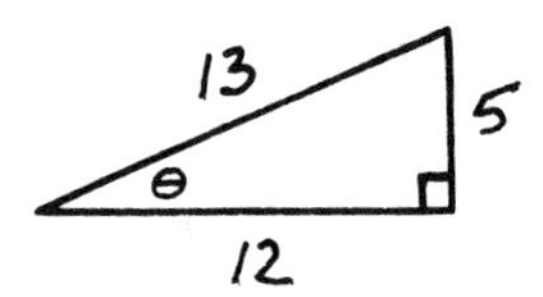

21. We use the arithmetic-geometric mean inequality (see 6.2): $\frac{x^2 + 4y^2}{2} \geq \sqrt{4x^2y^2} = 2xy$, so $x^2 + 4y^2 \geq 4xy$, or $4xy \leq 50$. Hence $(x + 2y)^2 = x^2 + 4xy + 4y^2 \leq 50 + 50 = 100$, and $x + 2y \leq 10$. Equality occurs when $x = 2y = 5$. See problems 16, Spring 1983 and 9, Fall 1982.

22. Let $AE = kx$, $AC = ky$, and suppose $AD = 4$ and $DB = 3$. Then, using absolute value for area, $|ADE| = \frac{1}{2} \cdot 4 \cdot kx \cdot \sin A$, $|ABC| = \frac{1}{2} \cdot 7 \cdot ky \cdot \sin A$ (see 6.34) and the ratio of the areas is $\frac{4x}{7y} = \frac{1}{2}$. Hence $\frac{x}{y} = \frac{7}{4} \cdot \frac{1}{2} = \frac{7}{8}$, and $\frac{AE}{EC} = 7$.

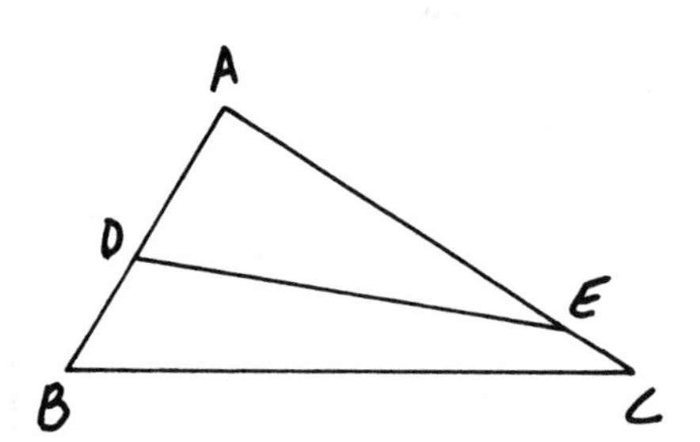

23. If they bought g sticks of gum, r rolls, and c candy bars, then $g + r + c = 100$ and $g + 10r + 50c = 500$. Subtracting, $9r + 49c = 400$. Hence $1 \leq c < 8$ (since $9r = 400$ is impossible) and $r = \frac{400 - 49c}{9} = 44 - 5c + \frac{4 - 4c}{9}$. Since r is integral, 9 divides $4 - 4c = 4(1 - c)$. Only $c = 1$ works. Then $r = 39$ and $g = 60$.

24. If the given polynomial is $P(x)$, its roots are equal to those of $Q(x) = \frac{1}{3}P(x)$. Since $Q(x)$ can be factored as $(x - r_1)(x - r_2)\ldots(x - r_5)$, it is not hard to see that the given expression is merely $-Q(-1)$. This is, numerically, $-\left(-1 + \frac{8}{3} - 1 + \frac{1}{3} + \frac{4}{3} + \frac{1}{3}\right) = -\frac{8}{3}$.

25. The numbers x and x^3 will always have identical remainders upon division by A if and only if the number $x^3 - x = x(x + 1)(x - 1)$ is always a multiple of A. Since any set of three consecutive integers contains at least one even number and exactly one multiple of 3, $x(x + 1)(x - 1)$ is always a multiple of 6. Trying the first few possible values of x, we find:

$$6^3 - 6 = 6 \cdot 7 \cdot 5$$
$$7^3 - 7 = 7 \cdot 8 \cdot 6$$
$$8^3 - 8 = 8 \cdot 9 \cdot 7 = 6 \cdot 4 \cdot 3 \cdot 7$$
$$9^3 - 9 = 9 \cdot 10 \cdot 11 = 6 \cdot 3 \cdot 5 \cdot 11.$$

At this point it is clear that 6 is the largest possible A.

26. $[2ab + (a^2 + b^2 - c^2)][2ab - (a^2 + b^2 - c^2)]$

$= [(a + b)^2 - c^2][c^2 - (a - b)^2]$

$= (a + b + c)(a + b - c)(c - a + b)(c + a - b)$

$= 2s(2s - 2a)(2s - 2b)(2s - 2c)$

$= 16s(s - a)(s - b)(s - c) = 16 = k^2 = s(s - a)(s - b)(s - c)$

$= 1$ (see 3.61).

27. One of the squares must be larger than $90/3 = 30$, hence it must be 36, 49 or 64. Now $90 - 64 = 26 = 25 + 1$, $90 - 49 = 41 = 25 + 16$, $90 - 36 = 54$ is not a sum of distinct squares. The desired triples are (1,5,8) and (4,5,7).

28. Squaring the first two equations and adding, we have:

$\cos^2 x + \cos^2 y + \cos^2 z + 2 \cos x \cos y + 2 \cos y \cos z + 2 \cos x \cos z$

$+ \sin^2 x + \sin^2 y + \sin^2 z + 2 \sin x \sin y + 2 \sin y \sin z$

$+ 2 \sin x \sin z = 0$, or $3 + 2[\cos(x - y) + \cos(y - z) + \cos(z - x)]$

$= 0$ (see 4.11 and 4.314), so that the required sum is $-\frac{3}{2}$.

Note: If we think of $\vec{x} = (\cos x, \sin x)$ as a unit vector and define $\vec{y}$ and $\vec{z}$ similarly, the given equations state that the three vectors sum to 0. Since they each have unit length, this means that their directions must differ by 120°. Visualizing this leads to a quick solution of the problem.

29. Suppose a set of cubes displays all possible colorings exactly once. Place the cubes on a table so that their green sides are all face down. Then there are five possible colors showing on their top faces. For the "side" faces there are then, a priori, 24 colorings. But each cube can be rotated four ways to get another set of colorings on the sides, so there are actually only six distinct colorings for these faces. Hence there are $5 \times 6 = 30$ possible colorings altogether.

30. Let us suppose, quite generally, that the sides of the triangle are a,b,c with $a \leq b \leq c$. First we look at chords intersecting $\overline{AB}$ and $\overline{AC}$ at D and E, respectively. Suppose $AD = xc$, $AE = yb$. Then the area of $ADE = \frac{1}{2}xybc \sin A = \frac{1}{2} \cdot \frac{1}{2}bc \sin A$, so that $xy = \frac{1}{2}$ (see 3.64). By the law of cosines, $DE^2 = x^2c^2 + y^2b^2 - 2xybc \cos A = x^2c^2 + y^2b^2 - bc \cos A$. To minimize DE, we need to minimize $x^2c^2 + y^2b^2$, subject to the condition that $xy = \frac{1}{2}$. Since the condition involves multiplication and the function to be minimized involves addition, we can try the arithmetic-geometric mean inequality (see 6.1). It works: $\frac{x^2c^2 + y^2b^2}{2} \geq \sqrt{x^2y^2b^2c^2} = xybc = \frac{bc}{2}$, with equality occurring when $x^2c^2 = y^2b^2 = \frac{bc}{2}$. This gives the minimum length for DE as $bc - bc \cos A = bc(1 - \cos A) = bc\left(1 - \frac{b^2 + c^2 - a^2}{2bc}\right)$ $= \frac{a^2 - (b - c)^2}{2}$. A similar result holds if we assume the chord intersects two other sides. For the given values, the shortest is found to have length $\sqrt{\frac{15}{2}}$. See also problem 16, Spring 1983 or 9, Fall 1982.

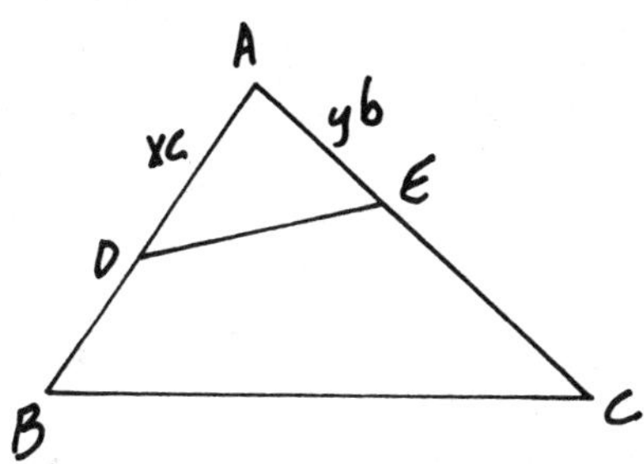

Solutions—Fall 1983

1. Writing out a few terms, we find that the sequence is $\underbrace{1,-1,-1},\underbrace{1,-1,-1},\underbrace{1,-1,-1},\ldots$ Since every term after the fifth depends only on the previous terms, a repetition of six terms will guarantee further repetitions. Hence every third term is -1. Since $1983 = 3 \cdot 661$, $a_{1983} = -1$.

2. We want to minimize $|24a - 15b|$, for positive integers a and b. Note that $N = |24a - 15b| = 3|8a - 5b|$. Hence N is always a multiple of 3 and is thus no smaller than 3. Letting $a = 2$, $b = 3$ shows that N in fact can equal 3. If Sam circled every pth number and Janet every qth, the minimal N would be the greatest common divisor of p and q (see Dudley, or any book on number theory).

3. Let the number be $a_1a_2a_3a_4a_5$. Clearly $a_1 > a_2 > a_3 > a_4 > a_5$. If $a_5 = 1$, $a_4 \geq 2$, $a_3 \geq 4$, $a_2 \geq 8$, and $a_1 \geq 16$, which is impossible. Hence $a_5 = 0$. If $a_4 \geq 2$, an argument like the one above shows $a_1 \geq 12$, which is again impossible. Hence $a_4 = 1$. Reasoning similarly, we find $a_3 \geq 3$ to be impossible, so $a_3 = 2$. The three solutions 94210, 95210, and 84210 follow quickly.

4. $2 \sin\left(\frac{A+B}{2}\right) \cos\left(\frac{A-B}{2}\right) = 2 \cos\left(\frac{A+B}{2}\right) \cos\left(\frac{A-B}{2}\right)$ (see 4.321 and 4.323), so $\sin \frac{A+B}{2} = \cos \frac{A+B}{2}$ OR $\cos \frac{A-B}{2} = 0$. Thus $\frac{A+B}{2} = \frac{\pi}{4}$, so $A + B = \frac{\pi}{2}$ or $A - B = \pi, 3\pi$, etc., which is not possible within the given limits.

5. Note that $\overline{OX} \perp \overline{XY}$ and $\overline{PY} \perp \overline{XY}$. If we draw $\overline{OZ} \| \overline{XY}$, with Z on $\overline{PY}$, then in triangle $\overline{OZP}$, $OP^2 = OZ^2 + PZ^2 = 7 + (PY - ZY)^2 = 7 + (PY - OX)^2 = 16$.

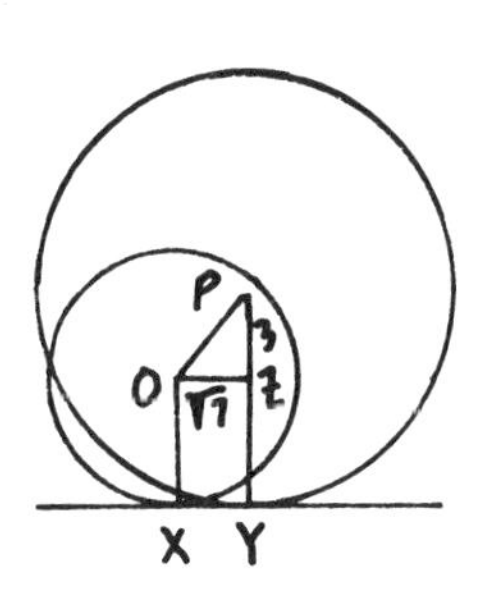

6. Clearly N must be a multiple of both 2 and 3. Let $N = 2^a \cdot 3^b \cdot k$. Then $\frac{N}{2} = 2^{a-1} \cdot 3^b \cdot k$. If this is a perfect square, $a - 1$ and b must be even, and k must be a square. Also, $\frac{N}{3} = 2^a \cdot 3^{b-1} \cdot k$.

Hence a and $b - 1$ must be multiples of 3, and k must be a cube. The smallest a satisfying both of these conditions is 3, and the smallest b is 4. For N to be smallest, we choose $k = 1$, so $N = 2^3 \cdot 3^4 = 8 \cdot 81 = 648$.

7. The shortest distance will be the smallest number of the form $75a + 330b$ (for integers a,b with $a > 0$). This is the greatest common divisor of 330 and 75, or 15.

8. Clearly, the center of the given circle is on the perpendicular bisector of line segment $\overline{AB}$. If this center is O, then (see diagram) $OX = OA = OB = r$, and $XY = 3 = XO + OY$. From right triangle AOY, $(3 - r)^2 + 2^2 = r^2$, so $9 - 6r + 4 = 0$ and $r = \frac{13}{6}$.

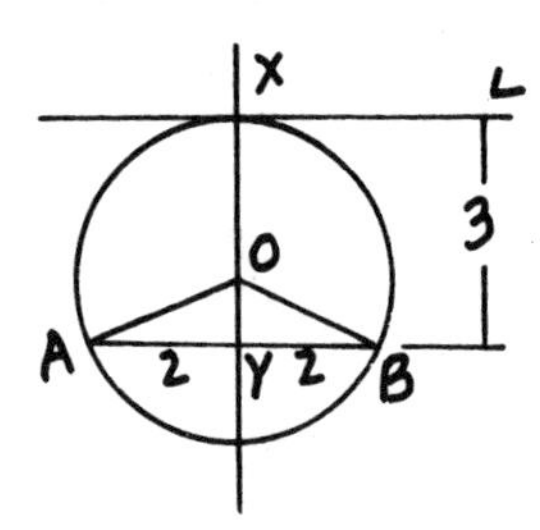

9. The diagram shows triangles drawn on the four sides of $ABXY$ that are congruent to triangle ABC. These form a new large square (since angle XBZ = angle CAB is complementary to angle CBA and C, B, and Z are collinear), whose side is 12. OC is half its diagonal, or $6\sqrt{2}$.

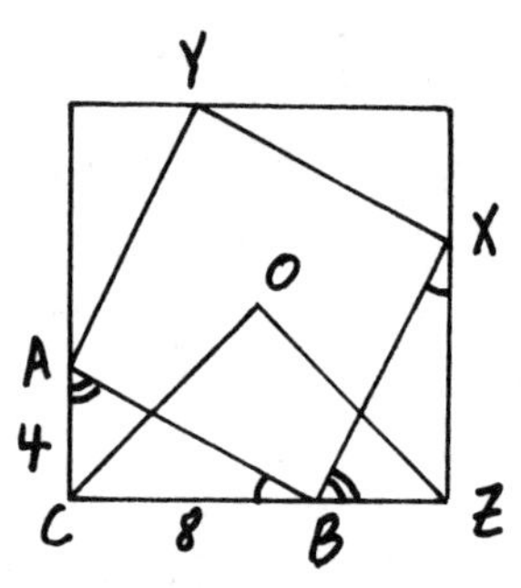

10. A number is a multiple of 9 if and only if the sum of its digits is a multiple of 9. Since $3^{1983} = 9 \cdot 3^{1981}$, each a_i is a multiple of 9. Now $3^{1983} = \frac{1}{3} \cdot 3^{1984} = \frac{1}{3} \cdot 9^{992} < \frac{1}{3} \cdot 10^{992}$, so a_1 has at most 992 digits. Then $a_2 < 9 \cdot 992 < 10000$, $a_3 \leq 4 \cdot 9 = 36$, and $a_4 \leq 9$. Since a_4 is a multiple of 9, we must have $a_4 = 9$. For $i > 4$, $a_i = 9$.

11. A comparison of triangle ABC with a 3-4-5 triangle shows that angle B is obtuse, so angle ACB is acute. Since angle ACD is the supplement of angle ACB, cos angle ACD = -cos angle ACB. Using the law of cosines,

$AB^2 = BC^2 + AC^2 - 2BC \cdot AC \cdot \cos \text{angle } ACB$

$AD^2 = CD^2 + AC^2 + 2CD \cdot AC \cdot \cos \text{angle } ACB$

so $AB^2 + AD^2 = 16 + 36 + 16 + 36 = 104$, and $AD^2 = 104 - 9 = 95$. Note that this method shows that if $\overline{AC}$ is a median of triangle ABC, then $4AC^2 = 2AB^2 + 2AD^2 - BD^2$.

11. (Continued)

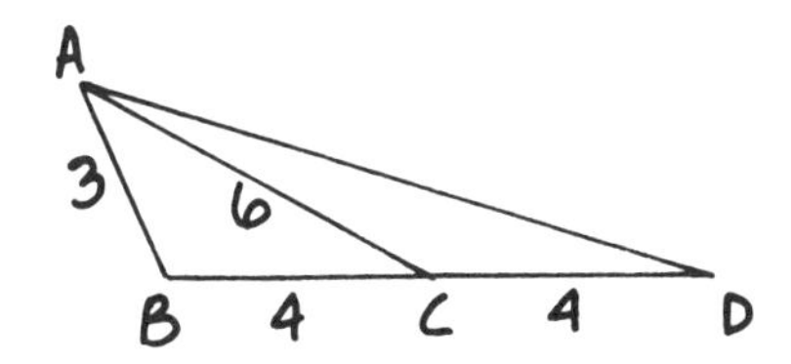

12. We need $xy = 10^k$, k a nonnegative integer. Since $x,y < 100$, $k < 4$. If $k = 0$, $(x,y) = (1,1)$. For $k = 1$, $xy = 10$. There are four integral divisors of 10, none of them greater than 100, so there are four ordered pairs. For $k = 2$, we have (2,50), (4,25), (5,20), (10,10), (20,5), (25,4), (50,2). For $k = 3$, we have (20,50), (25,40), (40,25), (50,20). Altogether, there are 1 + 4 + 7 + 4 = 16 possible ordered pairs.

13. Since $\sin^2 A + \cos^2 A = 1$ (see 4.11), $\cos^2 A = \sin^2 B$, so $\cos A = \pm \sin B$. If $\cos A = \sin B$, $A + B = 90°$, so $C = 90°$. If $\cos A = -\sin B$, $A + B = 180°$, which is impossible.

14. The number 1440 (base q) must be a multiple of 11; that is, 11 must divide $q^3 + 4q^2 + 4q$, or 11 divides $q(q + 2)^2$. Since 11 is prime, this means that 11 divides q or $q + 2$, and since $q \leq 10$, $q + 2 = 11$, and $q = 9$. Checking, $1441_9 = 1090_{10} = 11 \cdot 99 + 1$.

15. Any figure remains fixed upon rotation (about any point) of 360°. The figure in the problem must remain fixed under any combination of rotations about P by multiples of 48° or of 360°; that is, under rotations by angles whose degree-measure is of the form $48a + 360b$ (for positive or negative integers a,b). The smallest such number is $24 = 48 \cdot 8 - 360 \cdot 1$.

16. Triangle BXY is similar to triangle ABC, and its area is $\frac{1}{2}$ that of triangle ABC. Since the ratio of the areas of similar figures is the square of the ratio of the sides, the ratio of sides here must be $1:\sqrt{2}$. Hence $BX = BC/\sqrt{2} = 8/\sqrt{2}$ or $4\sqrt{2}$.

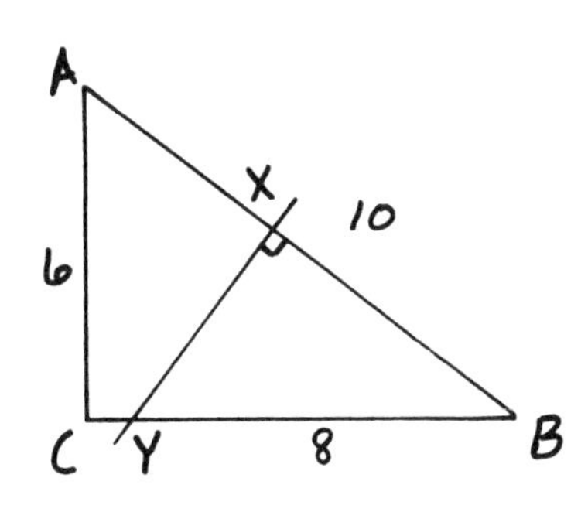

17. Since $\log_p q = \dfrac{1}{\log_q p}$ (see 5.12), we have $\dfrac{1}{\log_x n} = \dfrac{3}{\log_x m}$, or $3 \log_x n = \log_x n^3 = \log_x m$, and $m = n^3$.

18. Draw $\overline{QM} \perp \overline{PR}$. Since $\overline{BC}$ connects two midpoints of triangle PQR, it bisects any line segment from Q to line PR. Hence (see diagram) $QM = MH = BX$. Now $AX = YD = \frac{1}{2}(AD - BC) = 1$, so $BX = \sqrt{36 - 1} = \sqrt{35} = QM$. Since $BM = 3$, $BQ = \sqrt{35 + 9} = \sqrt{44}$ and $PQ = 2\sqrt{44} = 4\sqrt{11}$.

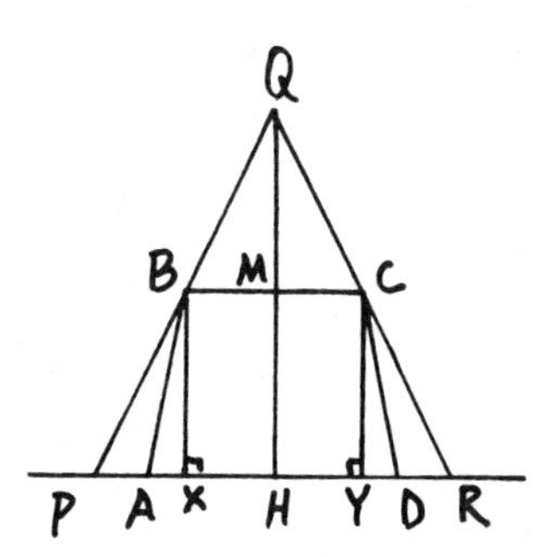

19. We have: $a_1{}^2 = 49$, so $a_2 = 4 + 9 = 13$

$a_2{}^2 = 169$, so $a_3 = 1 + 6 + 9 = 16$

$a_3{}^2 = 256$, so $a_4 = 2 + 5 + 6 = 13$.

Since a_i depends only on a_{i-1}, the sequence is 7,13,16,13,16, 13,16,... and $a_{1983} = 16$.

20. Call the given property $\emptyset$. Clearly, if a set of m (> 4) points has $\emptyset$, so must any subset with $(m - 1)$ points, so we can "build up" the set one point at a time. Suppose we have 4 points, A, B, C, D having $\emptyset$. Then 3 form an equilateral triangle, say triangle ABC, and D could be anywhere. Adding a 5th point E, we see that D and E must form an equilateral triangle with at least 2 of A, B, C; hence with exactly 2 (say, A and B), since in a plane 2 points can form an equilateral triangle with at most 2 other points. We now have A, B, C, D, E as in the diagram (note that the positions of D and E are determined only relative to each other). Try to add a 6th point F. Now F must form an equilateral triangle with 2 of D, E, C, hence must be an X or an O; but $\{X,A,B,D\}$ and $\{O,A,B,D\}$ don't have $\emptyset$. Thus $n = 5$.

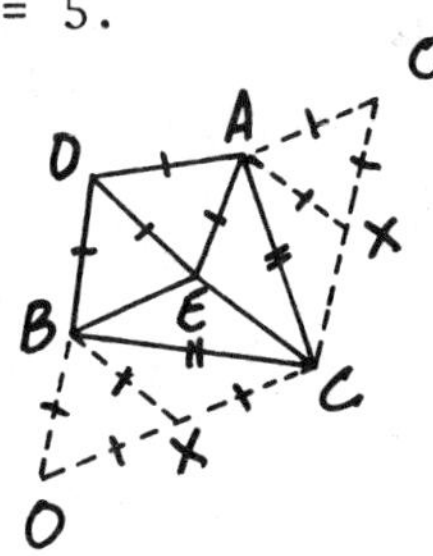

21. Adding, then subtracting, the last two equations gives $x = \frac{1}{2}(2 + a)$, $y = \frac{1}{2}(2 - a)$. Substitution in the first equation shows that $2a + a^2 + 2 - a = 2$, or $a^2 + a = 0$, and $a = 0, -1$. Both solutions work.

22. Triangles AMP, ANQ, ACB are similar, and the ratio of their areas is 1:2:3. Hence the ratio of their sides is $1:\sqrt{2}:\sqrt{3}$, so $\frac{NQ}{BC} = \frac{\sqrt{2}}{\sqrt{3}} = \frac{\sqrt{6}}{3}$. If $NQ = 4$, then $BC = 2\sqrt{6}$.

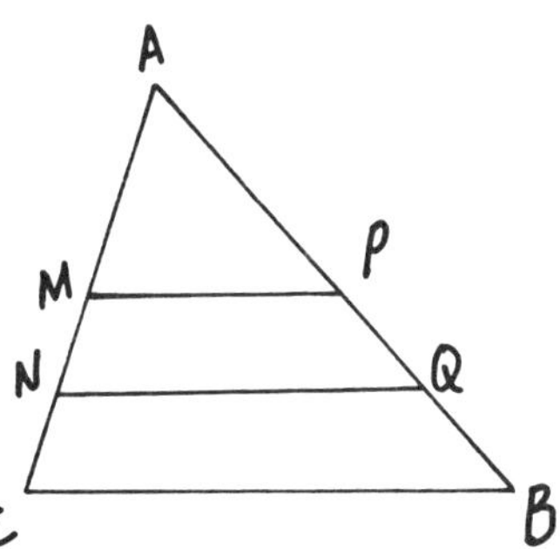

23. Draw $\overline{ZQ} \parallel \overline{BC}$. Since $AZ:ZB = AQ:QC$, $AQ = \frac{1}{3}AC$, and $QC = YC = \frac{2}{3}AC$. Also, $ZQ = AQ = \frac{1}{3}AC = XC$, so triangles ZQC and YXC are congruent. Thus angle YXC = angle QZC is complementary to angle QCZ and angle $XPC = 90$. Note that this remains true as long as the ratios $AX:XC$, $CY:YB$, $BZ:ZA$ are equal.

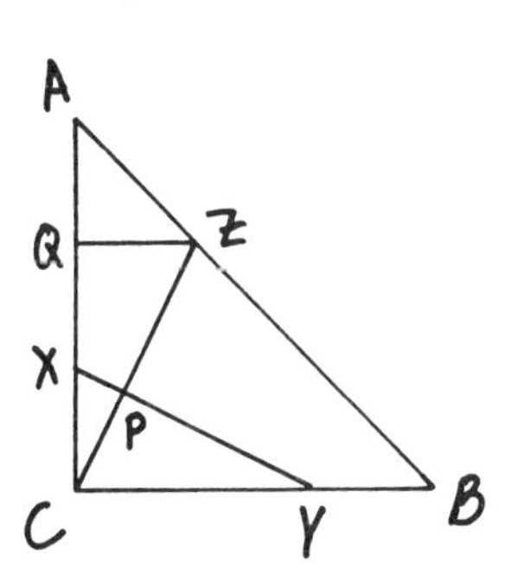

24. Write $n = 3k + r$ where $k \geq 0$ is a nonnegative integer and $r = 0, \pm 1$. $n^2 = 9k^2 + 6kr + r^2$, $\frac{n^2}{3} = 3k^2 + 2kr + \frac{r^2}{3}$, and since $\frac{r^2}{3} = 0$ or $\frac{1}{3}$, $\left[\frac{n^2}{3}\right] = 3k^2 + 2kr = k(3k + 2r)$. This is prime only if $k = 1$ and $r = 0$ or 1. These produce $n = 3k + r = 3$ or 4 (both check).

25. The two-digit multiples of 17 are 17,34,51,68,85, and those of 23 are 23,46,69,92. None of these numbers begins with the digit 7. Every other digit begins only one number, except 6, which begins two. Hence the sequence can "branch" whenever a 6 occurs but must end if a 7 occurs. Hence all such finite sequences end in 7. The shortest such, for example, is 9,2,3,4,6,8,5,1,7.

26. Draw $\overline{DY} \perp \overline{BC}$. Since arcs AB and XD are equal, $AB = XD = CD$, so angle YXD = angle BCD = angle BAD = 30°, and $XY = (\sqrt{3}/2)\ XD$ $= 4(\sqrt{3}/2) = 2\sqrt{3}$. Hence $XC = 4\sqrt{3}$.

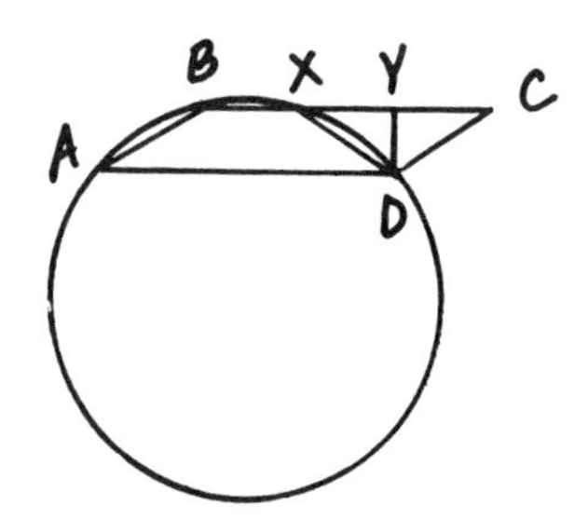

27. The original cone is similar to the cone formed by one of the new volumes. Since the ratio of the volumes of similar figures is the cube of the ratio of linear dimensions,

$$\frac{1}{2} = \frac{x^3}{12^3}, \text{ and } x = \frac{12}{\sqrt[3]{2}}.$$

28. Adding the three equations and noting that $[x] + \{x\} = x$, we have $x + y + z = 3.3$. Subtracting from this the sum of the first two equations shows that $[y] + \{x\} = 0$. The sum of an integer and a fraction $(0 \leq \{x\} < 1)$ can be 0 only if both are 0, so x is an integer, and $0 \leq y < 1$. Hence the first equation can be written as $x + \{z\} = 1.1$, so that $x = 1$, $\{z\} = .1$. The second equation shows that $y + [z] = 2.2$, so $[z] = 2$ and $y = .2$. Hence (x,y,z) $= (1,.2,2.1)$.

29. We have $2^{\tan x} = 2^{3\sin x}$, so $\tan x = 3 \sin x = \dfrac{\sin x}{\cos x}$.

Since $\sin x \neq 0$ in the given domain, $3 = \dfrac{1}{\cos x}$ and $\cos x = \dfrac{1}{3}$.

30. Let m angle $B = \beta$, and draw $\overline{CP}$. Then (see 3.93) $AC = 2R \sin \beta$. Let us find PQ in terms of β. We have angle $QCB = 90 - \beta$ = angle BAP (from right triangles ABM, CBN). And angle BAP = angle BCP, since they both intercept arc BP. Hence m angle $QCP = 180 - 2\beta$ and $PQ = 2R \sin(180 - 2\beta) = 2R \cos 2\beta$. Hence $2 \cdot 2R \cos 2\beta$ $= 7 \cdot 2R \sin \beta$. Using (4.22), we have: $2 \cos 2\beta = 2 - 4 \sin^2\beta$ $= 7 \sin \beta$. Thus $4 \sin^2\beta + 7 \sin \beta - 2 = 0$ and $\sin \beta = \dfrac{-7 \pm \sqrt{49 + 32}}{8}$ $= \dfrac{-7 \pm 9}{8} = -2, \dfrac{1}{4}$. The root -2 is extraneous.

30. (Continued)

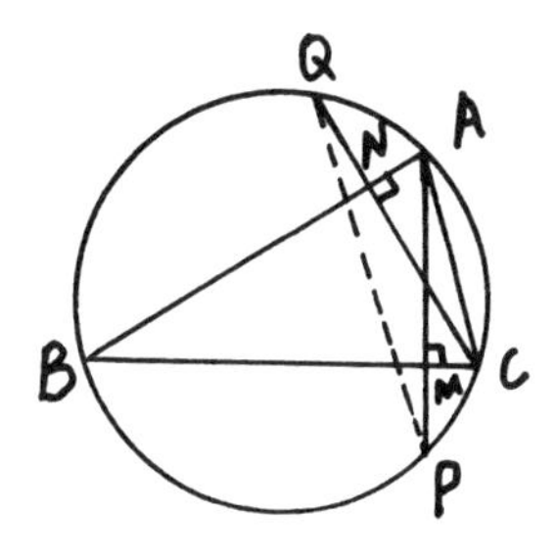

Solutions—Spring 1984

1. If the original number was N, then $2\sqrt{N} = 10$, $\sqrt{N} = 5$, and $N = 25$. She should have taken $2 \cdot 25^2 = 1250$.

2. Since $\overline{PT} \perp \overline{OT}$, angle $POT = 50°$, so arc $AX = 100°$ and arc $OX = 80°$.

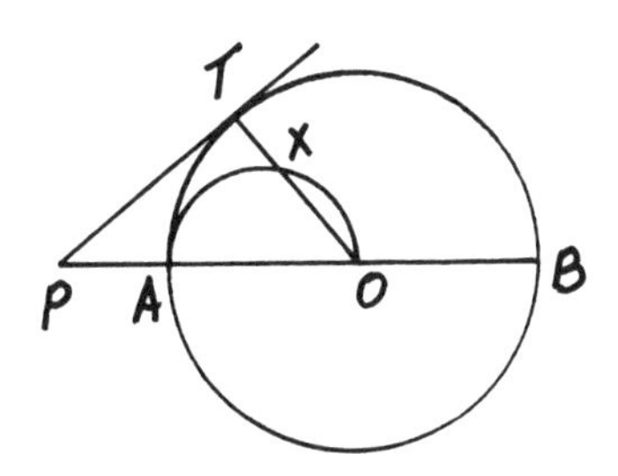

3. Draw $\overline{OQ} \| \overline{AC}$ and $\overline{PQ} \| \overline{BC}$ as shown. Since angle OCX + angle XCW = angle $WCP = 45° + 90° + 45° = 180°$, points O, C, and P are collinear. Then OQP is an isosceles right triangle, $OQ = OR + RQ = \frac{1}{2} AC + \frac{1}{2} CW = 7$, so $OP = 7\sqrt{2}$.

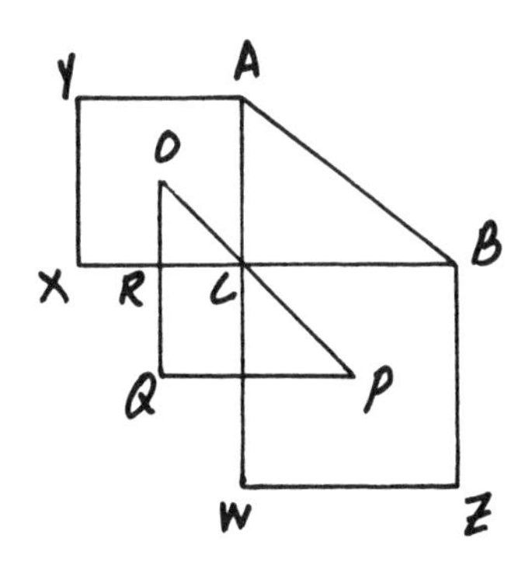

4. Since $(\sqrt{13} + \sqrt{3})(\sqrt{13} - \sqrt{3}) = 13 - 3 = 10$,

$\log_{10}(\sqrt{13} + \sqrt{3}) + \log_{10}(\sqrt{13} - \sqrt{3}) = \log_{10}10 = 1$,

and $\log_{10}(\sqrt{13} - \sqrt{3}) = 1 - a$.

5. We have: $a_n - a_{n-1} = 2n$

$$a_{n-1} - a_{n-2} = 2(n - 1)$$

$$a_{n-2} - a_{n-3} = 2(n - 2)$$

$$\vdots$$

$$a_2 - a_1 = 2 \cdot 2$$

$$a_1 = 1.$$

Add: $a_n = 1 + 2(2 + 3 + 4 + \ldots + n) = -1 + 2(1 + 2 + \ldots + n)$ (see 1.53) $= -1 + n(n + 1) = n^2 + n - 1$.

6. We can relate the given ratios to segments along diagonal $\overline{PR}$ by drawing $\overline{QA}||\overline{SB}||\overline{XY}$ (see diagram). Then, from properties of parallel lines, $PQ{:}XP = AP{:}ZP = 3$ and $SP{:}YP = BP{:}ZP = 4$. Since triangle $AQP \cong$ triangle RSB, we have $PR = PB + BR = PB + PA$, so $AP{:}ZP + BP{:}ZP = PR{:}ZP = 3 + 4 = 7$ and $\frac{PZ}{PR} = \frac{1}{7}$.

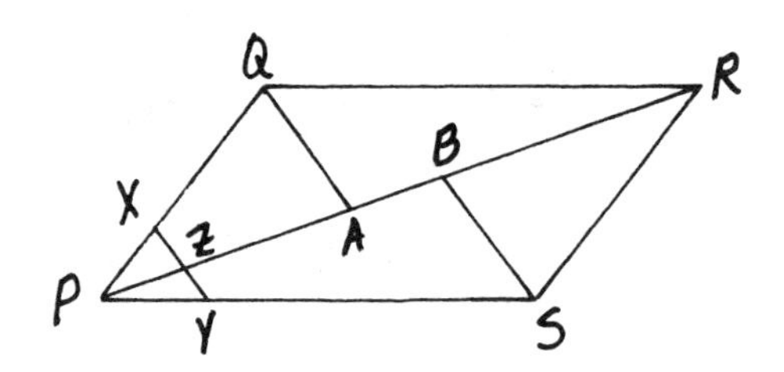

7. N must be a perfect cube, and its cube root is less than 10 (since $10^3 = 1000$). Hence N can only be $5^3 = 125$, $6^3 = 216$, $7^3 = 343$, $8^3 = 512$, or $9^3 = 729$. A quick check shows that only 512 works.

8. There are 5 trains per hour on the Jerome line and 10 trains on the White Plains line. Hence there are 15 trains per hour on the Lexington line, and the interval between them is 4 minutes. Actually, other constant intervals are possible, if we allow trains to wait before proceeding down the trunk line. Such a solution would allow a schedule in which trains travel down the trunk line at arbitrary constant intervals. But it would create a "backup" of trains at the beginning of the trunk line.

9. If the given expression were greater than 4, one of the addends would have to be greater than 2. But both are clearly less than 2. Hence the required integer is either 2 or 3. Note that $\log_4 9 > \log_4 8 = \frac{3}{2}$, $\log_9 28 > \log_9 27 = \frac{3}{2}$, so the sum is greater than 3.

10. First let us assume that the square is symmetric about the altitude of the triangle. Then, if $x = MD$ (see diagram), angle $AMD = 45°$, so $DE = EM = \dfrac{x}{\sqrt{2}}$, and $AE = \dfrac{x/\sqrt{2}}{\sqrt{3}} = \dfrac{x}{\sqrt{6}}$. But $AE + EM = \dfrac{x}{\sqrt{2}} + \dfrac{x}{\sqrt{6}} = 6$,

$$x = \frac{6}{\frac{1}{\sqrt{2}} + \frac{1}{\sqrt{6}}} = 9\sqrt{2} - 3\sqrt{6} \text{ and } (p,q) = (9,-3).$$

To prove that the position of the square must indeed be as indicated, it is enough to prove that the square is unique. This can be done by noting that point D is the image of point F when rotated about M by 90°. Hence D is on the image of line BC under the rotation. This image is a line and can intersect AC at only one point.

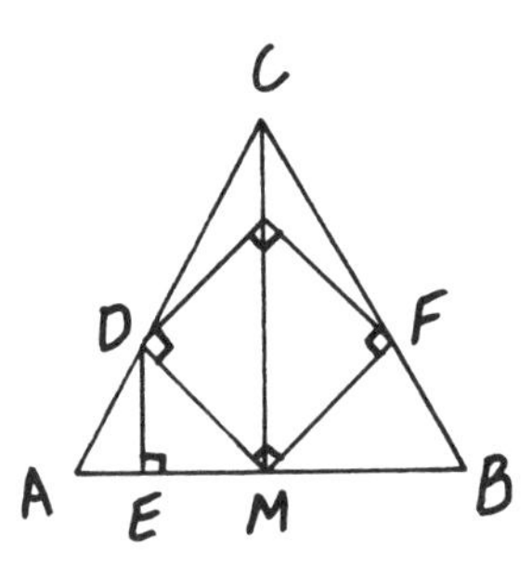

11. Note that in this range, $\cos x - \sin x$ is positive. Now $(\cos x + \sin x)^2 + (\cos x - \sin x)^2 = 2$ (see 4.11), so $(\cos x - \sin x)^2 = 2 - 25/16 = 7/16$, and $\cos x - \sin x = \sqrt{7}/4$.

12. We can take the sum in pairs. Let $R(x)$ = the remainder when x is divided by

x	$2^x + x^2$	$R(2^x + x^2)$
0	1 + 0	1
1	2 + 1	3
2	4 + 4	0
3	0 + 1	1
⋮	⋮	⋮
$4K$	0 + 0	0
$4K + 1$	0 + 1	1
$4K + 2$	0 + 4	4
$4K + 3$	0 + 1	1

The required sum is
$R(1 + 3 + 0 + 1 + 24(0 + 1 + 4 + 1) + 0)$
$= R(5 + 24 \cdot 6) = R(5) = 5$.

13. N must be a perfect sixth power. The only such number in the given range is 729. In fact, $729 = 9^3 = 27^2$.

14. Angle XCW is supplementary to angle ACB. By the law of cosines, $AB^2 = 5^2 + 7^2 - 2 \cdot 5 \cdot 7 \cos\theta$ and $XW^2 = 5^2 + 7^2 + 2 \cdot 5 \cdot 7 \cos\theta$. Hence $AB^2 + XW^2 = 2(5^2 + 7^2) = 148$.

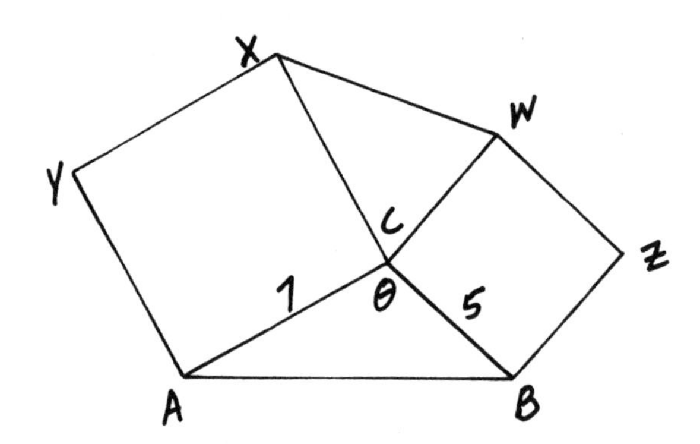

15. If the roots are $r_1, r_2, \ldots, r_{17}$, then their sum is 0, and $r_1^{17} = 3r_1 - 1$, $r_2^{17} = 3r_2 - 1$, $r_3^{17} = 3r_3 - 1$, ..., $r_{17}^{17} = 3r_{17} - 1$. Adding gives $r_1^{17} + r_2^{17} + \ldots + r_{17}^{17} = 3(r_1 + r_2 + \ldots + r_{17}) - 17 = -17$.

16. A train leaves Times Square at 10:30 and arrives at Main Street at noon. This is the first train met by the noon train out of Main Street. The noon train meets each train out of Times Square from 10:30 until 1:30 (when it arrives at Times Square). There are 13 such trains.

17. Since angle CED and angle CFD are both inscribed in semicircles, each is a right angle. Hence $CFDE$ is a rectangle and $EF = CD$. The area of the triangle is 54, which must be equal to $\frac{1}{2}AB \cdot CD = \frac{1}{2} \cdot 15 \cdot CD$. Hence $EF = CD = \dfrac{54}{(15/2)} = \dfrac{36}{5}$.

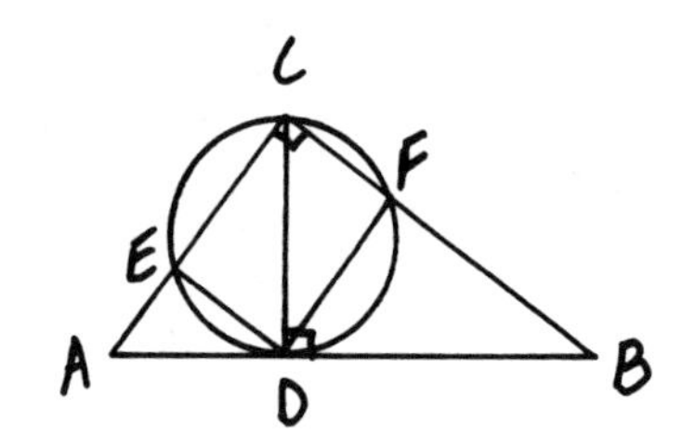

18. Write $\cos^6 x - \sin^6 x + \frac{(\sin^2 2x)(\cos 2x)}{4} = \cos^6 x - \sin^6 x$

$+ \frac{(4\sin^2 x \cos^2 x)(\cos^2 x - \sin^2 x)}{4}$ (see 4.21 and 4.22)

$= \cos^6 x - \sin^6 x + \sin^2 x \cos^4 x - \sin^4 x \cos^2 x$

$= (\cos^4 x - \sin^4 x)(\cos^2 x + \sin^2 x)$

$= (\cos^2 x - \sin^2 x)(\cos^2 x + \sin^2 x)(1)$ (see 4.11)

$= (\cos^2 x - \sin^2 x)(1) = \cos 2x.$

Now $\cos 2x = 0$ if and only if $2x = 90°$ or $270°$ and $x = 45°$ or $135°$.

19. Since $4(AB) < 100$, $A = 1$ or 2. Since CA is a multiple of 4, hence even, $A = 2$. Therefore, $B < 5$. Now since $4(AB)$ ends in 2, $B = 3$, so $C = 9$.

20. Let the rectangles be $ABCD$ and $AXCY$. Then triangles ADP, PYC, CBQ, QXA are congruent (see diagram). If $DP = x = PY$, then $AP = AY - PY = 8 - x$. Thus $AD^2 + DP^2 = 2^2 + x^2 = AP^2 = (8 - x)^2$, so $4 = 64 - 16x$ and $x = \frac{15}{4}$. Then $\overline{AX} \cong \overline{AD}$ is an altitude in parallelogram $APCQ$, whose area is thus $2(8 - x) = \frac{17}{2}$.

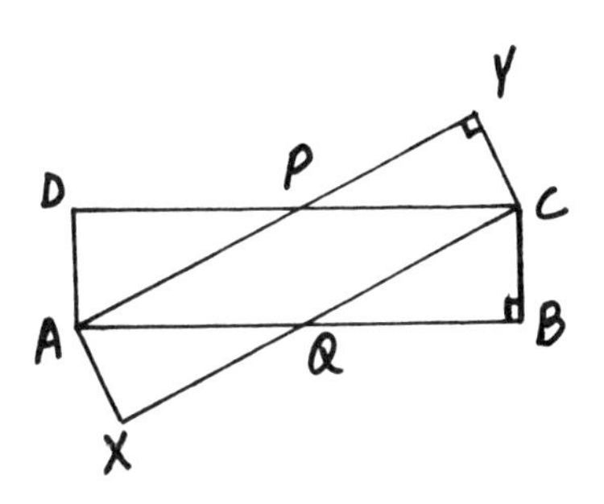

21. Let $n = \sqrt{a}$. Then $\sqrt{b} = kn$, and if $1000 \leq a < b < 10000$, then certainly $31 < n < kn < 100$. Hence k can be only 2 or 3. Suppose a, in decimal notation, is $ABCD$. If $k = 2$, then $4(ABCD) = DCBA$. Hence A is even and cannot be more than 2 (or $b > 10000$). If $A = 2$, then $4(ABCD) = DCBA > 8000$, so $D = 8$ or 9. D cannot be 8, as no square can end in 8 (in decimal notation). And if $D = 9$, $4(ABCD)$ ends in 6, not 2. Hence k cannot be 2. If $k = 3$, $9(ABCD) = DCBA$, so that $A = 1$ and $D = 9$. The only perfect square satisfying this is $33^2 = 1089$. Indeed, $9801 = 99^2$ and $99 = 3 \cdot 33$.

22. We use absolute value to denote area. Since triangle $ABX \sim$ triangle CDX, $AX:XC = AB:DC = 2:3$, and $AP:PD = AX:XC = 2:3$ as well. Since triangles APX, PXD have the same altitude from X, the ratio of their areas is also 2:3. Since triangle $PXD \sim$ triangle ABD and $DP:DA = 3:5$, then $|PXD|:|ABD| = 9:25$. Thus if $|APX| = 2k$, then $|PXD| = 3k$ and $|ABD| = (25/9) \cdot 3k = (25/3)k$, and the ratio $|APX|:|ABD| = 6:25$.

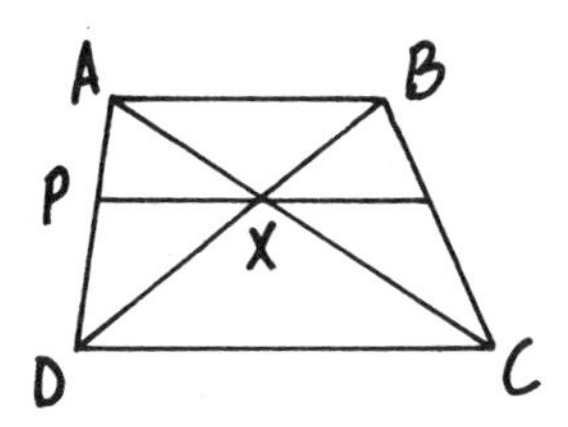

23. For any point P outside the unit circle, the length of the tangent (see diagram) is $\sqrt{x^2 + y^2 - 1}$. If (x,y) is on the given hyperbola, this distance equals $\sqrt{2y^2 - 4y + 8} = \sqrt{2(y - 1)^2 + 6} \geq \sqrt{6}$, with equality when $y = +1$, $x = \pm\sqrt{6}$.

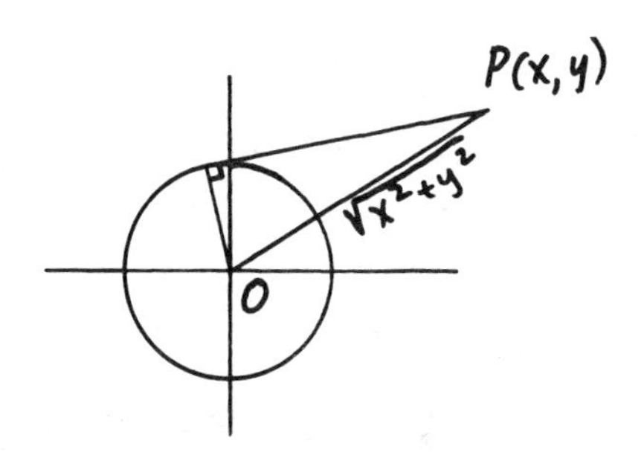

24. Let x = the required interval. If a bus passes the student at point A, then the next bus arrives at point A x-minutes later. It must then spend $12 - x$ minutes catching up, so it takes $12 - x$ minutes to travel the same distance that the student walks in 12 minutes. The ratio of the rates is then $\frac{12 - x}{12}$. If the student meets a bus at point B, then she passes the next bus 4 minutes later, so that the next bus has spent $x - 4$ minutes going the same distance that the student has gone in 4 minutes. Hence the ratio of the rates is $\frac{x - 4}{4}$. The two ratios must be equal (since the speeds are constant). Hence $\frac{12 - x}{12} = \frac{x - 4}{4}$, and $x = 6$.

25. If there are b boys and g girls, then $b = g + 1$ and $2(g - 1) = b$. Hence $2g - 2 = g + 1$, $g = 3$, and $b = 4$. There are 7 children in the family.

26. By symmetry, $\overline{OY}$ (the perpendicular bisector of $\overline{RS}$ in the diagram) bisects angle AOB. Then angle OQR = angle OQX + angle $XQR = 30° + 90° = 120°$. If $QR = s$, then $XQ = \frac{s}{2}$ and from 30-60-90 triangle OXQ, $OQ = \frac{s\sqrt{3}}{3}$. Using the law of cosines in triangle OQR, $13^2 = OR^2 = s^2 + \frac{s^2}{3} + \frac{s^2\sqrt{3}}{3} = s^2\left(\frac{4 + \sqrt{3}}{3}\right)$. Hence $s^2 = \frac{13^2 \cdot 3}{4 + \sqrt{3}} = 39(4 - \sqrt{3})$ $= 156 - 39\sqrt{3}$.

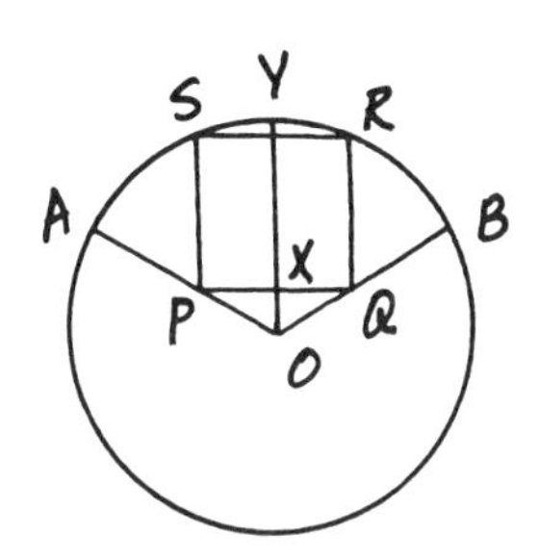

27. We have $x^7 - \frac{1}{2}x^6 - \frac{5}{2}x^5 + \frac{17}{2}x^4 + 0x^3 - \frac{419}{2}x^2 - \frac{372}{2}x - \frac{10}{2} = 0$.

Also, $(1 + r_1)(1 + r_2)\ldots(1 + r_7) = 1 + \sum_{i=1}^{7} r_i + \sum r_i r_j + \sum r_i r_j r_k$ $+ \ldots + r_1r_2r_3r_4r_5r_6r_7$. These are just the "symmetric functions" of (1.4), so this sum equals $1 - a_6 + a_5 - a_4 + a_3 - a_2 + a_1 - a_0$ $= (2 + 1 - 5 - 17 + 0 + 419 - 372 + 10)/2 = 32/8 = 19$. A more direct solution is accomplished by using the result from problem 24, Spring 1983. Using this insight, if $P(x)$ is the given polynomial, we need only calculate $(-1/2)P(-1)$.
For more information on the symmetric functions of the roots of a polynomial equation, see Borofsky, or Shklyarsky, Chentsov, and Yaglom (Chapter 9).

28. We can write: $y^2 = x - 1 + \frac{6}{x}$. Since x and y are integers, $\frac{6}{x}$ must be an integer. Hence $x = 1,2,3,6$, and $y^2 = 6,4,4,6$. But if y^2 is a perfect square, y^2 can only be 4. The possible solutions are thus (2,2) and (3,2).

29. The student can always choose from the three largest piles, whose sum must be at least 50. For suppose their sum were less. Then the smallest of the 3 could not have more than 15 pennies (if it had 16 or more, the sum of these 3 largest piles would be greater than $16 + 17 + 18 > 50$ pennies). Then the remaining piles could not total more than $14 + 13 + 12 + 11 = 50$ pennies, and there could not be 100 pennies altogether. A set of piles of 19, 16, 15, 14, 13, 12, and 11 coins shows that disturbing three piles may in fact be necessary.

30. There are 2 cases to consider:

<u>Case 1</u>: The right angle intersects two different sides of the triangle. Moving from the symmetrical situation $\overline{BP} \cong \overline{BQ}$ to P' and Q' we see Area I > Area II, since $\overline{OP} \cong \overline{OQ}$, angle $POP' \cong$ angle QOQ', and $P'O > Q'O$ ($P'O$ is further from the perpendicular). So we lose more area than we gain and area $OPBQ$ > area $OP'BQ'$ (with limiting case $P' = B$, Q' between B and C).

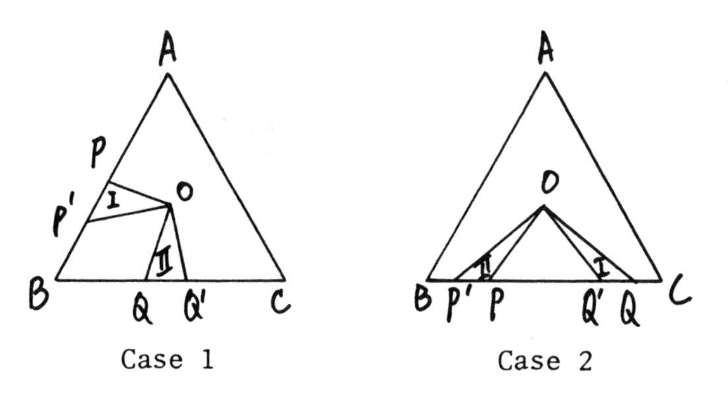

Case 1 Case 2

<u>Case 2</u>: The right angle intersects the same side of the triangle twice. We can start with $\overline{BP} \cong \overline{CQ}$ and "rotate" the angle towards vertex B. By an argument similar to that of Case 1, area triangle POQ < area triangle $P'OQ'$ (with the same limiting case as above). Hence the maximum area = area $OPBQ$ as in case 1.

Note $OB = \frac{2}{3}\left(\frac{\sqrt{3}}{2}\right) = \frac{\sqrt{3}}{3}$.

In case 1, from triangle BPR,

$$\frac{\frac{\sqrt{3}}{3} - x}{x} = \sqrt{3}, \text{ or } x = \frac{\sqrt{3}}{3(1 + \sqrt{3})} \text{ and the area } OPBQ = \frac{1}{2}(PQ)(OB)$$

$$= \frac{1}{2}(2x)\left(\frac{\sqrt{3}}{3}\right) = \frac{1}{3(1 + \sqrt{3})}.$$

30. (Continued)

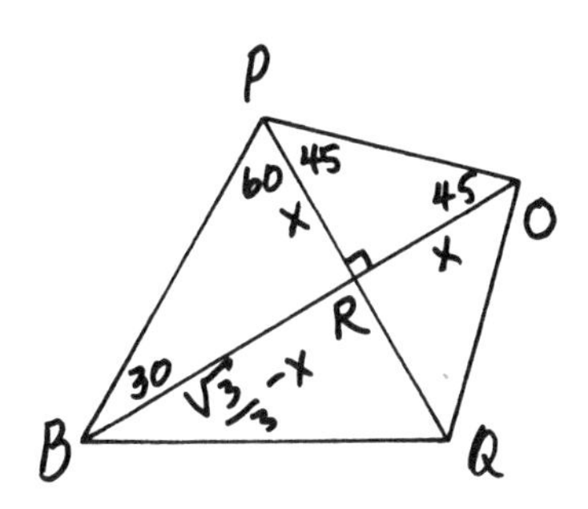

A short calculation will show that the maximal area in case 2 is only $\frac{1}{12}$.

Solutions—Fall 1984

1. Let $AP = x$ and $PB = 3x$. Then triangles APS, PBQ, QCR, RDS are congruent, and the sum of their areas is $6x^2$. The area of the large square is $16x^2$, so the area of the smaller square is $10x^2$, and the required ratio is 10:16 = 5:8.

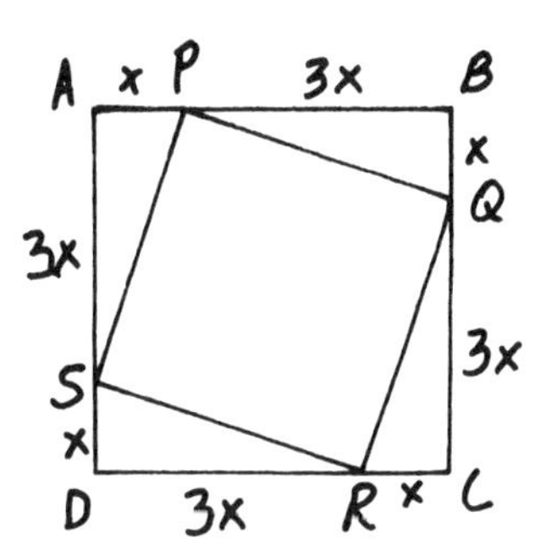

2. She will get all three questions correct $\frac{1}{8}$ of the time. Failing this, she will get only the first problem wrong $\frac{1}{8}$ of the time, only the second $\frac{1}{8}$ of the time, and only the third $\frac{1}{8}$ of the time. Hence she will get at least two right $\frac{1}{8} + \frac{1}{8} + \frac{1}{8} + \frac{1}{8} = \frac{1}{2}$ of the time.

3. Using (3.5), we have $CX = OX = 2$, $XD = TD = 4$. If M is the intersection of $\overline{BD}$ and $\overline{OP}$, then $BM = MD$ (by symmetry), so $MD = 5$ and $MT = MD - TD = 1$. Hence in right triangle OMT, $OM = \sqrt{5}$ and $OP = 2\sqrt{5}$.

3. (Continued)

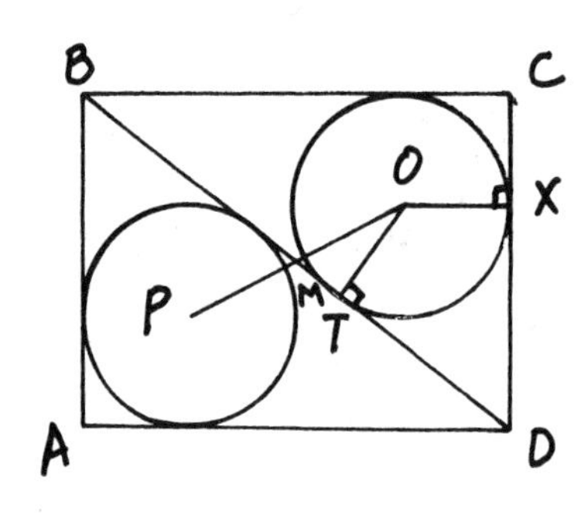

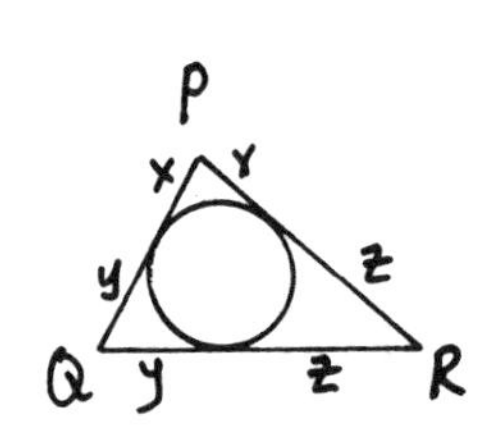

4. The given cube is slightly less than $8 \cdot 10^9$, so the integer is slightly less than 2000. Also the rightmost digit must be a 3 (otherwise the cube would not end with a 7). The possible choices are 1993, 1983, 1973, etc. Trial and error can finish the solution. Or, we can note that $1993^3 > 1990^3 = 10^3(200 - 1)^3 = 10^3(8 \cdot 10^6 - 3 \cdot 4 \cdot 10^4 + 6 \cdot 10^2 - 1) > 10^7(800 - 12) = 788 \cdot 10^7$, so 1993 is too big. Writing 1973^3 as $(2 \cdot 10^3 - 3^3)^3$ and proceeding similarly, we can see that this number is too small.

5. We have: (i) $x^2 - 2xy = 0$ and (ii) $x^2 = y + 3$.
From (i), $x(x - 2y) = 0$, so $x = 0$ or $x = 2y$. If $x = 0$, $\log_3 x$ is undefined. If $x = 2y$, we have $4y^2 - y - 3 = 0$, so $(4y + 3)(y - 1) = 0$ and $y = \frac{3}{4}$, 1 and $x = \frac{-3}{2}$, 2. Only the positive values will satisfy (ii).

6. Expanding and rearranging the terms yields:

$(\cos 15 \sin 75 + \sin 15 \cos 75) + (\cos 15 \cos 45 - \sin 15 \sin 45)$
$= \sin(15 + 75) + \cos(45 + 15) = 1 + \frac{1}{2} = \frac{3}{2}$ (see 4.311 and 4.313).

7. $h(x) = f(1 - x) = 1 - \frac{1}{1 - x}$. When $h(x) = 8$, we solve for x to yield $1 - \frac{1}{1 - x} = 8$ so $x = \frac{8}{7}$.

8. Let $N \frac{\cos 15° + \sin 15°}{\cos 15° - \sin 15°}$.

Then $N^2 = \frac{\cos^2 15° + \sin^2 15° + 2 \cos 15° \sin 15°}{\cos^2 15° + \sin^2 15° - 2 \cos 15° \sin 15°}$

$= \frac{1 + \sin 30°}{1 - \sin 30°} = \frac{\frac{3}{2}}{\frac{1}{2}} = 3.$ (See 4.11, 4.21.)

Since $\cos 15° > \frac{1}{2} > \sin 15°$, N is positive and so $N = \sqrt{3}$.

9. Draw $\overline{XA}$, $\overline{YB}$, $\overline{ZC}$. Since $BC = CX$ and triangles ZBC, ZCX have a common altitude, $|\text{triangle } ZBC| = |\text{triangle } ZCX|$ (using absolute value for area). Also, since $AB = BZ$, $|\text{triangle } ACB| = |\text{triangle } CBZ|$. Similarly, we find that $|\text{triangle } AXC| = |\text{triangle } AXY| = |\text{triangle } ABC|$ and $|\text{triangle } AYB| = |\text{triangle } BYZ| = |\text{triangle } ABC|$. Hence the required ratio is 7:1.

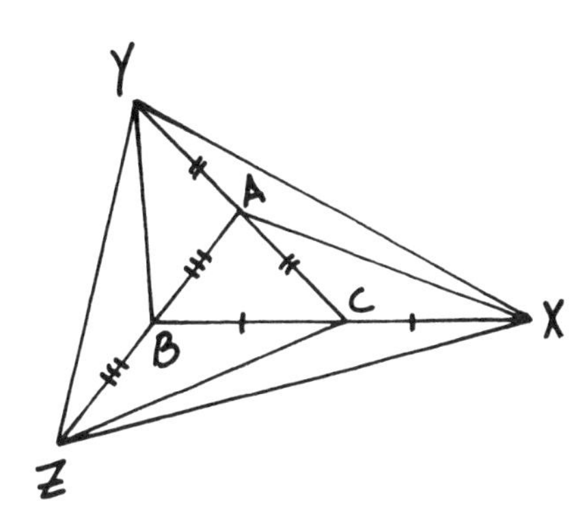

10. If the roots are $r - d$, r, and $r + d$, then the sum of the roots is $3r = 6$, so $r = 2$ is a root. Substitution in the original equation gives $8 - 24 - 48 + c = 0$, and $c = 64$.

11. A round trip for A takes 40 minutes. A round trip for B takes 30 minutes. A round trip for C takes 64 minutes. We need the least common multiple of these three. Since $30 = 2 \cdot 5 \cdot 3$, $40 = 2^3 \cdot 5$, $64 = 2^6$, the least common multiple is $2^6 \cdot 3 \cdot 5 = 64 \cdot 15$ minutes, or $64 \cdot \frac{1}{4} = 16$ hours.

12. From the Pythagorean theorem, $AB = 1$. Also, the median to the hypotenuse is half as long as the hypotenuse. Therefore, $\tan\theta = \frac{\sin\theta}{\cos\theta} = \frac{1}{2}$. From this equation, $\cos\theta$ is clearly the longer leg, and by the Pythagorean theorem: $\left(\frac{\cos\theta}{2}\right)^2 + \cos^2\theta = 1$ and solving yields $\cos\theta = \frac{2\sqrt{5}}{5}$.

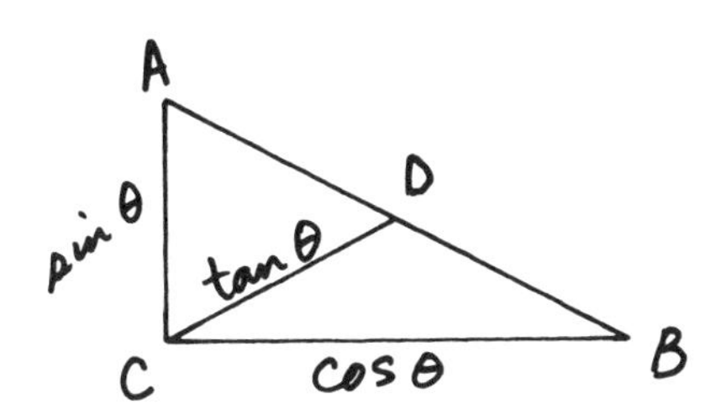

13. A line connecting two midpoints of a triangle is equal to half the third side. Hence $MP = \frac{1}{2}FD$ and $MP = \frac{1}{2}AC$, so $AC = DF$. Similarly, $BC = EF$ and $AB = DE$, so that the required ratio is 1.

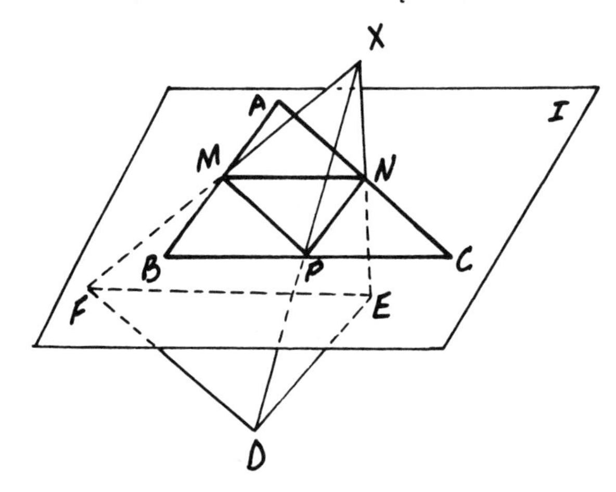

14. Letting $x = 5$, we have $2f(5) + f(-4) = 25$. Letting $x = -4$, we have $2f(-4) + f(5) = 16$. These are two equations in two unknowns. Solving, we find $f(5) = \frac{34}{3}$, $f(-4) = \frac{7}{3}$. In general, $f(x) = \frac{x^2 + 2x - 1}{3}$.

15. We use 3.93. Let angle $ABC = \beta$, angle $ACB = \gamma$, and the radii of the circles be r and R (the circle of radius r is through A, B, and P). Then $AP = 2r \sin \beta = 2R \sin \gamma$, so $r{:}R = \sin \gamma{:}\sin \beta = 5{:}8$ (see 3.4).

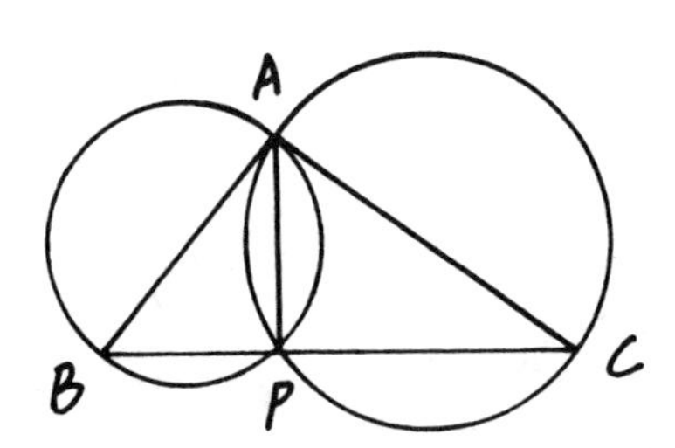

16. We have $\frac{x^2 + y^2}{x - y} = \frac{x^2 - 2xy + y^2 + 2xy}{x - y} = \frac{(x - y)^2 + 2}{x - y}$ $= (x - y) + \frac{2}{x - y}$. Let $z = x - y$. Then $z > 0$ and $z \cdot \frac{2}{z} = 2$, a constant, so $z + \frac{2}{z}$ is minimal when $z = \frac{2}{z}$, or $z = \sqrt{2}$ (see 6.4). This gives $x = \frac{\sqrt{6} + \sqrt{2}}{2}$, $y = \frac{\sqrt{6} - \sqrt{2}}{2}$, and $\frac{x^2 - y^2}{x - y} = 2\sqrt{2}$. Except to check that both are positive, it is not necessary to solve explicitly for x and y. Refer also to problem 9, Fall 1982.

17. We need only consider $n > 0$. Let $n = 3k + \ell$ where $\ell = 0, 1, 2$ and $k > 0$. If $\ell = 0$, $n^2 = 9k^2$, $\left[\frac{n^2}{3}\right] = 3k^2$, prime only for $k = 1$. If $\ell = 1$, $\left[\frac{n^2}{3}\right] = 3k^2 + 2k = k(3k + 2)$, prime only for $k = 1$. If $\ell = 2$, $\left[\frac{n^2}{3}\right] = 3k^2 + 4k + 1 = (3k + 1)(k + 1)$, which cannot be prime. Letting $n = 3 \cdot 1 + 1 = 4$, $\left[\frac{n^2}{3}\right] = 5$.

18. Let P' be the image of P when rotated about a typical point Q. Triangle $PP'Q$ is always right and isosceles, so angle $QPP' = 45°$ and $PP' = QP\sqrt{2}$. Hence we can obtain P' from Q by rotating $45°$ about P and dilating the resulting figure (performing a homothecy) about P by a factor of $\sqrt{2}$. This means that the locus of P' is another square, whose side is $\sqrt{2}$ and whose area is thus 2. For more on homothecy, see references cited in problem 21, Spring 1975.

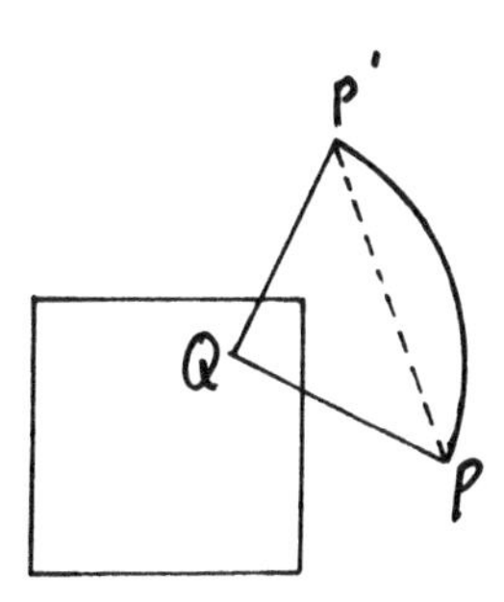

19. Let circle O be tangent to $\overline{AB}$ at X and to $\overline{BC}$ at Y. By symmetry, point O is on diagonal $\overline{BD}$. If $OD = OX = OY = r$, then $OXBY$ is a square, so $OB = r\sqrt{2}$. Thus $OD + OB = r + r\sqrt{2} = DB = \sqrt{2}$, so

$$r = \frac{\sqrt{2}}{1 + \sqrt{2}} = -2(1 - \sqrt{2}) = 2 - \sqrt{2}, \text{ and } (p,q) = (2,-1).$$

20. Adding, we have $x^2 + 2xy + y^2 + x + y = 42$, $(x + y)^2 + (x + y) - 42 = 0$. Let $x + y = z$. Then $z^2 + z - 42 = 0$, or $z = -7, 6$. Subtracting the original equations, we have $x^2 - y^2 + x - y = (x + y)(x - y) + (x - y) = (x + y + 1)(x - y) = -14$,

so $x - y = \frac{7}{3}$ or $x - y = -2$

$x + y = -7$ $\qquad$ $x + y = 6$

$(x,y) = (-7/3, -14/3)$ $\qquad$ $(x,y) = (2,4)$

21. Clearly $OT = 20$. Since $OQ = 24$, $QP = 7$, we know $OP = 25$. Then, if angle $POQ = \theta$, angle $TOP = \emptyset$, we can write angle $TOQ = (\theta + \emptyset)$ and $OX = a = OT \cos(\theta + \emptyset)$. Now $\cos\theta = \frac{OQ}{OP} = \frac{24}{25}$, $\cos\emptyset = \frac{OT}{OP} = \frac{20}{25}$, so $\cos(\theta + \emptyset) = \frac{24}{25} \cdot \frac{20}{25} - \frac{7}{25} \cdot \frac{15}{25} = \frac{375}{625} = \frac{3}{5}$ and $OX = 20 \cdot \frac{3}{5} = 12$ (see 4.313).

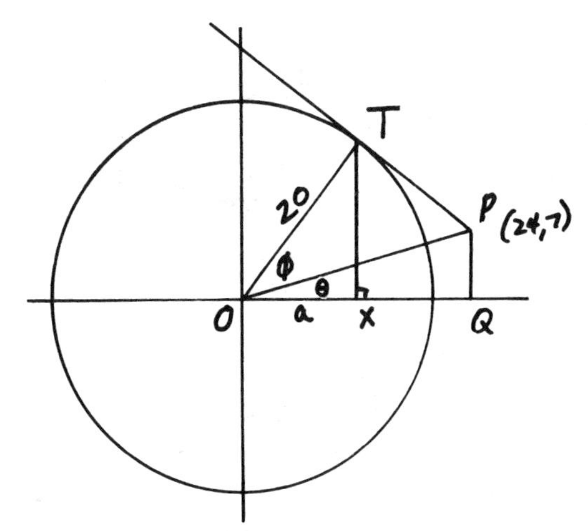

22. If we draw the circle through D and C which is tangent to $\overrightarrow{BA}$ at E, angle DEC is $\frac{1}{2}$ arc CD. For any other point E' on $\overrightarrow{BA}$, angle $DE'C$ is less than $\frac{1}{2}$ arc CD, since other points are outside the circle. Hence point E is the point we want. To find BE, let O be the center of the circle and M the midpoint of $\overline{CD}$. Then $\overline{OM} \perp \overline{CD}$, $\overline{OE} \perp \overline{BE}$, and $OE = BM = 2$. Then $EB = MO$, and $EB^2 = MO^2 = OD^2 - MD^2 = OE^2 - MD^2 = 4 - 1 = 3$. Thus $BE = \sqrt{3}$.

22. (Continued)

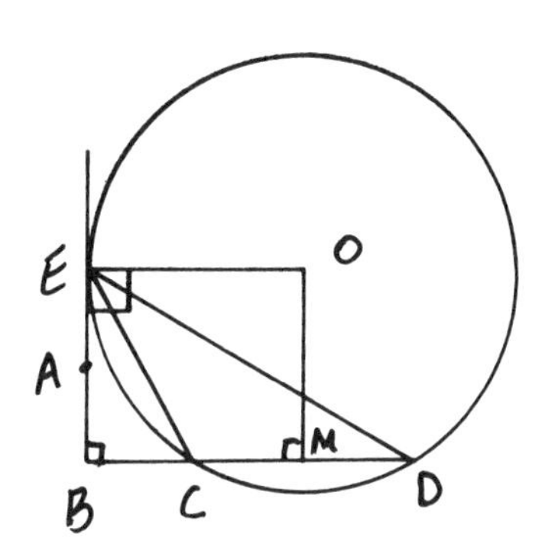

23. The probability of neither being a spade is $\frac{39}{52} \cdot \frac{38}{51} = \frac{19}{34}$. Hence the probability of getting at least one spade is $1 - \frac{19}{34} = \frac{15}{34}$.

24. We compute the ratios of areas of triangles with equal altitudes by comparing their bases. Using absolute value for area, let $|QRT| = A$. Then $|TRX| = A$, $|RXS| = \frac{3A}{2} = |XSY|$, $|YTX| = |TRX| = A$ (all from ratios along line QS). Also, since $|PTS|:|RTS| = 5:4$, $|PTS|:\left(A + \frac{3}{2}A\right) = 5:4$, so $|PTS| = \frac{25}{8}A$. From ratios along PR: $|PYT| = \frac{5}{2}A$. Thus $|PTSY| = |PTY| + |TYX| + |YXS| = 5A$, and $|PYS| = |PTSY| - |PTS| = 5A - \frac{25}{8}A = \frac{15}{8}A$.

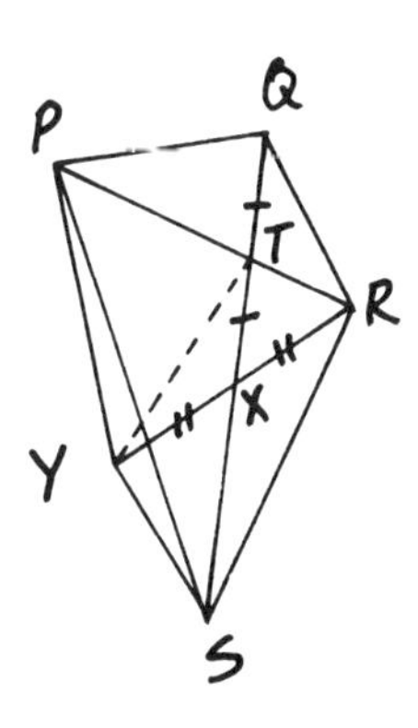

25. If $PA = x = PB$, then $PC = 10 - x$, and $PC^2 + CA^2 = PA^2$, or $(10 - x)^2 + 4^2 = x^2$, $100 - 20x + x^2 + 16 = x^2$, $116 - 20x = 0$, $x = \frac{116}{20} = \frac{29}{5}$.

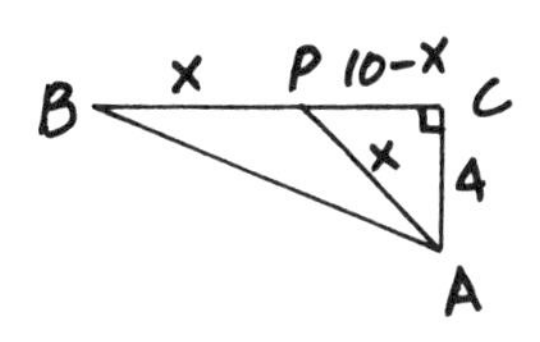

26. After the first sock is picked, the chance of the second sock matching is 4/9.

27. $\frac{f(x)}{g(x)} = 0$ if and only if $f(x) = 0$ while $g(x) \neq 0$. This happens for 2 and 9.

28. Suppose there are n numbers in the set. Clearly, n is more than 2. We will first show that n cannot be 3, 4, or 5. For example, if $n = 3$, the numbers can be represented by a, b, and b/a, so that $b/a = ab$. Hence $|a| = |b| = |b/a| = 1$. But then at least two of the three numbers must be equal, contrary to the assumptions of the problem. Similar arguments will show that n cannot be 4 or 5. Next we can show that n cannot be greater than 6. Indeed, if the first two numbers are a and b, then the next five are b/a, $1/a$, $1/b$, a/b, and a, so that the seventh number is equal to the first. Hence $n = 6$. The sequence 2,3,3/2,1/2,1/3,2/3 (among others) will satisfy the conditions of the problem.

29. Drop altitude $\overline{AE}$ onto $\overline{BC}$. Then $\overline{AD}$ is the bisector of angle EAB, so $BD:DE = AB:AE = 2:\sqrt{3}$ (see 3.31). This leads to $DB = 1/(2 + \sqrt{3})$ and $DC = \frac{1 + \sqrt{3}}{2 + \sqrt{3}}$. Then the area of triangle ACD is $(1/2)(DC)(AE)$ $= (3 - \sqrt{3})/4$.

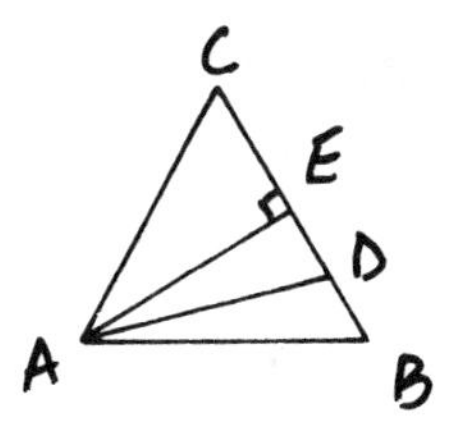

30. Label any vertex of the 11-gon as point A. Then we can label the next five vertices going clockwise around the circle as $B1$, $B2$, $B3$, $B4$, $B5$, and the five vertices going counterclockwise as $C1$, $C2$, $C3$, $C4$, $C5$. Let us first find out how many of the triangles described have a vertex at A. Clearly, of the other two vertices of such a triangle, one must be some Bi and another some Cj for $1 \leq i, j \leq 5$. And in fact, it is not hard to see that $i + j \geq 6$ if the triangle is to contain the center of the circle. Now it is not hard to count the triangles: each corresponds to the solution of the above inequality for i and j natural numbers less than 6. There is one solution if $i = 1$, two if $i = 2$, and so on. Hence there are $1 + 2 + 3 + 4 + 5 = 15$ triangles with one vertex at A. Letting each vertex of the 11-gon play the role of A, we now have $11 \cdot 15 = 165$ triangles. But we have then counted each triangle three times (having let each vertex of each triangle play the role of A), so that the number of triangles is actually $165/3 = 55$. This argument can be generalized for a regular polygon with an odd number of sides.

APPENDIX A

ALGEBRA

1.1 If $a:b = c:d$, then:

1.11 $(a + b):b = (c + d):d$

1.12 $(a + b):(a - b) = (c + d):(c - d)$

1.13 $a:b = c:d = (a + c):(b + d)$

1.21 Binomial Theorem: If $\binom{n}{k} = \frac{n!}{k!(n-k)!}$, (for $k \leq n$),

then: $(a + b)^n = \sum_{k=0}^{n} \binom{n}{k} a^k b^{n-k}$

1.22 $(a + b)^3 = a^3 + b^3$ if and only if $a = 0$ or $b = 0$ or $a + b = 0$.

1.311 Factor Theorem: For $P(x)$ a polynomial in x and r a real or complex number, $x - r$ divides $P(x)$ if and only if $P(r) = 0$.

1.312 (Corollary) The polynomial $P(x)$ divides the polynomial $Q(x)$ if and only if every (complex) root of $P(x) = 0$ is also a root of $Q(x) = 0$ (including multiplicities of distinct roots).

1.321 $a^3 + b^3 = (a + b)(a^2 - ab + b^2)$

1.322 $a^3 - b^3 = (a - b)(a^2 + ab + b^2)$

1.33 $x^n - y^n = (x - y)(x^{n-1} + x^{n-2}y + x^{n-3}y^2 + \ldots + xy^{n-2} + y^{n-1})$

1.11
S81-#1, p. 34
F82-#16, p. 42

1.12
S75-#22, p. 3
F77-#8, p. 16

1.21
F75-#30, p. 3
S76-#4, p. 7
S77-#30, p. 15

1.22
S75-#2, p. 1
S76-#30, p. 9
F77-#10, p. 16

1.311
S75-#18, p. 2
F75-#8, p. 4
F75-#10, p. 4
F76-#2, p. 10
F77-#20, p. 17
F78-#18, p. 22
S80-#10, p. 29

1.312
F77-#20, p. 17

1.321
S77-#6, p. 13
S77-#8, p. 13
F77-#26, p. 17
S79-#21, p. 25
F81-#4, p. 37
F81-#9, p. 37
F81-#18, p. 38

1.322
S80-#26, p. 30
F81-#4, p. 37
F81-#18, p. 38

1.33
S77-#25, p. 14
F77-#7, p. 16
F78-#8, p. 21
S80-#19, p. 30
F80-#14, p. 32

1.34 For odd n, $x^n + y^n = (x + y)(x^{n-1} - x^{n-2}y + x^{n-3}y^2 - \dots - xy^{n-2} + y^{n-1})$.

1.34
S79-#10, p. 24

1.4 If $P(x) = x^n - a_{n-1}x^{n-1} + a_{n-2}x^{n-2} - \dots + (-1)^{n-1}a_1x + (-1)^n a_o = 0$ has roots $r_1, r_2, r_3, \dots, r_n$, then:

$a_{n-1} = \sum_{i=1}^{n} r_i$ (the sum of the roots)

$a_{n-2} = \sum_{1\leq i<j\leq n} r_i r_j$ (the sum of the roots "taken two at a time")

$a_{n-3} = \sum_{1\leq i<j<k\leq n} r_i r_j r_k$ (the sum of the roots "taken three at a time")

$\vdots$

$a_o = r_1 r_2 r_3 \dots r_n$

1.4
S75-#30, p. 3
F75-#6, p. 4
F75-#8, p. 4
F75-#9, p. 4
F77-#13, p. 16
F77-#21, p. 17
F78-#24, p. 22
S79-#17, p. 24
S79-#20, p. 25
S84-#27, p. 51

1.51 Arithmetic progression: If $a_1, a_2, \dots, a_n$ are in arithmetic progression with common difference d, then:

1.511 $a_n = a_1 + (n - 1)d$

1.511
S76-#7, p. 7
F80-#10, p. 32

1.512 $\sum_{i=1}^{n} a_i = \frac{n}{2}[2a_1 + (n - 1)d] = \frac{n}{2}[a_1 + a_n]$

1.512
F75-#11, p. 4
S79-#29, p. 26
F79-#7, p. 26

1.52 Geometric progression: If $a_1, a_2, \dots, a_n$ are in geometric progression with common ratio r, then:

1.521 $a_n = a_1 r^{n-1}$

1.522 $\sum_{i=1}^{n} a_i = \frac{a_1(1 - r^n)}{1 - r}$

1.522
S82-#8, p. 40
S82-#18, p. 40

1.523 $\sum_{i=1}^{\infty} a_i = \frac{a_1}{1 - r}$ if $|r| < 1$

1.523
S76-#1, p. 6
S76-#16, p. 8
S76-#15, p. 11
F76-#19, p. 11
S77-#12, p. 13
S78-#24, p. 20
S79-#25, p. 28
F81-#10, p. 37
F81-#14, p. 37
F82-#26, p. 43

1.53 $\sum_{i=1}^{n} i = n(n+1)/2$

1.53
S78-#6, p.
S78-#8, p.
S78-#20, p.
S78-#28, p.
S78-#23, p.

1.6 De Moivre's Theorem: For any complex number $r \text{ cis } \theta$ (or $r \cos\theta + ir \sin\theta$),

$$(r \text{ cis } \theta)^n = r^n \text{ cis } (n\theta).$$

1.6
F78-#18, p. 22
F78-#30, p. 23

1.7 Rational Root Theorem: If $P(x) = a_n x^n + a_{n-1} x^{n-1} + a_{n-2} x^{n-2} + \ldots + a_1 x + a_o$ is a polynomial with integer coefficients and p/q is a rational root of the equation $P(x) = 0$ (where p and q are relatively prime), then p must divide a_o, while q must divide a_n.

1.7
F75-#6, p. 4
F75-#10, p. 4
S76-#6, p. 7
S80-#30, p. 31

1.81 If $P(x)$ is a polynomial with real coefficients and $P(a + bi) = 0$ (where i is the imaginary unit and a,b are real), then $P(a - bi) = 0$ as well.

1.82 If $P(x)$ is a polynomial with rational coefficients and $P(a + b\sqrt{c}) = 0$ (where a, b, and c are rational and c is positive but not a perfect square), then $P(a - b\sqrt{c}) = 0$.

1.81
S80-#10, p. 29
S80-#20, p. 30

1.82
F77-#28, p. 17
S80-#10, p. 29

NUMBER THEORY

2.1 Pythagorean Triples: All positive integer solutions to the equation $a^2 + b^2 = c^2$ are given by:

$$a = k(m^2 - n^2)$$
$$b = k(2mn)$$
$$c = k(m^2 + n^2)$$

where m, n, and k are arbitrary positive integers, with $m > n$. For relatively prime triples (a,b,c), we need $k = 1$, and m and n relatively prime and of opposite parity.

2.1
F75-#25, p. 6
S76-#9, p. 7
F76-#30, p. 12
S78-#10, p. 18

2.2 Fermat's "Last Theorem" for $n = 3$: The equation $a^3 + b^3 = c^3$ has no solutions in integers unless $abc = 0$.

2.2
S78-#9, p. 18

2.3 Fermat's "Little" Theorem: If p is prime and a is relatively prime to p, then $a^{p-1} - 1$ is a multiple of p.

2.3
F82-#22, p. 43
F82-#29, p. 43

GEOMETRY

3.1 The medians of a triangle are concurrent at a point (the centroid) that divides each median in the ratio 2:1.

3.1
F77-#23, p. 17

3.2 For median m_c to side c, $m_c^2 = \frac{1}{2}(a^2 + b^2) - \frac{1}{4}c^2$.

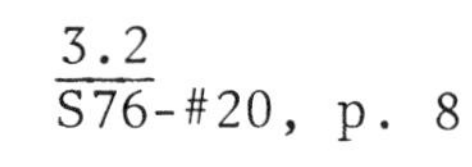
3.2
S76-#20, p. 8

3.3 If angle bisector t_c divides side c into segments of lengths m and n, then:

3.31 $a:b = m:n$

3.3
F75-#5, p. 4
S76-#2, p. 6
F78-#12, p. 21
F81-#26, p. 39
S83-#12, p. 45
F84-#29, p. 54

3.32 $ab - mn = t_c^2$

3.32
S76-#2, p. 6

3.4 The "Extended Law of Sines": If R is the circumradius of triangle ABC, $\frac{a}{\sin A} = \frac{b}{\sin B} = \frac{c}{\sin C} = 2R$

3.4
F78-#16, p. 21
F79-#24, p. 28
F79-#30, p. 28
F81-#16, p. 38
F81-#30, p. 39
F84-#15, p. 53

3.5 If X, Y, and Z are the points of contact of the sides of triangle ABC with its inscribed circle and s is the triangle's semiperimeter, then:

$CX = CY = s - c$
$BX = BZ = s - b$
$AY = AZ = s - a$

3.5
S77-#22, p. 14
S78-#30, p. 20
F82-#18, p. 43
F84-#3, p. 52

3.6 If a, b, c are the sides of triangle ABC, k its area, s its semiperimeter, R its circumradius, and r its inradius, then:

3.61 $k = \sqrt{s(s - a)(s - b)(s - c)}$ (Hero's Formula)

3.62 $k = rs$

3.63 $k = abc/4R$

3.64 $k = (1/2)ab \sin C$

3.7 In an equilateral triangle, the sum of the distances from any interior point to the three sides is always equal to the altitude of the triangle.

3.81 If $AX \| BY \| CZ$, then $\frac{1}{BY} = \frac{1}{AX} + \frac{1}{CZ}$.

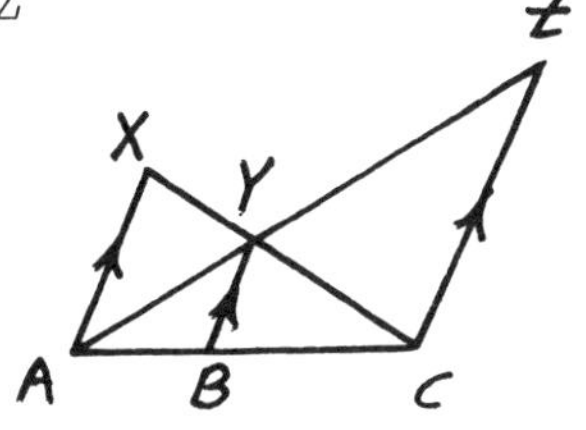

3.82 Ceva's Theorem: If AP, BQ, CR are concurrent at X (as shown), then:

3.821 $\frac{AR}{RB} \cdot \frac{BP}{PC} \cdot \frac{CQ}{QA} = 1$

3.822 $\frac{XP}{AP} + \frac{XQ}{RQ} + \frac{XR}{CR} = 1$

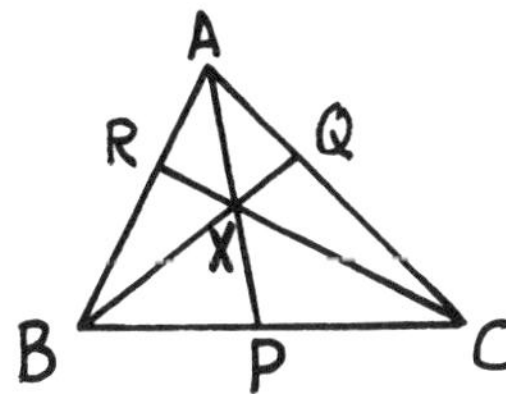

3.83 Menelaus' Theorem: If "transversal" PR intersects the sides of triangle ABC in P, Q, and R (as shown), then:

$$\frac{PP}{PB} \cdot \frac{BQ}{QC} \cdot \frac{CR}{RA} = 1$$

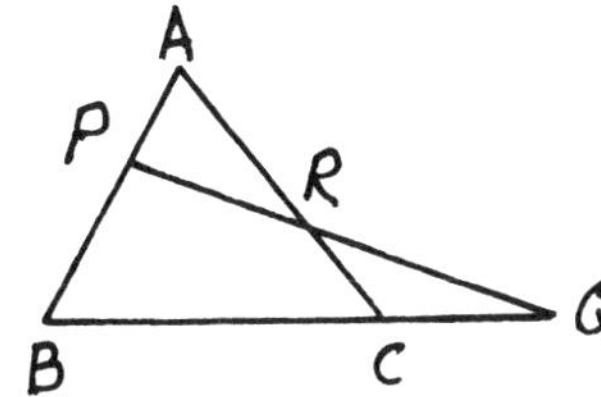

3.84 In parallelogram $ABCD$, $AC^2 + BD^2 = AB^2 + BC^2 + CD^2 + DA^2$.

3.6
S75-#3, p. 1
F76-#7, p. 10
F77-#25, p. 17
S78-#3, p. 18
S80-#18, p. 30
S81-#21, p. 35
S83-#26, p. 46

3.62
S75-#3, p. 1
S78-#2, p. 18
S80-#18, p. 30
F82-#18, p. 43

3.63, 3.64
S78-#3, p. 18
S79-#2, p. 23
S79-#15, p. 24
S79-#18, p. 24
F79-#4, p. 26
F79-#15, p. 27
S80-#12, p. 29
S82-#4, p. 39
S83-#22, p. 45
S83-#30, p. 46

3.7
S76-#14, p. 7

3.81
S77-#14, p. 14

3.83
S77-#1, p. 13

3.84
F80-#4, p. 31
F80-#12, p. 32

3.91 Ptolemy's Theorem: In any quadrilateral *ABCD*, $AC \cdot BD \leq AB \cdot CD + AD \cdot BC$, with equality holding if and only if quadrilateral *ABCD* can be inscribed in a circle.

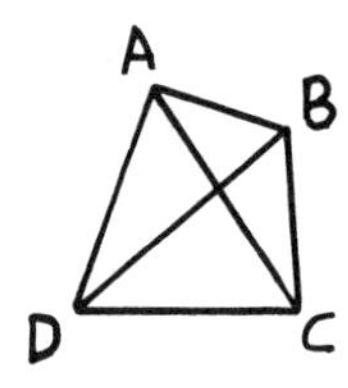

3.92 For a quadrilateral to be cyclic (to possess a circumscribing circle), it is necessary and sufficient that its opposite angles be supplementary.

3.93 In a circle of radius r, the length of a chord which subtends an inscribed angle θ is $2r \sin \theta$.

TRIGONOMETRY

The "Pythagorean" Identities:

4.11 $\sin^2 x + \cos^2 x = 1$

4.12 $\tan^2 x + 1 = \sec^2 x$

4.13 $\cot^2 x + 1 = \csc^2 x$

Double and Half-Angle Formulas:

4.21 $\sin 2x = 2 \sin x \cos x$

4.22 $\cos 2x = \cos^2 x - \sin^2 x = 2 \cos^2 x - 1 = 1 - 2 \sin^2 x$

3.91
F75-#9, p. 4
F75-#15, p. 5
F77-#3, p. 15
F77-#12, p. 16
S81-#20, p. 35
S81-#26, p. 36

3.92
F77-#12, p. 16
F81-#30, p. 39

3.93
F83-#30, p. 49
F84-#15, p. 53

4.11
F76-#12, p. 11
F76-#14, p. 11
S79-#8, p. 23
S79-#21, p. 25
F79-#5, p. 26
S80-#17, p. 30
F80-#19, p. 32
F81-#13, p. 37
S83-#10, p. 44
S83-#28, p. 46
F83-#13, p. 47
S84-#11, p. 50
S84-#18, p. 50
F84-#7, p. 52

4.12
F82-#19, p. 43

4.21
F76-#14, p. 11
S78-#13, p. 19
S78-#22, p. 19
F78-#30, p. 23
S79-#8, p. 23
S79-#18, p. 24
F79-#22, p. 28
F81-#29, p. 39
S82-#11, p. 40
F82-#30, p. 43
S84-#18, p. 50
F84-#7, p. 52

4.22
F78-#30, p. 23
S81-#8, p. 34
F81-#29, p. 39
S82-#11, p. 40
S84-#18, p. 50

4.23 $\tan x = (\sin 2x)/(1 + \cos 2x) = (1 - \cos 2x)/\sin 2x$

4.24 $\sin^2 x = \frac{1}{2}(1 - \cos 2x)$; $\sin \frac{x}{2} = \pm \sqrt{\frac{1 - \cos x}{2}}$

4.25 $\cos^2 x = \frac{1}{2}(1 + \cos 2x)$; $\cos \frac{x}{2} = \pm \sqrt{\frac{1 + \cos x}{2}}$

Addition Formulas:

4.311 $\sin (x + y) = \sin x \cos y + \cos x \sin y$

4.312 $\sin (x - y) = \sin x \cos y - \cos x \sin y$

4.313 $\cos (x + y) = \cos x \cos y - \sin x \sin y$

4.314 $\cos (x - y) = \cos x \cos y + \sin x \sin y$

4.321 $\sin x + \sin y = 2 \sin[(x + y)/2]\cos[(x - y)/2]$

4.322 $\sin x - \sin y = 2 \cos[(x + y)/2]\sin[(x - y)/2]$

4.323 $\cos x + \cos y = 2 \cos[(x + y)/2]\cos[(x - y)/2]$

4.324 $\cos x - \cos y = -2 \sin[(x + y)/2]\sin[(x - y)/2]$

4.331 $\sin x \sin y = (1/2)[\cos (x - y) - \cos (x + y)]$

4.332 $\cos x \sin y = (1/2)[\sin (x + y) - \sin (x - y)]$

4.333 $\cos x \cos y = (1/2)[\cos (x + y) + \cos (x - y)]$

4.334 $\sin x \cos y = (1/2)[\sin (x + y) - \sin (x - y)]$

4.341 $\tan (x + y) = \dfrac{\tan x + \tan y}{1 - \tan x \tan y}$

4.23
F76-#14, p. 11

4.24
F76-#14, p. 11
S77-#4, p. 13
F79-#5, p. 26
F79-#10, p. 27

4.311
S77-#23, p. 14
S77-#30, p. 15
F84-#6, p. 52

4.312
S77-#30, p. 15
F82-#30, p. 43

4.313
F76-#25, p. 9
F84-#6, p. 52
F84-#21, p. 53

4.314
F80-#27, p. 33
F82-#24, p. 43
S83-#28, p. 46

4.321
S77-#30, p. 15
F77-#17, p. 16
S79-#4, p. 23
F82-#30, p. 43
F83-#4, p. 46

4.322
S82-#27, p. 41
F82-#30, p. 43

4.323
F77-#9, p. 16
F77-#17, p. 16
F78-#10, p. 21
F83-#4, p. 46

4.331
F78-#10, p. 21

4.341
F76-#3, p. 10
S78-#12, p. 19
F79-#10, p. 27
F79-#14, p. 27
F81-#22, p. 38
S82-#24, p. 41

4.342 $\tan(x - y) = \dfrac{\tan x - \tan y}{1 + \tan x \tan y}$

4.4 $\tan^{-1}A + \tan^{-1}B = \tan^{-1}\left(\dfrac{A + B}{1 - AB}\right)$

4.51 $\sin 3x = 3 \sin x - 4 \sin^3 x$

4.52 $\cos 3x = 4 \cos^3 x - 3 \cos x$

MISCELLANY

For $a, b, c > 0$, $a, b, c \neq 1$:

5.11 $(\log_a b)(\log_b c) = \log_a c$ (the "chain rule" for logarithms)

5.12 $\log_a b = \dfrac{1}{\log_b a}$

If $[x]$ denotes the greatest integer no larger than x, then:

5.21 $[(x + m)/n] = [([x] + m)/n]$, for m, n integers, and $n > 0$.

5.22 For nonintegral $x > 0$, if $[x] = n$, then $[-x] = -n - 1$.

INEQUALITIES

6.1 If $x > 0$, $f(x) = x + 1/x$ attains its minimum value when $x = 1$.

6.2 The "arithmetic-geometric mean inequality:" for $a, b > 0$, $(a + b)/2 \geq \sqrt{ab}$, with equality holding if and only if $a = b$. In general, for $a_1, a_2 \ldots a_n \geq 0$, $\frac{1}{n}(a_1 + a_2 + \ldots + a_n) \geq \sqrt[n]{a_1 a_2 \ldots a_n}$, with equality holding if and only if $a_1 = a_2 = \ldots = a_n$.

6.3 If $a + b$ remains constant, the product ab is largest when $a = b$.

4.342
F79-#10, p. 27
F79-#11, p. 27

4.4
S77-#13, p. 13
F81-#17, p. 38

4.51
S79-#28, p. 25

5.11
S78-#7, p. 18
S78-#17, p. 19
F78-#11, p. 21

5.12
F79-#28, p. 28
S80-#13, p. 29
F82-#14, p. 42
F83-#17, p. 48

5.22
F75-#1, p. 3
S77-#29, p. 15

6.1
S76-#6, p. 7
F81-#2, p. 36
F82-#27, p. 43
S83-#30, p. 46

6.3
F76-#7, p. 10
S78-#29, p. 20
F80-#21, p. 33

6.4 If ab remains constant, the sum $a + b$ is smallest when $a = b$.

7. The Inclusion-Exclusion Principle: If we use $|A|$ to denote the number of elements in the set A, then:

7.1 $$\left|\bigcup_{i=1}^{n} A_i\right| = \sum_{i=1}^{n} |A_i| - \sum_{1 \le i < j \le n} |A_i \cap A_j| + \sum_{1 \le i < j < k \le n} |A_i \cap A_j \cap A_k| - \ldots + (-1)^{n+1} \left|\bigcap_{i=1}^{n} A_k\right|$$

In particular,

7.2 $|A \cup B| = |A| + |B| - |A \cap B|$

7.3 $|A \cup B \cup C| = |A| + |B| + |C| - |A \cap B| - |A \cap C| - |B \cap C| + |A \cap B \cap C|$

6.4
F82-#9, p. 42
F82-#15, p. 42
F82-#21, p. 43
F84-#16, p. 53

7.1
S82-#25, p. 41

7.2
S77-#2, p. 13

7.3
F82-#6, p. 42

APPENDIX B

The powerful method of mass points is a way to intuit various geometric results, in particular Ceva's and Menelaus' theorems (see Appendix A, 3.821, 3.822, 3.83). A more rigorous treatment could be based on properties of similar triangles, on addition of vectors, or on barycentric coordinates.

I. Definitions

1. A mass point is a set consisting of a positive number n (called the *weight*) and a point P. Notation for a mass point is (n,P) or simply nP.

2. Two mass points (a,A) and (b,B) are equal if and only if $a = b$ and $A = B$.

3. The *sum* of two mass points (a,A) and (b,B) is the mass point $(a + b, C)$, where C is on line segment AB and $AC:CB = b:a$ (note the reversal in the order of the numbers). The sum of (a,A) and (b,B) is sometimes called the *centroid* or *center of mass* of the system comprising the two mass points.

II. Properties of Addition

1. The sum of two mass points is unique. Hence, subtraction is well defined if we take $(c,C) - (b,B)$ to be $(c - b, A)$, where A is the point on ray BC such that $AC:CB = b:c - b$.

2. Addition of mass points is commutative: $(a,A) + (b,B) = (b,B) + (a,A)$.

3. Addition of mass points is associative: $(a,A) + (b,B) + (c,C) = [(a,A) + (b,B)] + (c,C)$ $= (a,A) + [(b,B) + (c,C)]$. This sum is called the *centroid* of the system comprising the three points.

   The first two properties above follow directly from the definition of the sum of two mass points. The third property can be proved using, for example, Ceva's theorem. Taken as an assumption, it is logically equivalent to Ceva's theorem.

III. Examples

1. In triangle ABC, D is the midpoint of BC and E is the trisection point of AC closer to A. Find the ratios $BG:GE$ and $AG:GD$.

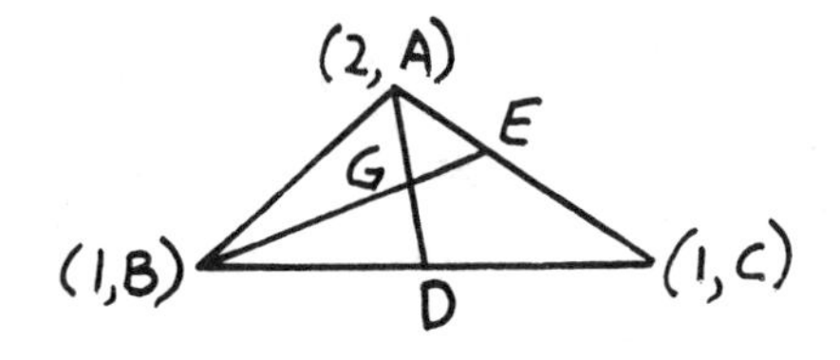

Solution: Assign weights of 1 to each of B and C and a weight of 2 to A. Then $(1, B) + (1,C) = (2,D)$ and $(2,A) + (1,C) = (3,E)$. By associativity of addition, $(1,B) + (1,C) + (2,A) = (2,D) + (2,A)$ $= (1,B) + (3,E) = (4,G)$, where G is on line segment AD such that $AG:GD = 1:1$ and G is also on line segment BE such that $BG:GE = 3:1$. Hence $BG:GE = 3:1$ and $AG:GD = 1:1$.

If CG is extended to intersect AB in X, a similar argument will show that $CG:GX = 3:1$.

2. Prove that the medians of a triangle are concurrent and divide each other in the ratio 2:1 (see Appendix A, 3.1).

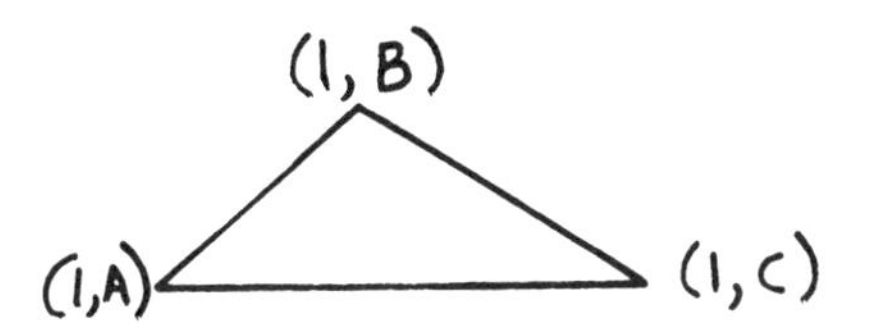

Solution: Assign weight 1 to each of the vertices of the triangle. Then $1A + (1B + 1C) = (1A + 1B) + 1C = (1B + 1C) + 1A$ (by the associative and commutative properties), and the three medians will all pass through the mass-point which is this sum. The 2:1 ratio follows immediately.

3. In triangle ABC, angle bisector BE intersects median AD at K. Find the ratios $BK:KE$ and $AK:KD$ in terms of the lengths of the sides of the triangle.

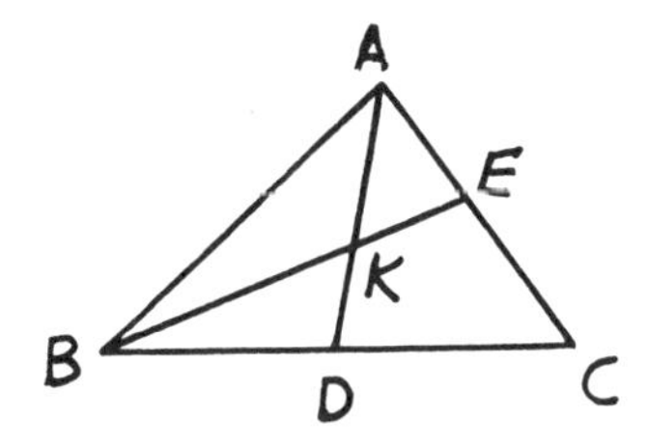

Solution: If the lengths of the sides of the triangle are (as usual) a, b, and c, we can assign a weight of a to point A and a weight of c to points B and C. Then $(a,A) + (c,C) = (a + c, E)$ (see 3.31 in Appendix A) and $(c,C) + (c,B) = (2c,D)$. The sum $(a,A) + (c,B) + (c,C) = (a + 2c, K)$, and we find $BK:KE = (a + c):c$, while $AK:KD = 2c:a$.

4. In quadrilateral $ABCD$, points E, F, G, and H are the trisection points of AB, BC, CD, and DA closer to A, C, C, and A, respectively. Show that $EFGH$ is a parallelogram.

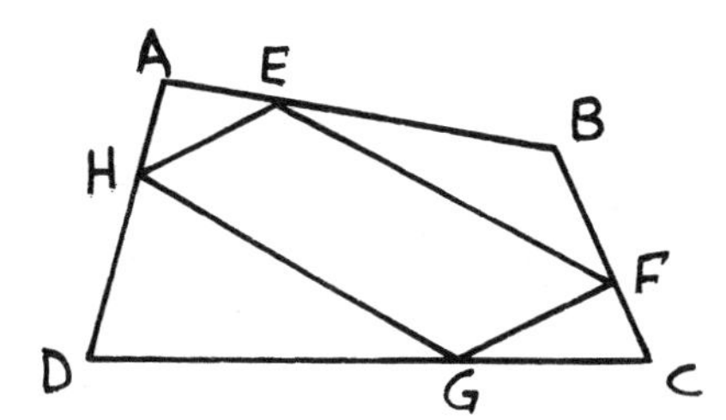

Solution: Assign a weight of 2 to points A and C and a weight of 1 to points B and D. Then the centroid of the system may be found in two ways: $(2A + 1B) + (2C + 1D) = 3E + 3G = (2A + 1D) + (2C + 1B) = 3H + 3F$. Since this centroid is unique, it follows that EG and FH bisect each other, and $EFGH$ is a parallelogram.

5. In trapezoid $ABCD$, base AB is half as long as base CD. Diagonals AC and BD intersect at F, and point E is chosen on line segment BC such that $BE:EC = 1:3$. If EF is extended to intersect AD at G, find the ratio $EF:FG$.

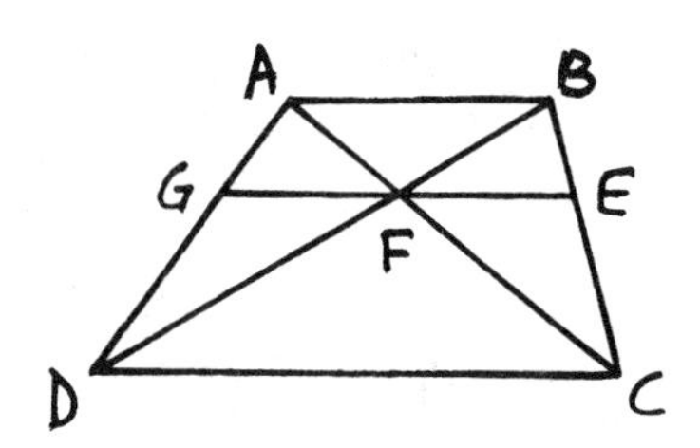

Solution: We want to assign weights so that the centroid of the system will be at point F. From similar triangles ABF, CDF, we find that $AB:CD = AF:FC = BF:FD = 1:2$. Hence the centroid will be at F if we assign any weight x to D, $2x$ to B, y to C, and $2y$ to A. But we would also like the sum of $(2x,B) + (y,C)$ to be at E. This requires that $2x:y = 3:1$. This can be conveniently arranged, for example, by letting $x = 3$, $y = 2$. Then $6B + 2C = 8E$, $4A + 6B + 2C + 3D = 15F$, and the difference $12F - (6B + 2C)$, which is the sum $4A + 3D$, must be collinear with E and F and also with A and D. Hence this difference is $7G$, and $EF:FG = 7:8$.

6. For this problem, we have the quite reasonable assumption that mass points in space behave the same way as mass points in the plane. With this assumption, show that the line segments joining each vertex of a tetrahedron to the centroid of the opposite face are concurrent and divide each other in the ratio 3:1.

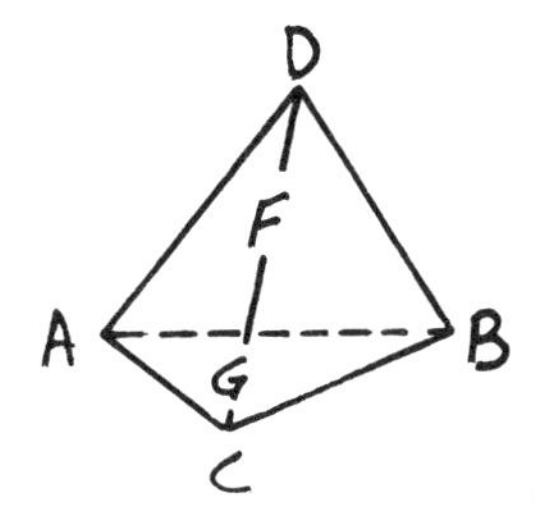

Solution: Assign a weight of 1 to each vertex of the tetrahedron. Then the centroid of each face is the mass-point sum of its three vertices, with a weight of 3. The sum of the four mass points at the vertices of the tetrahedron thus lies along the line segment connecting the centroid of each face with the vertex opposite, and this point divides each such segment in the required ratio.

For problems using mass points, refer to the index. For a more formal treatment, see Hausner.

ANNOTATED BIBLIOGRAPHY

Altshiller-Court, Nathan. College Geometry. New York: Barnes and Noble, 1952.

Extensive, thorough treatment of major topics of advanced Euclidean geometry, including similitude and homothecy, triangles and their associated circles, the Simson line, circles of Apollonius, harmonic division, theorems of Steward, Ceva and Menelaus, and many more topics. Hundreds of challenging and enlightening exercises. For the serious geometry student. Originally published in 1925.

Apostol, Tom M. Calculus. 2 vols. New York: Blaisdell Publishing Co., 1962.

An especially interesting calculus text, approaching the subject from an historical point of view. Integration precedes differentiation. A variety of intriguing and valuable problems are scattered throughout the exercise sets. A standard text used at many colleges and universities.

Aref, M. N., and William Wernick. Problems and Solutions in Euclidean Geometry. New York: Dover Publications, 1968.

A vast collection of intriguing geometrical problems (many with complete solutions), including a chapter devoted to solid geometry. Captivating and challenging.

Beckenbach, E. F., and R. Bellman. An Introduction to Inequalities. New Mathematical Library, Vol. 3. Washington, D.C.: Mathematical Association of America.

A solid introduction to the classical inequalities and their applications to problem-solving. Exercises with answers.

Beiler, Albert H. Recreations in the Theory of Numbers. 2nd ed. New York: Dover Publications, 1964.

A truly entertaining book overflowing with number theoretical facts, charts, tables, theorems, conjectures, nice exercises and solutions. Excellent bibliographical references.

Berman, Gerald., and K. D. Fryer. Introduction to Combinatorics. New York: Academic Press, 1972.

The authors maintain that math can be interesting and fun as well as useful and that learning math is best accomplished by doing math. They provide an abundance of particularly interesting problems.

Blank, Albert A. Problems in Calculus and Analysis. New York: John Wiley and Sons, 1966.

A companion book to Introduction to Calculus and Analysis by R. Courant and F. John. Hundreds of problems and solutions. Most will challenge and enlighten the first-year calculus student. Many excellent problems require no calculus.

Borofsky, S. Elementary Theory of Equations. New York: Macmillan Co., 1963.

A good introduction to the subject, although well beyond traditional secondary school syllabus. Excellent and abundant problems.

Boyer, Carl B. A History of Mathematics. New York: John Wiley and Sons, 1968.

An outstanding overview of the history of mathematics by one of the foremost scholars in the field. Boyer provides delightful sets of truly enlightening exercises at the end of each chapter. A surprising wealth of fascinating material for any problemist.

Brother Alfred Brousseau. Mathematics Contest Problems. Palo Alto, Calif.: Dale Seymour Publications, 1984.

A selection from the St. Mary's College competitions. Two levels of problems (7-9) and (10-12). Complete solutions.

Charosh, Mannis. Mathematical Challenges. Washington, D.C.: National Council of Teachers of Mathematics.

One hundred forty problems from the Mathematics Student Journal. Suitable for Grades 7-12. Good source of intersting, diversified, challenging problems and solutions.

Chrystal, G. Algebra (An Elementary Textbook. 6th ed. 2 vols. New York: Chelsea Publications, 1959.

A classic work first published in 1886. Extensive treatment of a wealth of topics. Superb collection of problems.

The Contest Problem Books (I, II, III, and IV). New Mathematical Library, Vols. 5, 17, 25, and 29. Washington, D.C.: Mathematical Association of America.

Collections of problems and solutions to the Annual High School Mathematics Examination, which began in 1950. A must for any serious problemist preparing for these competitions.

Coxeter, H. S. M. Introduction to Geometry. 2nd ed. New York: John Wiley and Sons, 1969.

In his introduction to the first edition, the author notes that "Americans have somehow lost interest in geometry. The present book constitutes an attempt to revitalize this sadly neglected subject." The text provides an extensive tour of the subject and goes well beyond basic work. A marvelous array of fascinating topics in geometry by one of the recognized masters. Exercises and answers.

Coxeter, H. S. M., and S. L. Greitzer. Geometry Revisited. New Mathematical Library, Vol. 19. Washington, D.C.: Mathematical Association of America.

Many classic topics from advanced Euclidean geometry. Clear, enjoyable presentations. Numerous exercises.

Dickson, L. E. History of the Theory of Numbers. 3 vols. New York: Chelsea Publications, 1959.

Extensive collection of number theoretic facts, theorems, and results. A valuable source book.

Dorofeev, G., M. Potapon, and N. Rozov. Elementary Mathematics--Selected Topics and Problem Solving. Moscow: Mir Publishers, 1973.
A business-like treatment of essential secondary school mathematics. Greater depth than the current (American) curriculum. Excellent and abundant sets of problems. A separate chapter on nonstandard problems.

Dorrie, Heinrich. One Hundred Great Problems of Elementary Mathematics. New York: Dover Publications, 1965.
Classic problems with solutions. Wide range of topics. Many historically significant problems and solutions collected in one book.

Dudley, Underwood. Elementary Number Theory. San Francisco: W. H. Freeman and Co., 1969.
An especially good text for talented secondary school students. Rich in problems. Straightforward presentation of essential topics.

Edwards, Harold M. "Fermat's Last Theorem," in Scientific American, Oct. 1978, Vol. 239, No. 4.
A complete discussion of the many results spawned by centuries of research into this elusive problem.

Feller, William. An Introduction to Probability Theory and its Applications. 2nd ed. 2 vols. New York: John Wiley and Sons, 1957.
Extensive treatise on the subject. Wealth of applications and treatment of the significant problems. A vast amount of material.

Fisher, L., B. Kennedy, and W. Medigovich. The Brother Alfred Brousseau Problem-Solving and Mathematics Competition. 2 vols. Palo Alto, Calif.: Dale Seymour Publications, 1984.
A nice, diverse collection, presented on reproducible worksheets. Introductory (7-10) and Senior Division (10-12). Includes complete solutions.

Gleason, A. M., R. E. Greenwood, and L. M. Kelly. The William Lowell Putnam Mathematical Competition, Problems and Solutions 1938-1964. Washington, D.C.: Mathematical Association of America, 1980.

Greger, Karl. "Square Divisors and Square-Free Numbers," in Mathematics Magazine, Sept. 1978, Vol. 51, No. 4.
Some fascinating results on this seemingly prosaic topic.

Greitzer, Samuel L., editor. International Mathematical Olympiads 1959-1972. New Mathematical Library, Vol. 27. Washington, D.C.: Mathematical Association of America.

Guy, Richard K. Unsolved Problems in Number Theory. New York: Springer-Verlag, 1981.

Hall, H. S., and S. R. Knight. Elementary Algebra. London: Macmillan and Co., 1959; New York: St. Martin's Press, 1959.
A classic work first published in 1885. Excellent in all respects. Unfortunately out of print and difficult to obtain.

Hall, H. S., and S. R. Knight. Elementary Trigonometry. London: Macmillan and Co., 1959; New York: St. Martin's Press, 1959.
Another out-of-print classic. Extraordinary wealth of material and superb problem sets.

-----. Higher Algebra. London: Macmillan and Co.; New York: St. Martin's Press, 1957.
A stunning work! A treasure chest of mathematical problems and a grand tour of the techniques of higher algebra. Unfortunately out of print.

-----. Key to Higher Algebra. London: Macmillan and Co.; New York: St. Martin's Press, 1957.
Every problem solved in detail. 374 pages. A common practice with 19th century texts was publication of a companion "key."

Hall, H. S., and F. H. Stevens. First Lessons in Geometry. London: Macmillan and Co., 1932.

Hausner, Melvin. "The Center of Mass and Affine Geometry," in American Mathematical Monthly, Oct. 1962, Vol. 69, No. 8.
A formal treatment of the concepts behind the method of mass points (see Appendix B).

Herstein, I. N. Topics in Algebra. Waltham, Mass.: Xerox College Publishing, 1964.
This introduction to the essential notions of abstract algebra is fast becoming a classic. It includes discussion of groups, rings, vector spaces, and fields.

Honsberger, Ross. Ingenuity in Mathematics. New Mathematical Library, Vol. 23. Washington, D.C.: Mathematical Association of America.
Nineteen captivating essays treating a variety of topics including number theory, combinatorics, and geometry, all rounded out with very worthwhile exercises.

-----. Mathematical Gems I, II.
-----. Mathematical Morsels.
-----. Mathematical Plums.
Dolciani Mathematical Exposition Series, 1973-1979. Washington, D.C.: Mathematical Association of America.
Gems I is an intriguing collection of 13 brief expositions on such topics as Reuleaux triangles, Hamiltonian circuits, recursion, Morley's theorem, and other not-so-well-known "mathematical gems." Gems II contains another 14 fascinating essays. Both volumes include problems and solutions. Morsels presents 91 ingenious problems and solutions that have appeared in various journals over the years. Plums contains ten more essays, edited by Honsberger, in which he shares with us "the wonder and excitement of ingenious mathematical work at the elementary level." Exercises and solutions included.

The Hungarian Problem Book, I and II. New Mathematical Library, Vols. 11 and 12. Washington, D.C.: Mathematical Association of America.
Essay-style questions from the Eo"tvos competitions from 1894 through 1928 provide challenging and thought-provoking problems. Complete solutions, often accompanied by generalizations and related theorems. Fine preparatory material for Olympiad hopefuls.

Huntley, H. E. *The Divine Proportion.* New York: Dover Publications, 1970.
Many results, both algebraic and geometric.

Kasner, Edward, and James R. Newman. *Mathematics and the Imagination.* New York: Simon and Schuster, 1958.
A popular classic, containing many important and entertaining results.

Kay, David C. *College Geometry.* New York: Holt, Rinehart and Winston, 1969.
Part I (famous theorems of geometry), Part III (absolute geometry and concepts of parallelism), and several chapters of Part IV (development of geometry from models) are brimming with stunning theorems and problems.

Kazarinoff, N. D. *Geometric Inequalities.* New Mathematical Library, Vol. 4. Washington, D.C., Mathematical Association of America.
Applications of inequalities to isoperimetric problems and other classic geometric situations. Some very essential information and important techniques. Valuable exercises with solutions.

Khinchin, A. Ya. *Continued Fractions.* Chicago: University of Chicago Press, 1964.
Exposition suitable for more advanced students having some knowledge of the calculus. No exercises.

Klamkin, M., E. Barbeau, and W. Moser. *1001 Problems in High School Mathematics.* Montreal: Canadian Mathematical Congress.
A diverse set of problems and solutions chosen from a variety of sources. Appendices include "Mathematical Toolchest"--a helpful set of techniques and facts. Problems are similar in style to "Olympiad-type" questions, and range in difficulty from medium to quite challenging.

Knoebel, R. Arthur. "Exponentials Revisited," in *American Mathematical Monthly*, April 1981, Vol. 88, No. 4.
Resolves a number of convergence problems associated with "towers of exponents."

Knuth, Donald E. *The Art of Computer Programming.* 3 vols. Reading, Mass., Addison-Wesley, 1973.
An encyclopedic reference work covering an enormous area. It is sometimes surprising how useful our mathematics can be, in the most unexpected places.

Korovkin, P. P. *Inequalities.* Little Mathematical Library. Moscow: Mir Publishers, 1975.
A concise introduction to some basic inequalities and their applications to problems of extremes, approximations, limits, etc. Good exercises and complete solutions.

Krechmar, V. A. *A Problem Book in Algebra*. Moscow: Mir Publishers, 1974.
A valuable collection of challenging problems with complete solutions. Problems range in difficulty from moderate to Olympiad level.

Larsen, Loren C. *Problem-Solving Through Problems*. New York: Springer-Verlag, 1983.
A manual of problem-solving. Over 700 superb problems arranged by subject drawn from major competitions, problem sections of journals, and a variety of other sources. Complete solutions for nearly one-third of the problems. Essentially Olympiad-Putnam level.

Lidsky, V., editor. *Problems in Elementary Mathematics*. Moscow: Mir Publishers, 1973.
Six hundred fifty-eight exceptionally fine problems with complete solutions. Covers algebra, geometry (plane and solid), and trigonometry. Good source of practice problems for all levels of competitions.

Liff, Allan I. "On Solutions to the Equation $x^a + y^b = z^c$," in *Mathematics Magazine*, Sept. 1968, Vol. 41, No. 4.

Markushevich, A. I. *Recursion Sequences*. Little Mathematical Library. Moscow: Mir Publishers. 1975.
A concise, straightforward introduction to recursion sequences. Some good worked examples.

Moore, Charles G. *An Introduction to Continued Fractions*. Washington, D.C.: National Council of Teachers of Mathematics, 1964.
A good book to introduce the basic concepts at a gentle pace. Twenty-one problem sets, with solutions.

National Council of Teachers of Mathematics. *Enrichment Mathematics for High School*. 28th Yearbook. Washington, D.C.: NCTM, 1963.
Many articles suitable for students and instructors. Some nice articles on problem-solving, and collections of contest-style problems. Annotated bibliography, although dated, provides some good references.

Newman, Donald J. *A Problem Seminar*. New York: Springer-Verlag, 1982.
One hundred nine well-selected problems. Hints and solutions. Mostly on the undergraduate level, but many suitable for secondary school students.

Niven, Ivan. *The Mathematics of Choice*. New Mathematical Library, Vol. 15. Washington, D.C.: Mathematical Association of America.
An introduction to some of the classic notions from the realm of permutations and combinations together with numerous well-selected problems.

Niven, Ivan. *Numbers: Rational and Irrational*. New Mathematical Library, Vol. 1. Washington, D.C.: Mathematical Association of America.
Here is a clear, interesting, and informative introduction to the structure of the real numbers, including Cantor's proof of the existence of transcendental numbers. Twenty-nine problem sets provide ample practice and are well selected and always interesting.

Niven, Ivan, and Herbert S. Zuckerman. An Introduction to the Theory of Numbers. 3rd ed. New York: John Wiley and Sons, 1972.
A standard number theory text. Thorough and rich in fine exercises and problems.

Olds, C. D. Continued Fractions. New Mathematical Library, Vol. 9. Washington, D.C.: Mathematical Association of America.
A clear exposition of the subject, including classic problems, historical results, and many fine exercises (with solutions).

Polya, George. How to Solve It. Princeton, N.J.: Prince University Press, 1945. (2nd ed.: Garden City, N.Y.: Doubleday, 1956.)
Not a source of problems but a guide to problem-solving techniques and heuristics. A classic in the field.

-----. Mathematical Discovery. 2 vols. New York: John Wiley and Sons, 1962.
One of the classic works on problem-solving. Numerous problems, examples, and demonstrations illustrate various techniques.

-----. Mathematics and Plausible Reasoning. 2 vols. Princeton, N.J.: Princeton University Press. Vol. 1: 1954. Vol. 2 (rev. ed.): 1969.

-----, and J. Kilpatrick. The Stanford Mathematics Problem Book. New York: Teachers College Press, Columbia University, 1974.
Twenty essay-style contests. Hints and complete solutions. Many problems provide those essential problem-solving skills needed to bridge the gap from the short-answer contest problems to the longer, essay style problems that require a more sustained effort of thought.

Posamentier, A. Excursions in Advanced Euclidean Geometry. Portland, Maine: J. Weston Walsh, 1979.
The classical theorems and applications. A nice chapter on the Golden Section and Fibonacci numbers. Interesting exercises.

-----, and William Wernick. Geometric Constructions. Portland, Maine: J. Weston Walsh, 1973.
Five chapters covering this very neglected but extremely worthwhile segment of geometry. Thought-provoking exercises. Some good ideas for student research topics.

-----, and C. Salkind. Challenging Problems in Algebra. 2 vols. New York: Macmillan Co., 1970. (O.P.)
Excellent source of problems organized by topic. Many problems are extended or generalized. Valuable practice material for short-answer contests such as AHSME, AIME, ARML, etc.

-----. Challenging Problems in Geometry. 2 vols. New York: Macmillan Co., 1970. (O.P.)
Similar to "Algebra" set but this time covering such geometric topics as the theorems of Ptolemy, Ceva, Menelaus, and Stewart, the Simson line, and a lively selection of challenging, nonstandard problems. Many classics, such as the Butterfly theorem and the Steiner-Lehmus theorem, are solved in several different ways.

Rademacher, Hans, and Otto Toeplitz. The Enjoyment of Mathematics. Princeton, N.J.: Princeton University Press, 1970.

Twenty-eight wonderful essays. Exciting, always interesting, and informative. A favorite of Honsberger.

Rawson, Hugh. A Dictionary of Euphemisms and Other Doubletalk. New York: Crown Publications, Inc., 1981.

Ruderman, Harry, editor. NYSML-ARML Contests 1973-1982. Norman, Okla.: Mu Alpha Theta, 1983.

The complete collection of the problems from the American Regions Math League and the New York State Math League. The annual contests consist of team, individual, relay and "power" (essay) questions. Complete solutions. Excellent source of problems.

Rudin, Walter. Principles of Mathematical Analysis. New York: McGraw-Hill, 1976.

A standard text, particularly rich in advanced problems.

Salmon, G. A Treatise on Conic Sections. 6th ed. New York: Chelsea Publications (n.d.).

The classic, standard work on conic sections. Modern readers may be quite amazed by the depth of the coverage, variety of stunning theorems, and challenging exercises with which math students of a century ago were familiar.

Shklyarksy, D. O., N. N. Chentsov, and I. M. Yaglom. Selected Problems and Theorems in Elementary Mathematics. Moscow: Mir Publishers, 1979.

A superb source of 350 excellent problems at all levels of difficulty, with complete solutions.

-----. The USSR Olympiad Problem Book. San Francisco: W. H. Freeman and Co., 1962. (O.P)

Exceptionally fine diverse collection of problems, many from the USSR Olympiads. Complete solutions are especially rich in detail.

Sierpinski, W. A Selection of Problems in the Theory of Numbers. New York. Macmillan Co. 1963.

A collection of exciting problems and results from one of the old masters.

Solow, Daniel. How to Read and Do Proofs. New York: John Wiley and Sons, 1982.

A short book in which the author describes "some of the rules by which the game of theoretical mathematics is played." An introduction to the techniques of creating, doing, and writing proofs. Many illustrative examples and exercises. Complete solutions. Especially valuable for preparing for essay-style competitions or polishing research projects and papers.

Spiegel, Murray R. Finite Differences and Difference Equations. Schaum's Outline Series. New York: McGraw-Hill, 1971.

Teaches the theory of finite differences through presentation of 420 well-selected and completely solved problems. Particularly valuable sections on summation of series and recursion.

Thomas, G. B. Calculus with Analytic Geometry. 4th ed. Reading, Mass.: Addison-Wesley, 1969.
A standard text, especially rich in good problems.

Trigg, Charles. Mathematical Quickies. New York: McGraw-Hill, 1967.
A collection of 270 problems from the "Quickies" section of Mathematics Magazine.

Yaglom, A. M., and I. M. Yaglom. Challenging Mathematical Problems with Elementary Solutions. 2 vols. San Francisco: Holden Day, 1967.
Exceptionally fine collection of 174 problems and complete, detailed solutions. Exceptionally powerful methodology and very elegant solutions to challenging classic problems. Volume I is especially rich in problems in combinatorics. Difficulty ranges from moderate to extremely challenging.

Yaglom, I. M. Geometric Transformations I, II and III. New Mathematical Library, Vols. 8, 21, and 24. Washington, D.C.: Mathematical Association of America.
A comprehensive introduction to the subject. Quite rich in detail. Numerous applications of theory including solutions to classic problems of extremes.

INDEX

Algebra

EQUATIONS

Geometry

Miscellaneous